# Recent Developments in Precision Actuation Technologies

# Recent Developments in Precision Actuation Technologies

Guest Editors

**Bin-tang Yang**
**Yikun Yang**
**Xiaoqing Sun**

MDPI

Basel • Beijing • Wuhan • Barcelona • Belgrade • Novi Sad • Cluj • Manchester

*Guest Editors*

Bin-tang Yang
State Key Laboratory of Mechanical System and Vibration
Shanghai Jiao Tong University
Shanghai
China

Yikun Yang
State Key Laboratory of Mechanical System and Vibration
Shanghai Jiao Tong University
Shanghai
China

Xiaoqing Sun
College of Mechanical Engineering
Donghua University
Shanghai
China

*Editorial Office*

MDPI AG
Grosspeteranlage 5
4052 Basel, Switzerland

This is a reprint of the Special Issue, published open access by the journal *Actuators* (ISSN 2076-0825), freely accessible at: https://www.mdpi.com/journal/actuators/special_issues/9FVW646D55.

For citation purposes, cite each article independently as indicated on the article page online and as indicated below:

| Lastname, A.A.; Lastname, B.B. Article Title. *Journal Name* **Year**, *Volume Number*, Page Range. |
|---|

**ISBN 978-3-7258-6095-1 (Hbk)**
**ISBN 978-3-7258-6096-8 (PDF)**
**https://doi.org/10.3390/books978-3-7258-6096-8**

# Contents

*actuators* MDPI

*Article*

# Research on the Vibration Reduction Mechanism of a New Tensioning Platform with an Embedded Superstructure

**Xiaoqing Sun [1,2,*], Zhengyin Yang [1], Ju Wang [1], Xiusong Hou [2] and Yikun Yang [3]**

1 College of Mechanical Engineering, Donghua University, Shanghai 201620, China
2 National Innovation Center of Advanced Dyeing & Finishing Technology, Tai'an 271000, China
3 State Key Laboratory of Mechanical System and Vibration, Shanghai Jiao Tong University, Shanghai 200240, China; yyk199164@sjtu.edu.cn
* Correspondence: sunxq@dhu.edu.cn

**Abstract:** Aiming at the problem of precision driving and vibration suppression for sensitive payloads on-orbit, this paper proposes a new compliant platform based on an embedded superstructure and a smart material actuator. Firstly, the main structure of the platform is designed and optimized to achieve the expected indicators via the response surface method. Then, the vibration reduction mechanism of the platform with the embedded superstructure is studied by establishing an equivalent model. Following that, a four-phase superstructure is matched and designed with a compact space, and the results are verified by finite element modal analysis. Finally, both the tensioning performance and vibration reduction performance under fixed frequency harmonic disturbance are studied via transient dynamic simulation. Based on the obtained results, directions for future improvements are proposed. The relevant conclusions can provide a reference for function integration of precision tensioning and vibration suppression.

**Keywords:** sensitive payloads; embedded superstructure; compliant platform; optimization design; FEA

## 1. Introduction

With the rapid development of science and technology, the aerospace field must refine the design requirements of high precision and high integration for sensitive payloads on-orbit. Most sensitive loads are flexible and susceptible to micro-vibration on-orbit. Thus, many researchers are interested in finding ways to dampen on-orbit micro-vibration while also ensuring that sensitive payloads operate normally. Qiu [1] and other well-known scholars conducted an in-depth analysis of the vibration characteristics of a compliant mechanism with carefully crafted holes, which provided a new direction for the vibration control of compliant mechanisms such as solar arrays. Feng [2] and other scholars also studied the active vibration suppression and relative position control of large flexible antennas, which provided fast and accurate position adjustment capabilities for antennas. Song et al. [3] designed a multi-degree-of-freedom vibration isolation platform consisting of three units: a protected object, a nonlinear energy pool, and an x-shaped structure, and thus both effectively reduced the amplitude of resonance peaks and improved vibration isolation performance for practical engineering applications. In order to decrease the transmission of vibration and achieve the attenuation of the vibration magnitude of an isolated object, Zhai et al. [4] designed a new type of permanent and electromagnetic composite vibration isolation system based on the negative stiffness theory. To reduce disturbances generated by flywheels on board spacecraft, Xu et al. [5] proposed a new viscoelastic micro-vibration isolation and mitigation platform composed of four elements, and established an analytical model of a coupled system. The results showed that their platform effectively reduced micro-vibration disturbances induced by the flywheel. Song et al. [6] proposed an active-passive integrated controller based on piezoelectric ceramics; they also and established a feedforward feedback control system to solve the low-frequency vibration problem. Niu et al. [7] designed a bionic vibration isolator with a flexible structure for the vibration suppression of sensitive loads, thereby achieving a better vibration isolation effect. Wang et al. [8] designed a self-learning vibration absorber with negative electromagnetic stiffness, and provided a useful framework for the vibration control of sensitive loads. However, the design size and space of these sensitive loads on-orbit are usually limited, so it is necessary to integrate the vibration isolation and precision positioning devices for sensitive payloads, especially considering the high precision requirements during operation.

Precision actuation systems based on smart materials have undergone substantial research in response to the precision directional function required by sensitive loads on-orbit, particularly optical loads. Clark [9], Sun [10], Yang [11], and Herpe [12] designed different configurations of compliant platform schemes using piezoelectric and giant magnetostrictive materials, respectively, and all achieved excellent precision positioning and tracking capabilities. In order to integrate vibration reduction and precision positioning functions, Sun [13], Ciani [14], Yi [15] etc., performed excellent work on structure and control, but there is still room for improvement in both overall scheme design and miniaturization and integration. The embedded superstructure design, which is based on the principle of locally resonant phonon crystal, provides a new method for the functional integration of vibration reduction and precision positioning platform [16,17]. Jin [18], Fan [19] and other scholars have constructed superstructure models with different configurations and achieved the expected vibration suppression function. Throughout these studies, although precise superstructure design can provide reliable technology for on-orbit micro-vibration suppression, there is a relative lack of research on its functional integration with precision actuation platforms. An integration platform may not only effectively reduce the overall size of sensitive payloads, but it could also provide a convenient means for subsequent on-orbit maintenance.

Therefore, this work provides a compliant actuation platform approach for typical optical sensitive payloads in orbit by combining an embedded superstructure and smart material actuation technology. The micro-vibration suppression mechanism of the embedded superstructure, the design and optimization of the main structure of the platform, and the vibration suppression performance with integration of the embedded superstructure are all studied; subsequently, the relevant schemes and principles are verified by both theoretical modeling and finite element numerical simulation [20]. After additional research on the existing issues, a design improvement path for the next generation of sensitive payloads is suggested, and the conclusions can serve as a theoretical and technological basis for several practical on-orbit applications.

## 2. Design and Optimization of the Main Scheme of the Actuation Platform

### 2.1. Principal Scheme

The main scheme shown in Figure 1 is proposed based on the requirement for precision tensioning of optical sensitive loads on-orbit. The scheme is based on smart material actuation technology; in addition, it combines with a compliant mechanism design and an embedded superstructure to realize its required function.

As the sensitive load requires simultaneous external tensioning in eight directions around the circumference, the actuation platform is designed symmetrically, and its input is uniformly pushed by the intelligent material actuator, typically a piezoelectric actuator. To amplify and commutate the input displacement of the piezoelectric actuator, a Scott–Russell compliant amplification mechanism is constructed as the core of this scheme [21]. The upper support disc locks the upper fulcrum of the SR compliant amplification mechanism into place, and the lower support disc is used to secure the input end of the SR mechanism. The integral mounting base is mounted with three symmetrically arranged I-beam pillars, and its upper end is used to secure the upper support plate. The pretension bolt both dynamically adjusts the preload force of the piezoelectric actuator and makes contact with the actuator via the pad. In realizing the micro-nanometer tensioning adjustment, the precision actuation is achieved via the modified SR amplification mechanism, which drives the sensitive load at the end of the SR mounting point. To mount the embedded resonant superstructure, which can be used to achieve desired fixed-frequency vibration suppression of external disturbances in particular frequency bands and is described in Section 3, the mounting rectangular slots have been positioned on the eight modified SR mechanisms as well as the three pillars. The overall plan incorporates the embedded superstructure and makes maximum use of the design space to achieve the objectives of precise actuation adjustment and natural frequency vibration reduction.

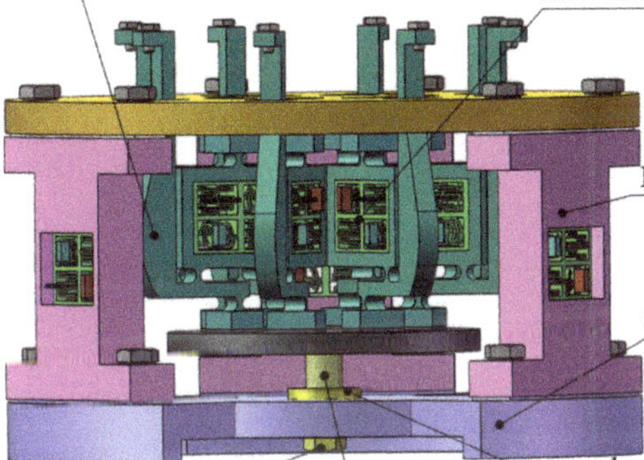


(**a**) front view

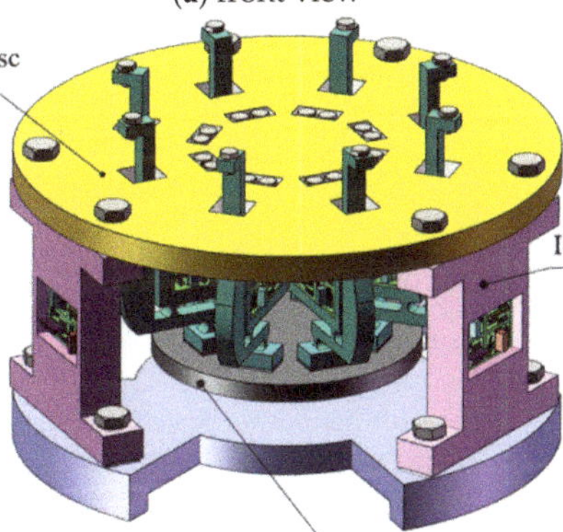


(**b**) Overhead view

**Figure 1.** Scheme diagram of the platform.

## *2.2. Optimization of Main Structure Parameters*

It is crucial to enhance performance indicators, such as the output displacement and natural frequency of the modified SR flexible amplification mechanism, because the primary platform must accomplish a specific operating range and bandwidth. The scheme adopts a unified parameter design for each flexible hinge and initially selects three parameters to be optimized, as shown in Figure 2a, which also illustrates the operation mechanism as follows: the length of the output end beam of the SR mechanism is $x_1$, the radius of the arc profile of the flexible hinge is $x_2$, and the width of the input end beam is $x_3$. In Figure 2b, $d_{in}$ is the input displacement, $d_{out1}$ is the displacement after primary amplification, and $d_{out2}$ is the displacement after secondary amplification. The improved Scott–Russell mechanism is equivalent to a series of lever and Scott–Russell mechanisms. In engineering applications, the magnification ratio of the entire compliant tensioning platform is mainly determined by the motion characteristics of Scott–Russell mechanisms via the velocity projection theorem. The total platform amplification ratio can be simply estimated as the product of the Scott–Russell mechanism and the lever mechanism amplification ratio: $(l_3/l_4) \times (l_2/l_1) = l_3 l_2 / l_4 l_1$.

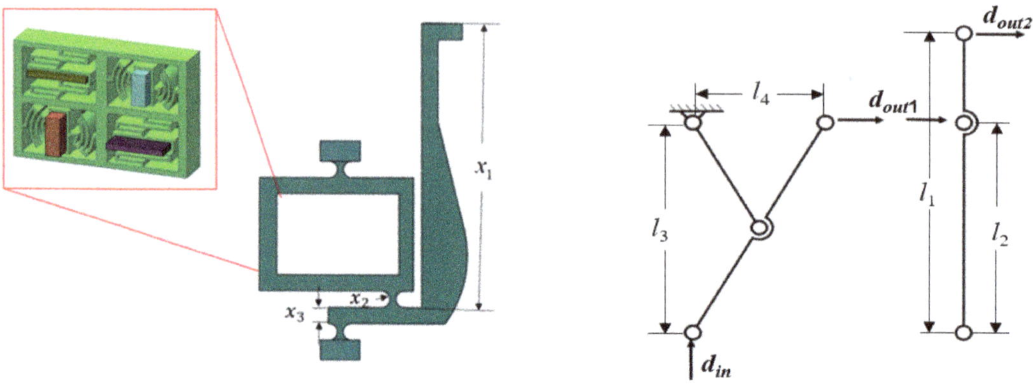


(**a**) Mechanism device diagram (**b**) Mechanism diagram

**Figure 2.** The key parameters of the SR compliant and operation mechanisms.

Overall, the three factors $x_1$, $x_2$, $x_3$ have little impact on the platform's basic structure; since they are not dependent on one another, a sensitivity analysis can be carried out. The starting value of the selected optimal size is defined by the size that, at the time of preliminary design, can achieve the performance of the entire flexible tensioning platform. As a result, the selected size range of the platform is mainly determined by the actual operation of the flexible tensioning platform. As can be observed from previous studies, the first-order natural frequency $Y_p$ and amplification ratio $Y_t$ of the flexible tension platform are chosen as the results of sensitivity analysis so that the larger motion output range and higher working frequency band are fulfilled. ANSYS software was used to generate 16 sets of orthogonal experimental parameters, followed by sensitivity analyses to obtain analysis results under different combinations; the specific values are shown in Table 1. Figure 3 shows the histogram of the effects of these three structural parameters on both the amplification ratio and the intrinsic frequency of the platform. It is evident that $x_3$ has a smaller effect on the overall platform performance compared to $x_1$ and $x_2$. In addition, $x_1$ shows a positive correlation with the deformation of the platform and a negative correlation with the frequency of the platform, while $x_2$ has the opposite effect.

**Table 1.** Results of platform sensitivity analysis.

| Serial Number | $x_1$/mm | $x_2$/mm | $x_3$/mm | $Y_t$/mm | $Y_p$/Hz |
|---|---|---|---|---|---|
| 1 | 75.78 | 0.700931 | 4.080975 | 0.033334 | 294.9327 |
| 2 | 70 | 0.75 | 4 | 0.029667 | 301.51 |
| 3 | 76.14604 | 0.776976 | 4.333236 | 0.031466 | 371.4842 |
| 4 | 71.55851 | 0.796421 | 4.175428 | 0.029454 | 303.9678 |
| 5 | 66.78863 | 0.711865 | 4.358513 | 0.029252 | 299.9477 |
| 6 | 72.04422 | 0.76038 | 4.321502 | 0.030315 | 361.7926 |

**Table 1.** *Cont.*

| Serial Number | $x_1$/mm | $x_2$/mm | $x_3$/mm | $Y_t$/mm | $Y_p$/Hz |
|---|---|---|---|---|---|
| 7 | 75.42766 | 0.712418 | 4.225901 | 0.032516 | 297.1097 |
| 8 | 73.88042 | 0.728615 | 4.121734 | 0.03181 | 298.5136 |
| 9 | 66.0245 | 0.789504 | 4.294115 | 0.027685 | 374.7236 |
| 10 | 76.78263 | 0.811695 | 4.146414 | 0.030889 | 302.6956 |
| 11 | 69.05379 | 0.791876 | 4.203257 | 0.02856 | 303.5848 |
| 12 | 70.41992 | 0.777609 | 4.398086 | 0.029325 | 363.1416 |
| 13 | 66.96664 | 0.700444 | 4.055081 | 0.029682 | 299.0751 |
| 14 | 65.94436 | 0.791094 | 4.278291 | 0.027621 | 374.4772 |
| 15 | 73.36935 | 0.716163 | 4.399461 | 0.031607 | 298.5916 |
| 16 | 70.58632 | 0.756065 | 4.155923 | 0.030033 | 380.6868 |

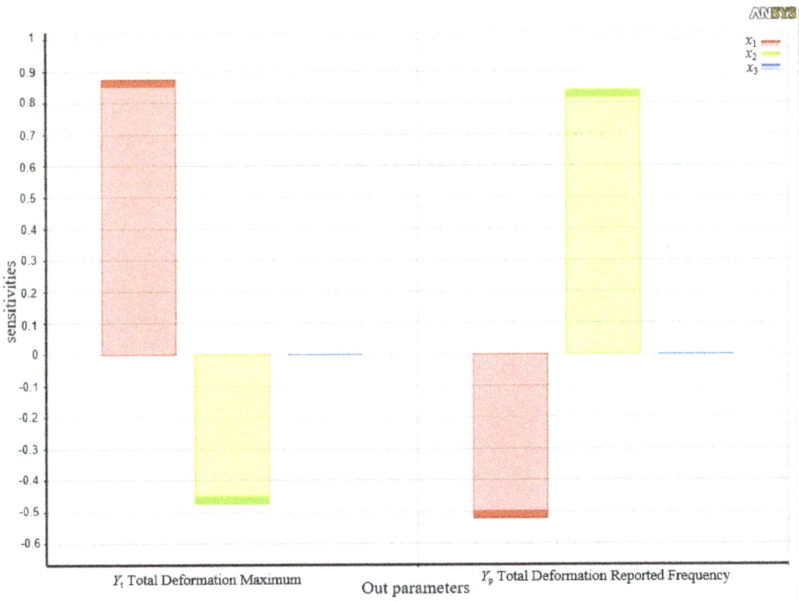


**Figure 3.** Sensitivity parameter influence relationships.

Given that the derivation of the theoretical model based on the principal structure is relatively complicated, the mapping model construction of the output performance, inherent frequency, and the three response surface method parameters [22] are adopted for optimizing the main structure with regard to multiple engineering objectives. After the orthogonal experiment, the regress function from MATLAB is used to fit the simulation results, and the mapping model of amplification ratio $Y_t$ and first-order natural frequency $Y_p$ with the three parameters is obtained [23]:

$$\begin{aligned} Y_t(x_1, x_2, x_3) = &-0.0329 + 0.0012x_1 - 0.0649x_2 + 0.0192x_3 - 0.0004x_1x_2 \\ &-0.0001x_1x_3 + 0.0177x_2x_3 + 0.0033x_2^2 - 0.0028x_3^2 \end{aligned} \tag{1}$$

$$\begin{aligned} Y_P(x_1, x_2, x_3) = &\ 509.3485 - 2.771x_1 - 198.8364x_2 - 95.4605x_3 + 0.9548x_1x_2 + \\ &0.5359x_1x_3 - 20.7123x_2x_3 - 0.0036x_1^2 - 76.7458x_2^2 + 8.8489x_3^2 \end{aligned} \tag{2}$$

Based on the above model, fitted response surfaces to deformation and intrinsic frequency are obtained as shown in Figures 4 and 5, respectively. The $x_1$ and $x_2$ parameters have a significant influence on the platform performance. The larger the $x_1$ and the smaller the $x_2$, the larger the amplification ratio of the tensioning platform. The smaller the $x_1$ and the larger the $x_2$, the greater the natural frequency of the tensioning platform. It is evident that the theoretical model fits well and better reflects the mapping relationship between relevant parameters and platform performance indicators.

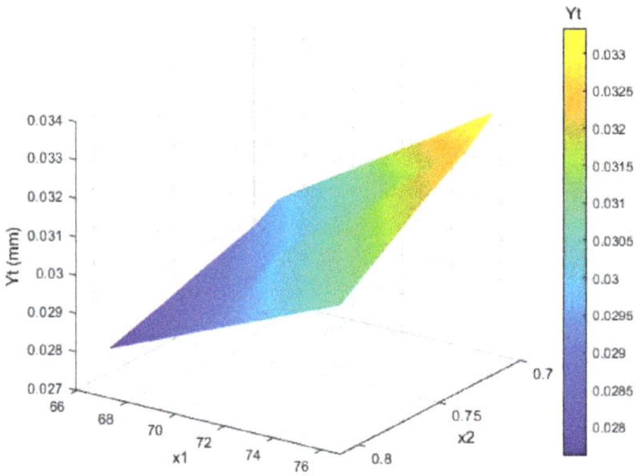


**Figure 4.** The deformation fitted response surface.

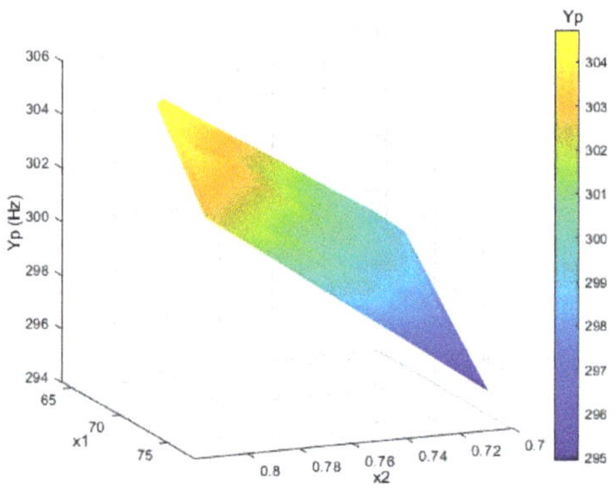


**Figure 5.** The frequency fitted response surface.

The following multi-objective optimization approach is proposed to balance the amplification ratio and natural frequency to obtain optimal performance characteristics. In addition, the constraints on the size parameters take into account the limitations of the actual processing of the flexible platform as well as the requirements of the overall envelope size.

$$\left\{ \begin{array}{l} Minimize\left(T = \frac{1}{aY_t + bY_p}\right) \\ s.t \left\{ \begin{array}{c} 65 < x_1 < 77 \\ 0.7 < x_2 < 0.825 \\ 4 < x_3 < 4.5 \end{array} \right\} \end{array} \right. \tag{3}$$

where $a$ and $b$ are weighted correction factors, respectively, which sum to 1. Then, using the fmincon function [24] the optimization results are corrected and rounded, and the key structural parameters are $x_1 = 75$, $x_2 = 0.8$, and $x_3 = 4.5$, respectively. The amplification ratio of the scheme is close to 3.4. The first-order solid frequency reached 365 Hz, compared with the finite element simulation results (as shown in Figures 6 and 7), and basically remained the same. This demonstrates the effectiveness of the optimization method; therefore, the presented scheme provides a basis for avoiding external disturbances in critical low-

and medium-frequency bands. Subsequently, as shown in Figure 8, the prototype is manufactured and assembled according to the above optimization results.

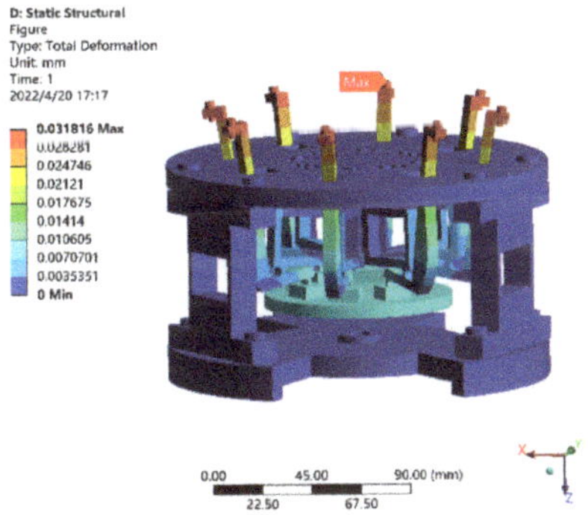


**Figure 6.** The displacement nephogram of the platform.

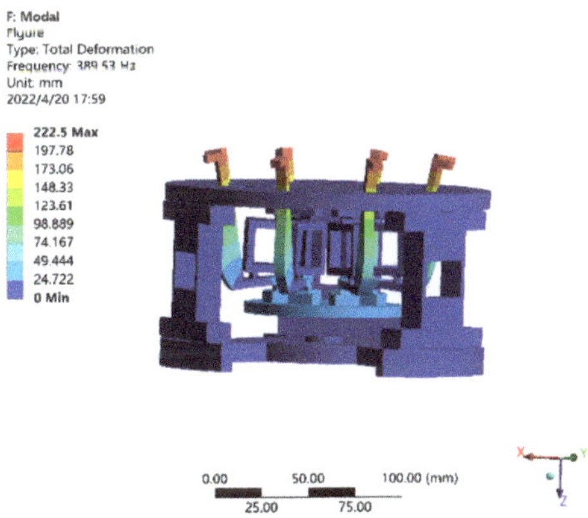


**Figure 7.** The stress nephogram of the platform.

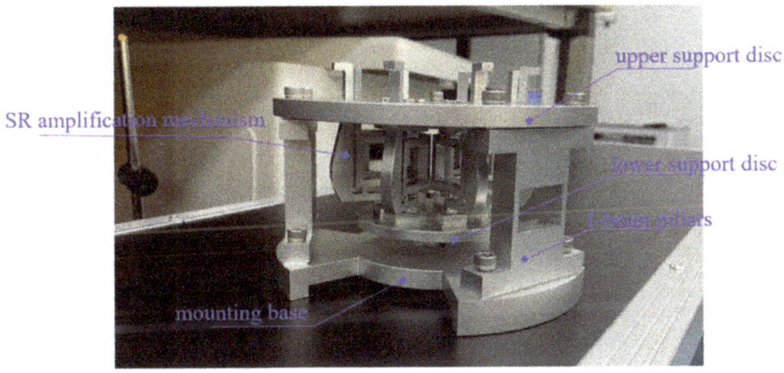


**Figure 8.** Prototype of the main platform.

## 3. Vibration Suppression Mechanism of the Embedded Superstructure

### *3.1. Equivalent Dynamic Modeling*

There are four major sources of vibration interference in the low-frequency region, which typically exhibits the harmonic disturbance type of interference, given that the target sensitive load is running on-orbit. Therefore, in this paper, an embedded four-phase superstructure scheme, as shown in Figure 9, is proposed for these four typical low-frequency operation disturbances.

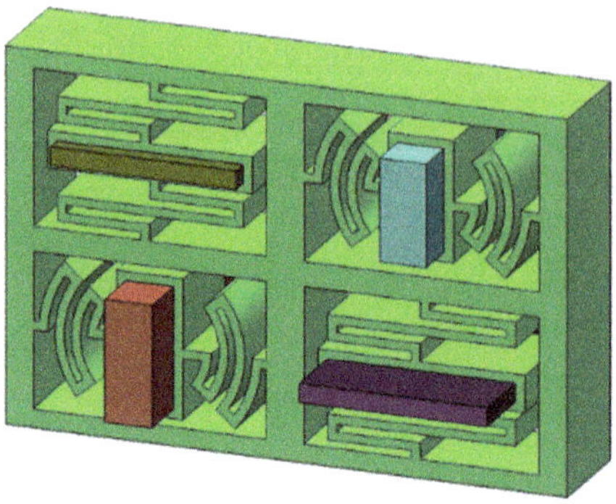

**Figure 9.** The proposed embedded superstructure scheme.

The equivalent vibration suppression model of the proposed scheme is simplified in accordance with the principle of local resonance; this allows further investigation of the impact of important structural characteristics of superstructure cells on the vibration suppression effect [25]. The equivalent model is shown in Figure 10. The main structure is subjected to an external harmonic perturbation $F(t)$ to produce a response $x_0$, where $m_0$, $k_0$, and $c_0$ represent the mass, stiffness, and damping of the main structure, respectively. $x_i$ ($I$ = 1, ... , 4), $m_i$ ($I$ = 1, ... , 4), $k_i$ ($I$ = 1, ... , 4), and $c_i$ ($I$ = 1, ... , 4) represent the displacement, mass, stiffness, and damping of each phase of the superstructure, respectively.

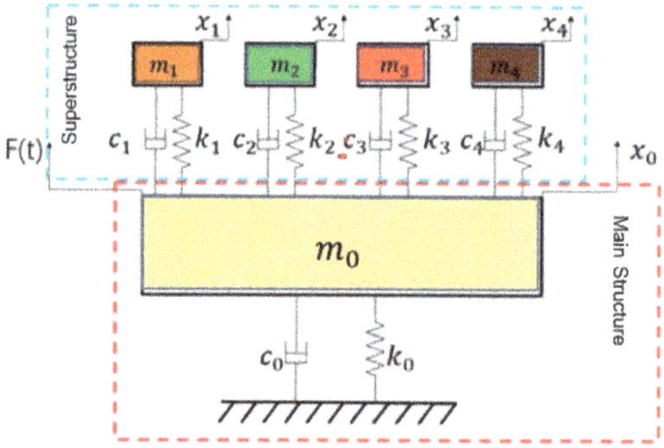


**Figure 10.** The equivalent vibration suppression model.

The dynamic equations of the system are established using the Newton–Euler method as:

$$M\ddot{x} + C\dot{x} + Kx = F(t) \tag{4}$$

where $x = \{x_0, x_1, x_2, x_3, x_4\}^T$; $M = diag\{m_0, m_1, m_2, m_3, m_4\}$;

$$C = \begin{pmatrix} \sum_{i=0}^{4} c_i & -c_1 & -c_2 & -c_3 & -c_4 \\ -c_1 & c_1 & 0 & 0 & 0 \\ -c_2 & 0 & c_2 & 0 & 0 \\ -c_3 & 0 & 0 & c_3 & 0 \\ -c_4 & 0 & 0 & 0 & c_4 \end{pmatrix} \tag{5}$$

$$K = \begin{pmatrix} \sum_{i=0}^{4} k_i & -k_1 & -k_2 & -k_3 & -k_4 \\ -k_1 & k_1 & 0 & 0 & 0 \\ -k_2 & 0 & k_2 & 0 & 0 \\ -k_3 & 0 & 0 & k_3 & 0 \\ -k_4 & 0 & 0 & 0 & k_4 \end{pmatrix} \quad (6)$$

Performing the Lasse transform on the above equation shows that the frequency response function transfer matrix of the overall system is:

$$H(\omega) = \begin{pmatrix} -\omega^2 m_0 + j\omega \sum_{i=0}^{4} c_i + \sum_{i=0}^{4} k_i & -j\omega c_1 - k_1 & \cdots & -j\omega c_4 - k_4 \\ -j\omega c_1 - k_1 & -\omega^2 m_1 + j\omega c_1 + k_1 & \cdots & 0 \\ \vdots & \vdots & \ddots & \vdots \\ -j\omega c_4 - k_4 & 0 & 0 & -\omega^2 m_4 + j\omega c_4 + k_4 \end{pmatrix} \quad (7)$$

The first term of this matrix is chosen as the primary system's frequency response function in order to study how the four-phase superstructure affects the vibration transfer characteristics of the main system:

$$H_M(\omega) = \frac{1}{k_0 \left[ 1 + 2\zeta_0 \lambda j - \lambda^2 - \lambda^2 \sum_{i=1}^{4} \frac{\mu_i (1 + 2\zeta_i a_i \lambda j)}{1 + 2\zeta_i a_i \lambda j - a_i^2 \lambda^2} \right]} \quad (8)$$

where the relevant parameters are set to:

$$\omega_n = \sqrt{\frac{k_0}{m_0}},\ \zeta_0 = \frac{c_0}{2m_0\omega_0},\ \lambda = \frac{\omega}{\omega_0};$$
$$\mu_i = \frac{m_i}{m_0},\ \omega_i = \sqrt{\frac{k_i}{m_i}},\ \zeta_i = \frac{c_i}{2m_i\omega_i},\ a_i = \frac{\omega_0}{\omega_i} (i = 1, 2, 3, 4)$$

The amplitude-frequency response of the main structure is mainly affected by the mass ratio $\mu_i$, damping ratio $\zeta_i$ and natural frequency $\omega_i$ of the superstructure element.

### 3.2. Vibration Suppression Mechanism

In MATLAB, the controlled variable approach is used to study the influence of the major parameters of the superstructure on the vibration characteristics of the main system. $m_0 + \sum_{i=1}^{4} m_i = m$ Order, $m_S = \sum_{i=1}^{4} m_i$ and $\mu = m_S / m_0$, where $m_s$ is the superstructure's overall mass. With other parameters unchanged, different mass ratios are selected to assess the influence of different mass ratios on vibration characteristics of the main system. The results are shown in Figure 11.

Studying the above trend shows that $\mu = 0$ indicates the main system response in the absence of superstructure. Compared with the main system without superstructure, when $\mu = 0.05$ the formant of the main system is greatly attenuated, the vibration is suppressed, and a new resonance peak is formed on the left side. As the mass ratio increases, the weakening amplitude of the main system increases, which demonstrates that while the vibration suppression ability is enhanced, the amplitude at the new natural frequency becomes gradually larger. Therefore, when designing the superstructure, the mass ratio of the superstructure should be considered comprehensively in order to balance its vibration suppression ability.

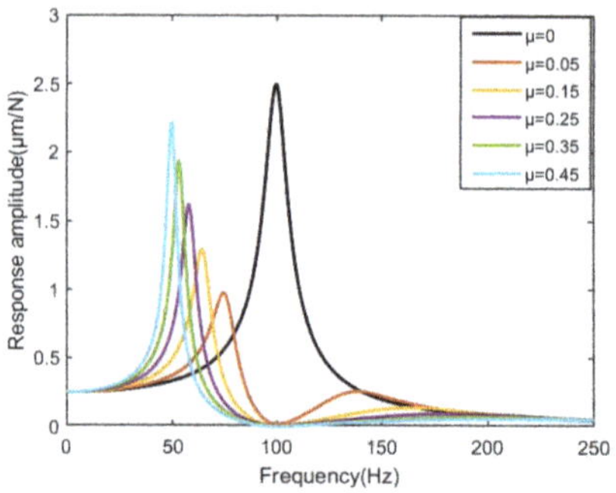


**Figure 11.** The frequency response with different mass ratio of superstructure.

Then, the influence of the superstructure and the natural frequency ratio of the main system on the vibration characteristics is explored. The simulation results shown in Figure 12 are obtained. $\varphi$ represents the ratio of the natural frequency of each unit of the superstructure to the natural frequency of the main structure. When the intrinsic frequency ratio is between 0 and 1, the resonance peak amplitude of the main structure decays as the frequency ratio increases, but the amplitude corresponding to the new natural frequency gradually increases. When the frequency ratio is greater than 1, the main structure's formant attenuation remains largely unchanged with the increase in the frequency ratio, but the new fixed frequency's amplitude increases on the left side while that on the right side gradually decreases. This demonstrates that, when the fixed frequency of the superstructure unit and the main structure are the same, the main structure's ability to dampen vibrations depends on how much vibration interference energy it can absorb from the environment.

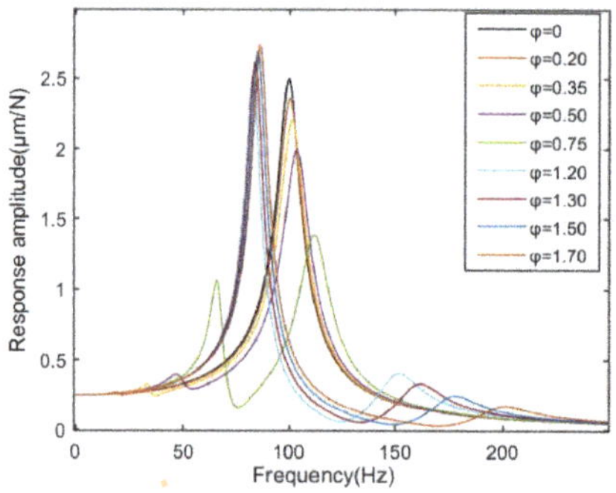


**Figure 12.** The frequency response with different natural frequency ratios of superstructure to main structure.

Finally, the effect of the relative damping coefficient of the superstructure on the vibration transmission characteristics of the main structure is studied; the simulation results are shown in Figure 13. The primary structure's attenuation amplitude increases as the relative damping ratio decreases, while additional natural frequencies start to emerge on the left and right sides of the original natural frequency. With the increase of the relative damping ratio, the attenuation amplitude of the resonance peak of the main structure decreases, indicating that the appropriate damping ratio is crucial to the main structure damping, and that the appropriate damping ratio should be selected according to different material properties in the future so as to enhance the vibration suppression effect. Therefore, based on the above conclusions, four-phase superstructure schemes with different natural frequencies of 35 Hz, 45 Hz, 55 Hz and 65 Hz are designed for typical low-frequency harmonic disturbances. The prototype can be 3D printed at a later stage to fabricate the external frame using TPU material [26] and the matching lead block injected by inlay to achieve the local resonance function. Considering the symmetry of the overall scheme, after embedding the superstructure a modal analysis of the single branch is carried out in order to reduce the simulation calculations. The first four orders of mode shape are shown in Figure 14. The findings demonstrate that the first four orders of inherent frequencies, 34.6 Hz, 44.4 Hz, 57.1 Hz, and 65.7 Hz, respectively, correspond to the local resonance of each phase of the four-phase superstructure and are essentially consistent with expectations. As a result, the simulation offers the foundation for vibration suppression.

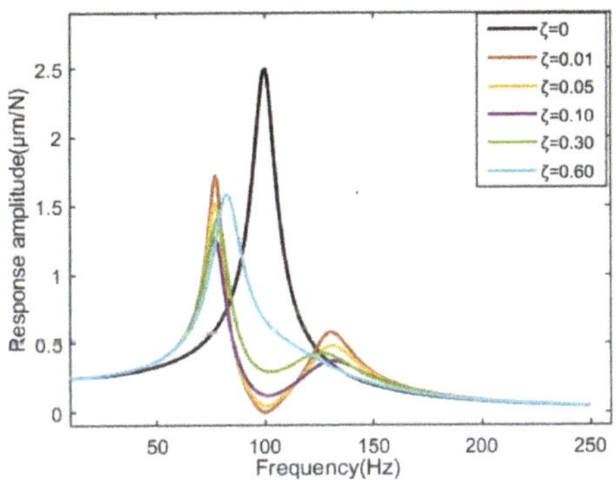


**Figure 13.** The frequency response with different damping ratios of superstructure to main structure.

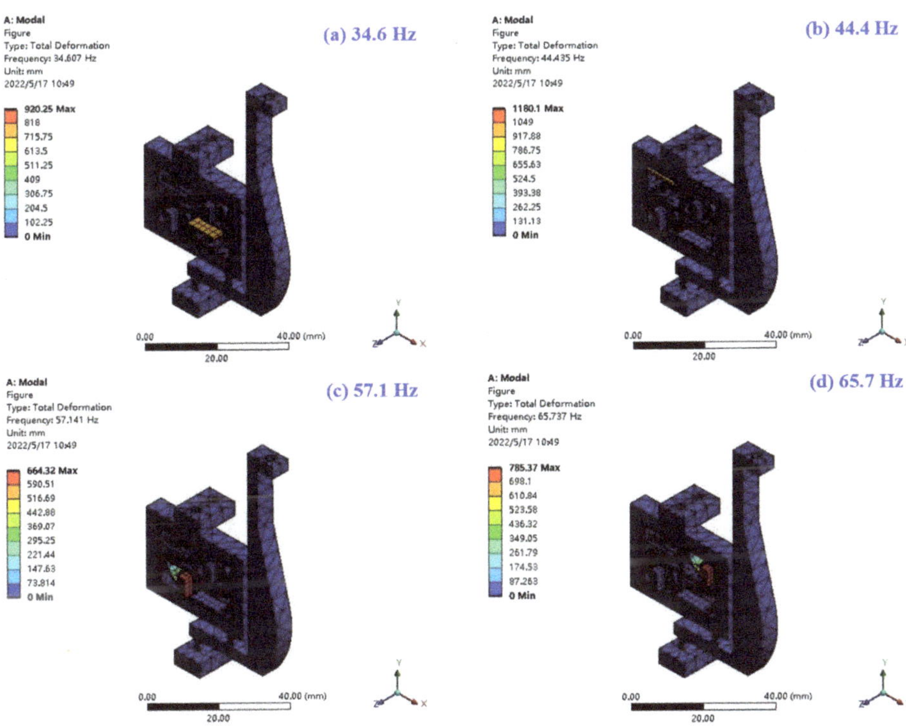


**Figure 14.** The mode shapes of the single system.

## 4. Numerical Research and Result Analysis

This section will use the finite element method to simulate the tensioning function of the platform and the vibration suppression effect on the common harmonic disturbance in accordance with the desired function of the compliant platform [27].

### 4.1. Static Performance

After integrating the superstructure, the different output displacement excitation of the smart material actuator is simulated and applied to the lower support disc's bottom surface in order to evaluate the overall platform's static performance. Static analysis is then used to determine the platform's input-output response results. The results are shown in Figure 15. The simulation amplification ratio is 3.2, slightly smaller than the theoretical value. The reason may be that the transverse deformation of the flexure hinges causes a reduction in output displacement.

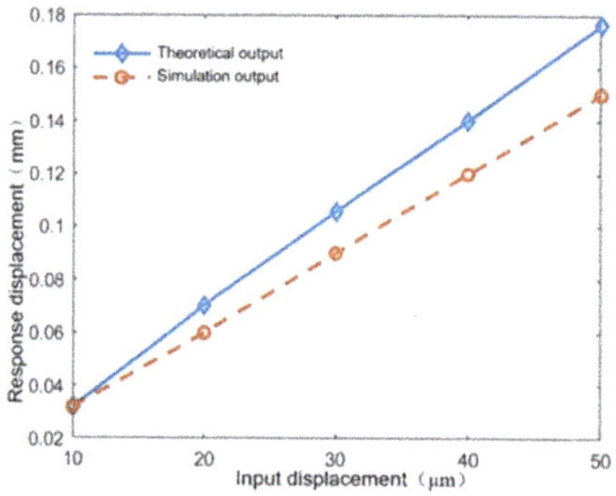


**Figure 15.** Tensioning platform displacement response results.

### 4.2. Dynamic Tensioning Performance

The lower support disc's bottom surface is exposed to sinusoidal displacement signals of varying amplitudes at a maximum operating frequency of 10 Hz in order to simulate the output displacement excitation of the piezoelectric actuator. This allows for evaluation of the tensioning output performance of the platform under certain input displacements by transient dynamics simulation [28]. During the simulation, the input displacement amplitude was set to several segments between 10 μm and 100 μm. The simulation time is set to 0.5 s, and the dynamic simulation results are shown in Figure 16.

Analyzing the above results, under the maximum operating frequency of 10 Hz and different amplitudes of excitation, the output displacement of the platform reaches 31 μm, 92 μm, 148 μm, 310 μm, respectively, and the output response amplitudes essentially conform to the 3 times amplification ratio. Meanwhile, from the simulation time domain curve it is evident that there is transient oscillation when the platform is initially excited; however, due to the existence of structural damping, within one cycle the oscillation gradually disappears and the platform reaches steady state. Meanwhile, the output matches the input sinusoidal signal well, which demonstrates that the platform can achieve the expected tension adjustment function.

### 4.3. Harmonic Response Tests of the Overall Platform

Harmonic disturbances with an amplitude of 10 m and a frequency ranging from 0 to 200 Hz are applied to the mounting base in order to examine the vibration suppression capabilities of the platforms with and without an embedded superstructure in the entire interference band. Then, harmonic response simulations are carried out; the vibration response curves are drawn in Figure 17. When compared with the platform without superstructure, it is obvious that the vibration suppression band gap is generated around the corresponding natural frequencies of the superstructure. When the external vibration is excited to the mounting base, the local resonance of the superstructure is generated. The vibration energy near the corresponding frequency is absorbed by the superstructure in the form of resonance, thus achieving the shielding and suppression of the corresponding vibration. The vibration with the corresponding frequency cannot pass through the band gap; in other words, the corresponding vibration has no impact on the main structure of the platform. Furthermore, according to the extracted displacement nephogram and stress nephogram of harmonic response shown in Figure 18, it is clear that external vibration causes local resonance of the embedded superstructure and results in obvious deformation, which conforms to the vibration reduction mechanism of the embedded superstructure.

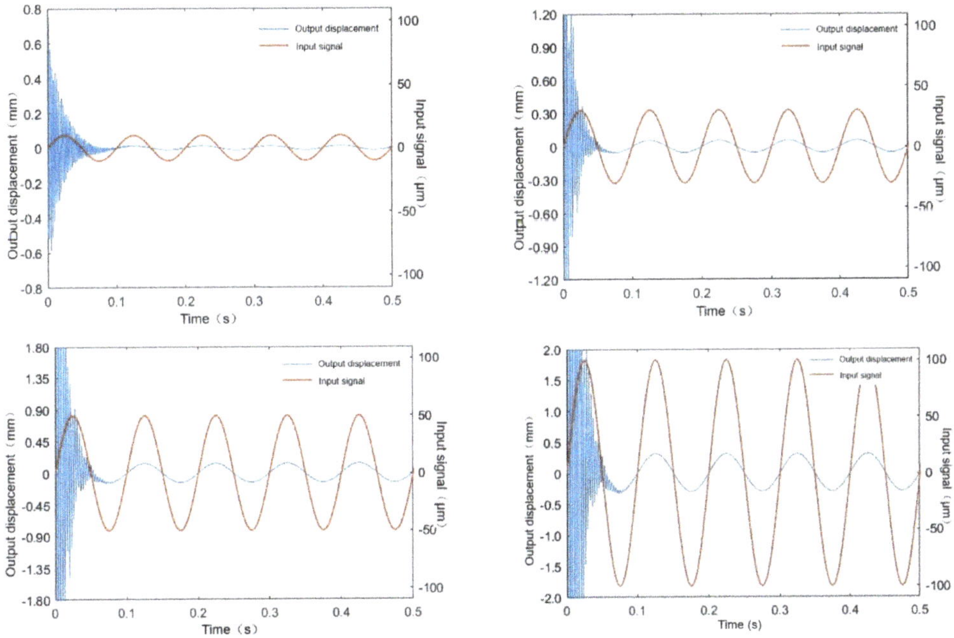


**Figure 16.** Output response of the actuation platform. (**a**) Output response at an input displacement amplitude of 10 μm. (**b**) Output response at an input displacement amplitude of 30 μm. (**c**) Output response at an input displacement amplitude of 50 μm. (**d**) Output response at an input displacement amplitude of 100 μm.

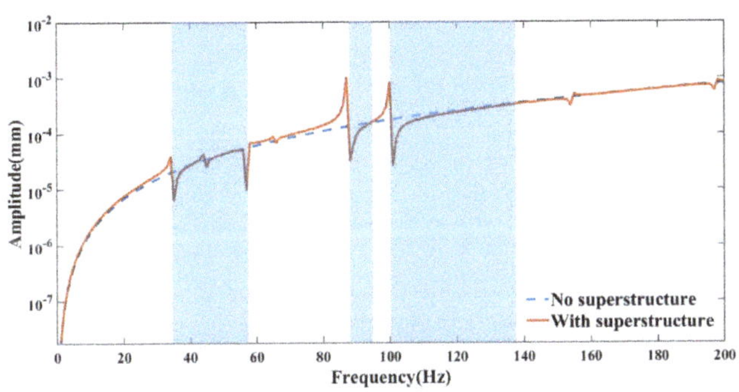


**Figure 17.** The harmonic response of the platform embedded with the optimized superstructure.

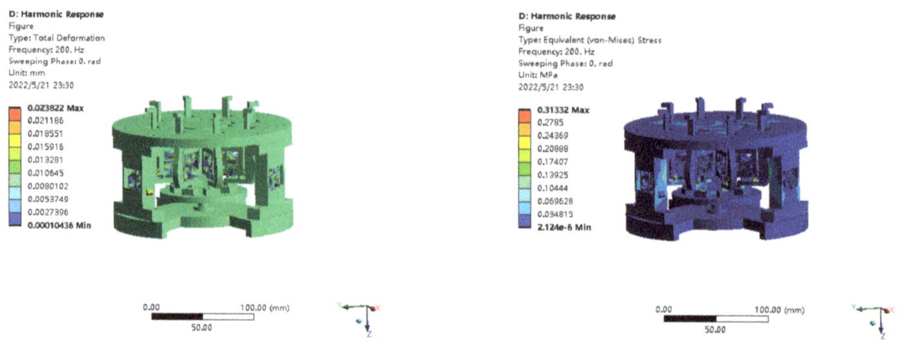


(**a**) Harmonic response displacement nephogram (**b**) Harmonic response stress nephogram

**Figure 18.** Platform harmonic response results.

### *4.4. Vibration Suppression Tests under Fixed Frequencies Disturbance*

A typical interference source is applied to the bottom of the mounting base in order to conduct a comparison study with or without the embedded superstructure to analyze the platform's performance in terms of vibration suppression under the desired fixed frequency. Since the main purpose of the designed four-phase superstructure is to achieve vibration suppression from external harmonic disturbances at the fixed operating frequency, the typical interference source can be set to

$$y = A_1 \sin(2\pi f_1 t) + A_2 \sin(2\pi f_2 t) + A_3 \sin(2\pi f_3 t) + A_4 \sin(2\pi f_4 t) \tag{9}$$

where $f_i$ ($i$ = 1, . . . , 4) is the corresponding resonant frequency of each phase of the superstructure, which can be determined via the previously mentioned modal simulation results. $A_i$ ($i$ = 1, . . . , 4) is the equivalent vibration amplitude, respectively, and can be set to 0.1 mm, 0.05 mm, 0.01 mm and 0.005 mm. After transient dynamic analysis [29], the simulation results in the time and frequency domains are processed using MATLAB, as plotted in Figure 19.

From Figure 19a,b, it is evident that the platform has a significant vibration suppression effect on the harmonic disturbances with frequencies of $f_1$ and $f_3$, and the amplitude is significantly reduced. Similarly, from Figure 19c,d, it is evident that the platform has a significant vibration suppression effect on the harmonic disturbances with frequencies of $f_1$ and $f_4$, and the amplitude is significantly reduced. Finally, from Figure 19g,h, it is evident that the platform also shows a significant vibration suppression effect for harmonic disturbances with frequencies of $f_1$, $f_2$, $f_3$ and $f_4$. The frequency domain results in Figure 19h show that four resonance peaks corresponding to the four excitation frequencies appear, and the amplitudes of these resonance peaks are obviously weakened, which demonstrates that the designed four-phase superstructure suppresses the vibration with desired frequency through local resonance function. The vibration suppression effects are further collated and listed in Table 2. It is evident that the vibration corresponding to the inherent frequency of each phase is suppressed to a certain extent, that the maximum vibration suppression effect appears at the first phase, and that the vibration suppression effect reaches −1.705 dB. Overall, although the designed superstructure achieves a certain vibration suppression capability, the suppression effect is limited. Because each branch of the planned superstructure does not form an array, only one single cell is used to perform vibration suppression, which has a restricted impact. Combined with the vibration suppression mechanism, future research should focus on the design of multiple embedded arrays in each branch, so as to improve the vibration suppression performance of the overall platform.

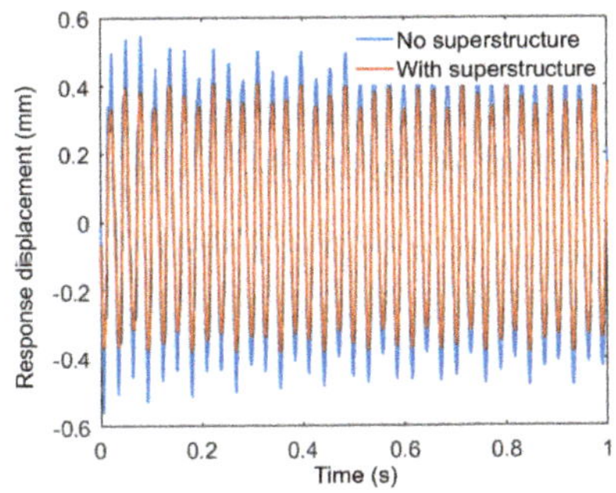


(**a**) Time domain response results under $A_1A_3$

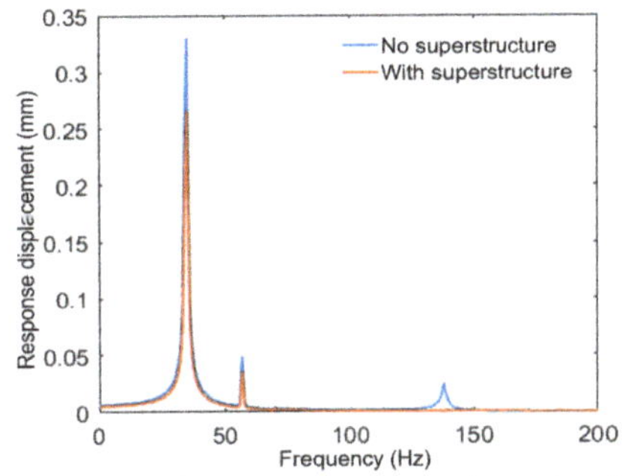


(**b**) Frequency domain response results under $A_1A_3$

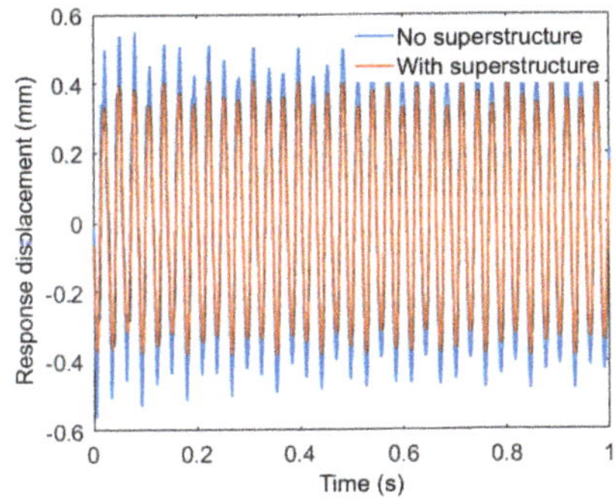


(**c**) Time domain response results under $A_1A_4$

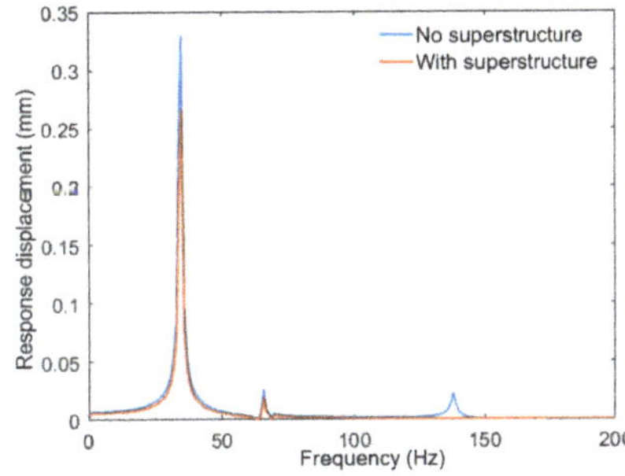


(**d**) Frequency domain response results under $A_1A_4$

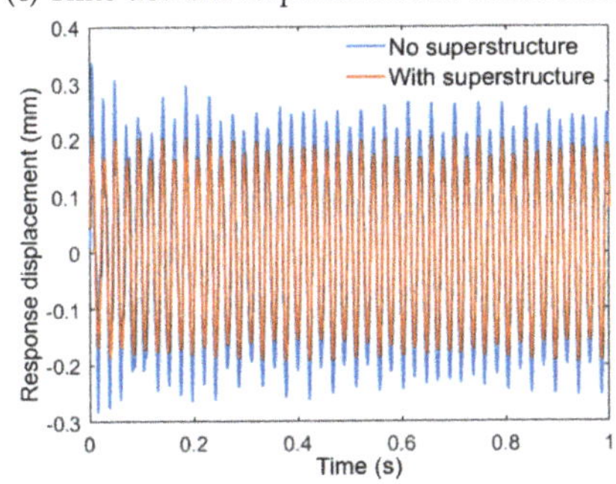


(**e**) Time domain response results under $A_2A_4$

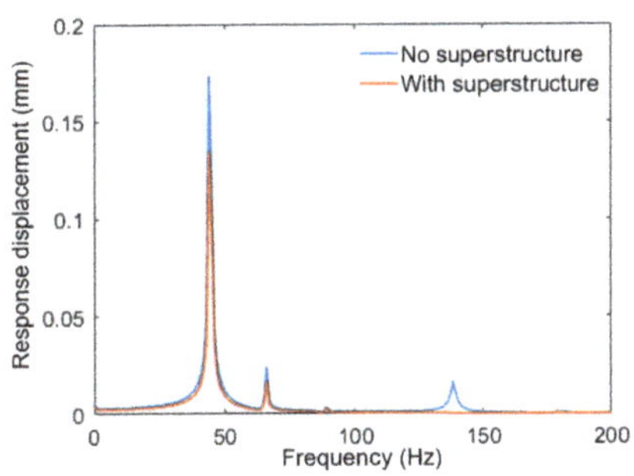


(**f**) Frequency domain response results under $A_2A_4$

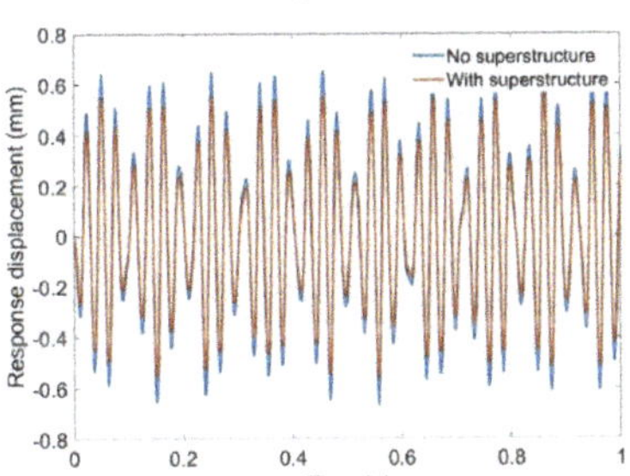


(**g**) Time domain response results under $A_1A_2A_3A_4$

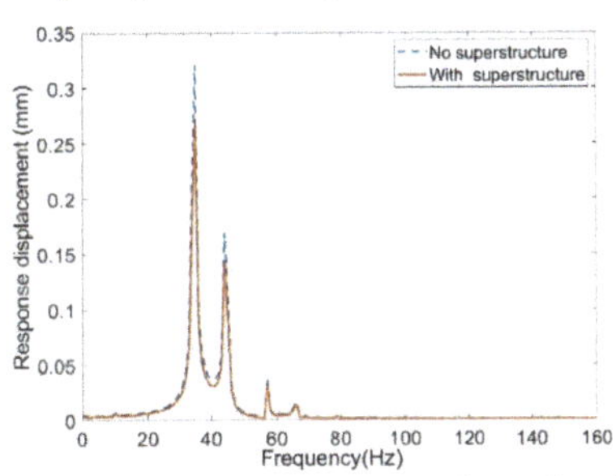


(**h**) Frequency domain response results under $A_1A_2A_3A_4$

**Figure 19.** The vibration suppression performance of the platform.

**Table 2.** The vibration suppression effects.

| Frequency/Hz | Vibration Amplitude without Superstructure/μm | Vibration Amplitude with Superstructure/μm | Attenuation/dB |
|---|---|---|---|
| 34.6 | 331 | 272 | −1.705 |
| 44.4 | 173 | 145 | −1.533 |
| 57.1 | 40 | 35 | −1.15 |
| 65.7 | 20 | 17 | −1.41 |

## 5. Conclusions

In this study, a compliant operating platform based on an embedded superstructure and piezoelectric actuators is proposed for dealing with ultra-precision tensioning driving and vibration suppression for on-orbit sensitive loads. The main structure of the scheme is optimized based on the response surface method. Based on the local resonance mechanism, a four-phase superstructure scheme is suggested for disturbances with natural frequencies. An equivalent vibration suppression model is established, and the evolution law between these key parameters and the vibration suppression performance is studied. Finally, the tensioning output performance and the vibration suppression performance of the platform under fixed frequencies is studied using finite element analysis. The findings demonstrate that the designed platform has effective tensioning output and vibration suppression abilities. However, the insufficient vibration suppression effect also suggests that in the future embedded superstructure cells need to be arranged in arrays to improve the vibration suppression effect.

**Author Contributions:** Conceptualization, X.S. and J.W.; methodology, X.S., J.W. and Z.Y.; software, J.W. and Z.Y.; validation, X.S., J.W. and X.H.; formal analysis, Z.Y. and Y.Y.; investigation, X.S.; resources, X.S.; data curation, J.W. and Z.Y.; writing—original draft preparation, X.S. and J.W.; writing—review and editing, X.S. and Z.Y.; visualization, J.W.; supervision, X.S.; project administration, X.S.; funding acquisition, X.S. All authors have read and agreed to the published version of the manuscript.

**Funding:** The authors would like to thank the handling editor and anonymous reviewers for their useful comments and constructive suggestions. This research was supported by National Natural Science Foundation of China (No. 51905087), the Research Project of State Key Laboratory of Mechanical System and Vibration (MSV202317), the Research Funds of National Innovation Center of Advanced Dyeing & Finishing Technology (2022GCJJ11), and the Fundamental Research Funds for the Central Universities (2232023D-16).

**Data Availability Statement:** The data that support the findings of this study are available on request from the corresponding author, upon reasonable request.

**Conflicts of Interest:** The authors declare no conflict of interest.

## References

1. Qiu, Z.; Zhu, X. Vibration modal characteristics analysis of compliant mechanism with opening. *Aerosp. Control. Appl.* **2022**, *48*, 1–11.
2. Feng, X.; Tang, L.; Guan, X.; Hao, R.; Zhang, K. Active damping and relative position control for large antenna. *Aerosp. Control. Appl.* **2022**, *48*, 17–26.
3. Song, Y.; Zhang, C.; Li, Z.; Li, Y.; Lian, J.; Shi, Q.; Yan, B. Study on dynamic characteristics of bio-inspired vibration isolation platform. *J. Vib. Control.* **2022**, *28*, 1470–1485. [CrossRef]
4. Zhai, M.; Zhang, B.; Li, X.; Long, Z. Design and Implementation of Permanent and Electromagnet Composite Vibration Isolation System Based on Negative Stiffness Theory. *Actuators* **2023**, *12*, 44. [CrossRef]
5. Xu, C.; Xu, Z.-D.; Huang, X.-H.; Xu, Y.-S.; Ge, T. Modeling and analysis of a viscoelastic micro-vibration isolation and mitigation platform for spacecraft. *J. Vib. Control.* **2018**, *24*, 4337–4352. [CrossRef]
6. Song, H.; Shan, X.; Hou, W.; Wang, C.; Sun, K.; Xie, T. A novel piezoelectric-based active-passive vibration isolator for low-frequency vibration sys-tem and experimental analysis of vibration isolation performance. *Energy* **2023**, *278*, 127870. [CrossRef]
7. Niu, M.; Chen, L. Analysis of a bio-inspired vibration isolator with a compliant limb-like structure. *Mech. Syst. Signal Process.* **2022**, *179*, 109348. [CrossRef]

8. Wang, X.; Wang, D.; Li, F.; Zhang, Y.; Xu, Z.; Wang, T.; Fu, G.; Lu, C. Self-learning vibration absorber with negative electromagnetic stiffness for variable vibration. *Int. J. Mech. Sci.* **2023**, *248*, 108225. [CrossRef]
9. Clark, L.; Shirinzadeh, B.; Zhong, Y.; Tian, Y.; Zhang, D. Design and analysis of a compact flexure-based precision pure rotation stage without actuator redundancy. *Mech. Mach. Theory* **2016**, *105*, 129–144. [CrossRef]
10. Sun, X.; Wang, Z.; Yang, Y. Design and Experimental Investigation of a Novel Compliant Positioning Stage with Low-frequency Vibration Isolation Capability. *Sens. Actuators A Phys.* **2019**, *295*, 439–449. [CrossRef]
11. Yang, X.; Wu, H.; Li, Y.; Chen, B. Dynamic Isotropic Design and Decentralized Active Control of a Six-axis Vibration Isolator via Stewart Platform. *Mech. Mach. Theory* **2017**, *117*, 244–252. [CrossRef]
12. Herpe, X.; Walker, R.; Dunnigan, M.; Kong, X. On a simplified nonlinear analytical model for the characterisation and design opti-misation of a compliant XY micro-motion stage. *Robot. Comput.-Integr. Manuf.* **2018**, *49*, 66–76. [CrossRef]
13. Sun, X.; Hu, W.; Bai, Z.; Yang, Y. Experimental investigation on a new sensitive payload platform with simultaneous positioning and vibration suppression capabilities. *J. Vib. Control.* **2022**, *28*, 1200–1213. [CrossRef]
14. Ciani, G.; Arain, M.A.; Aston, S.M.; Feldbaum, D.; Fulda, P.; Gleason, J.; Heintze, M.; Martin, R.M.; Mueller, C.L.; Kumar, D.M.N. Small optic suspensions for advanced LIGO input optics and other precision optical ex-periments. *Rev. Sci. Instrum.* **2016**, *87*, 114504. [CrossRef] [PubMed]
15. Yi, S.; Zhang, Q.; Sun, X.; Yang, B.; Meng, G. Simultaneous micropositioning and microvibration control of a magnetostrictive Stewart plat-form with synthesized strategy. *Mech. Syst. Signal Process.* **2023**, *187*, 109925. [CrossRef]
16. Yu, X.; Zhou, J.; Liang, H.; Jiang, Z.; Wu, L. Mechanical metamaterials associated with stiffness, rigidity and compressibility: A brief review. *Prog. Mater. Sci.* **2018**, *94*, 114–173. [CrossRef]
17. Zhuoqun, W.; Pengfei, W.; Yan, Z.; Jiang, Y. Vibration reduction technology of mechanical metamaterials presented to large scale structures. *Acta Aeronaut. Et Astronaut. Sin.* **2018**, *39*, 721651.
18. Jin, Y.; Shi, Y.; Yu, G.-C.; Wei, G.-T.; Hu, B.; Wu, L.-Z. A multifunctional honeycomb metastructure for vibration suppression. *Int. J. Mech. Sci.* **2020**, *188*, 105964. [CrossRef]
19. Fan, H.; Yang, L.; Tian, Y.; Wang, Z. Design of metastructures with quasi-zero dynamic stiffness for vibration isolation. *Compos. Struct.* **2020**, *243*, 112244. [CrossRef]
20. Gao, H.; He, W.; Zhang, Y.; Sun, C. Adaptive finite-time fault-tolerant control for uncertain flexible flapping wings based on rigid finite element method. *IEEE Trans. Cybern.* **2021**, *52*, 9036–9047. [CrossRef]
21. Wu, H.; Lai, L.; Zhang, L.; Zhu, L. A novel compliant XY micro-positioning stage using bridge-type displacement amplifier embed-ded with Scott-Russell mechanism. *Precis. Eng.* **2022**, *73*, 284–295. [CrossRef]
22. Lu, C.; Feng, Y.W.; Liem, R.P.; Fei, C.W. Improved Kriging with extremum response surface method for structural dynamic reliability and sensitivity analyses. *Aerosp. Sci. Technol.* **2018**, *76*, 164–175. [CrossRef]
23. Wu, Z.; Wang, S.; Zhang, Y.; Song, J.; Xue, B. Optimization of Process Parameters for Laser Cutting of AZ31B Magnesium Alloy Based on Orthogonal Experiment and BP Neural Network. *Trans. Indian Inst. Met.* **2023**. [CrossRef]
24. Le Luong, H.T.; Messine, F.; Henaux, C.; Mariani, G.B.; Voyer, N.; Mollov, S. Comparison between fmincon and NOMAD optimization codes to design wound rotor synchronous machines. *Int. J. Appl. Electromagn. Mech.* **2019**, *60*, S87–S100. [CrossRef]
25. Li, Z. Research on design and experiment of a XY-Decoupled compliant mechanism with high modal damping using embedded local resonance metastructure. *S. China Univ. Technol.* **2020**. [CrossRef]
26. Zu, R.; Fan, Q.; Chen, Y.; Chen, W.; Li, H.; Liu, H. Research progress and application of 3D Printing TPU materials. *Mod. Chem. Res.* **2023**, *130*, 20–24. [CrossRef]
27. Khushnood, M.A.; Wang, X.; Cui, N. Active vibration control of a slewing spacecraft's panel using H∞ control. *J. Vi-Broengineering* **2016**, *18*, 2959–2973. [CrossRef]
28. Harada, K.; Kinoshita, T.; Shiratori, N. The emergence of controllable transient behavior using an agent diversification strategy. *IEEE Trans. Syst. Man, Cybern.-Part A Syst. Hum.* **2003**, *33*, 589–596. [CrossRef]
29. Wang, Z.; Yin, B.; Yan, H. Weak Signal Detection Based on Pseudo Wigner Ville Distribution. In *Journal of Physics: Conference Series*; IOP Publishing: Bristol, UK, 2019; Volume 1176, p. 062040.

actuators

MDPI

Article

# A Novel Hierarchical Recursive Nonsingular Terminal Sliding Mode Control for Inverted Pendulum

**Hiep Dai Le** * **and Tamara Nestorović**

Mechanics of Adaptive Systems Department, Ruhr-Universität Bochum, 44801 Bochum, Germany; tamara.nestorovic@rub.de
* Correspondence: hiep.le-d9b@ruhr-uni-bochum.de

**Abstract:** This paper aims to develop a novel hierarchical recursive nonsingular terminal sliding mode controller (HRNTSMC), which is designed to stabilize the inverted pendulum (IP). In contrast to existing hierarchical sliding mode controllers (HSMC), the HRNTSMC significantly reduces the chattering problem in control input and improves the convergence speed of errors. In the HRNTSMC design, the IP system is first decoupled into pendulum and cart subsystems. Subsequently, a recursive nonsingular terminal sliding mode controller (RNTSMC) surface is devised for each subsystem to enhance the error convergence rate and attenuate chattering effects. Following this design, the HRNTSMC surface is constructed by the linear combination of the RNTSMC surfaces. Ultimately, the control law of the HRNTSMC is synthesized using the Lyapunov theorem to ensure that the system states converge to zero within a finite time. By invoking disturbances estimation, a linear extended state observer (LESO) is developed for the IP system. To validate the effectiveness, simulation results, including comparison with a conventional hierarchical sliding mode control (CHSMC) and a hierarchical nonsingular terminal sliding mode control (HNTSMC) are presented. These results clearly showcase the excellent performance of this approach, which is characterized by its strong robustness, fast convergence, high tracking accuracy, and reduced chattering in control input.

**Keywords:** adaptive sliding mode control; uncertainties; inverted pendulum; nonsingular terminal sliding mode; under-actuated system; hierarchical sliding mode

## 1. Introduction

The inverted pendulum (IP) system has been widely used in various industrial applications, such as balancing robots, rocket propellers, and humanoid robots [1–3], due to its cost-effectiveness and simple structure. However, the IP is an under-actuated system with fewer actuators than the degree of freedom to be controlled [4], which has strong nonlinearities, coupling properties, and is highly unstable [5]. Therefore, it provides many challenging problems with respect to linear and nonlinear control theories.

Since the 1950s, the IP has been employed as a platform for designing, evaluating, and comparing different control techniques [6]. Over the past decades, many controllers have been proposed for the IP to balance the pendulum at desired positions. One of the most straightforward controllers, a proportional–integral–derivative (PID) [7] controller, operates based on error values, which represent the difference between the actual values of the process and the desired ones. Thus, Ghosh [8] proposed the PID controllers to balance the IP at the desired cart position. This approach involved the use of two controllers: the first one for regulating the angular pendulum and the other for controlling the cart position. The robustness and performance of the PID controller were verified by simulations and practical experiments. However, a poorly tuned PID controller can drastically reduce system performance and lead to instability [9]. Thus, Wang [10] proposed a linear quadratic regulator (LQR) controller for the IP, with the response time and overshoot depending directly on the selection of the matrices **Q** and **R**. Although PID and LQR are known for their simple structures and easy implementation, they usually fail to effectively reject disturbances [11] such as system parameter uncertainties, friction forces, unmodeled dynamics, vibration, wind forces, and unpredictable collisions. In practice, these disturbances are present in most systems and can significantly degrade the control performance. To address this issue, numerous nonlinear control methods have been developed, such as fuzzy design [12], neural networks [13], and sliding mode control (SMC) [14]. Their performance outcomess have been investigated for the IP system in Ref. [15]. The authors of that study noted that none of these single controllers could meet the best comparative criteria, including simplicity, fast response, precision, disturbances rejection, adaptability, robustness, tracking capabilities, and reduced chattering. Among these approaches, the SMC has been considered to be an effective controller for the IP due to its ability to provide fast responses, precision, disturbances rejection, adaptability, and tracking capabilities. Conventional sliding mode control (CSMC), which was first designed by Utkin [16], is based on high-speed switching control laws to drive the system trajectory to a selected sliding surface [17]. Once the system states reach the selected sliding surface, the system's response becomes insensitive to disturbances and parametric uncertainties. Thanks to this property, the CSMC has been employed in many applications such as missiles [18], piezos [19], and grippers [20]. Nevertheless, the CSMC is not without its limitations, including slow response and chattering. Achieving fast convergence with the CSMC often requires high control input, which can drive actuators to saturation. In response to these limitations, the terminal sliding mode control [21] (TSMC) was developed to attain finite convergence without using high-control input. Compared to the CSMC, the TSMC has gained popularity in robust control because of its faster time convergence and diminished steady-state errors. Nonetheless, the TSMC suffers two major drawbacks: firstly, it exhibits slower performance than the CSMC when system states are far from equilibrium points $e(t) = 0$; secondly, it faces a singularity problem in its control law. To address this issue, Yu [22] proposed the fast terminal sliding mode control (FTSMC), which can increase the error convergence while conquering the first drawback of the TSMC with a faster response. Despite these improvements, the control design still brings a singularity problem. Therefore, Feng [23] introduced the nonsingular terminal sliding mode control (NTSMC) to increase the convergence speed and avoid the singularity problem. Subsequently, Shao [24] combined the NTSMC with the integral sliding mode to enhance tracking performance outcomes and reduce steady-state errors in the IP. Besides the convergence speed problems, an inherent challenge in the SMC is the presence of chattering caused by the signum function. When the chattering occurs, the control input oscillates at a high frequency, thereby reducing the life of the actuators, degenerating the control accuracy, causing high wear and tear in mechanical parts, and generating high power losses in electrical circuits [25]. A possible method to reduce the level of chattering is using the smooth function instead of the signum function [26,27]. However, it compromises the control's smoothness and overall system performance. Another feasible solution is the integration of a disturbance observer in the control design. Thanks to its ability to rapidly reject disturbances, disturbance observer-based control design has been widely used as an effective method to compensate for the disturbances and uncertainties stemming from both the environment and system [28]. Therefore, combining the SMC with the disturbance observers is a promising approach because the SMC only needs to address the disturbance estimation errors [29,30]. A comprehensive review of these disturbance observers [31] revealed that the linear extended state observer (LESO) is the most commonly used technique for disturbance estimation without requiring accurate modeling [32]. Moreover, the simplicity of the design process and parameter adjustments make the LESO easily implementable in engineering applications such as motor drive systems [33] and exoskeletons [34].

In previous studies, the SMCs were typically investigated and demonstrated for the fully actuated system, thus making them useless for direct applications in under-actuated systems [35]. This challenge arises from the inability to directly calculate control parameters for a sliding mode surface using Hurwitz conditions [4]. To deal with this limitation, Utkin [36] transformed the linearized equations of the system to regular forms and then employed them for SMC design. However, when the initial angular pendulum is far from an unstable equilibrium point, these control approaches may yield unsatisfactory results [37]. Another approach for designing the SMC without using a linear model for under-actuated systems is to use the SMC in a hierarchical structure, which not only retains the advantages of the SMC, but also has the ability to simultaneously control different outputs [38]. This method involves decoupling the under-actuated system into subsystems. For each subsystem, a sliding mode surface is meticulously designed based on the corresponding state system, thereby forming the initial level of a hierarchical structure. Then, the second level of the hierachical strucutere is constructed based on the first level. Finally, all the control laws of the different subsystems are interconnected to form the ultimate control law. By implementing this method, the high-dimensional system is effectively reduced to lower-dimensional subsystems, thereby reducing the complexity of the controller design. The conventional hierarchical sliding mode control (CHSMC) [39–42] was proposed for underactuated systems by using the CSMC in a hierarchical form. These results proved the effectiveness of the CHSMC controller for tracking the cart position while balancing the pendulum. However, using the CSMC to design the CHSMC leads to slow error convergences. Although adjusting controller coefficients can increase the error convergence speed, the errors of the sliding surface can not converge to equilibrium in a finite time. In response to this challenge, Refs. [43,44] designed the hierarchical nonsingular terminal sliding mode control (HNTSMC) for the IP to increase the error convergence speed. Nevertheless, these control laws contain signum structures that are premultiplied by the bound of the lumped disturbances, which means that chattering is not effectively reduced [45].

As aforementioned, both the CHSMC and the HNTSMC suffer from chattering problems in the control input and lack results related to the disturbance observer for the IP system. To the best of our knowledge, there is no research in the literature regarding the development of a hierarchical recursive nonsingular terminal sliding mode controller (HRNTSMC) for the IP. Compared to existing results, the main advantages of the proposed controller can be outlined as follows: (1) A new HRNTSMC has been introduced to the IP for reducing chattering phenomena in the control input, thereby increasing the error convergence speed and saving energy consumption. The RNTSMC has been developed for each subsystem, including the integration of the NTSMC to enhance the convergence and tracking precision of the subsystem states [46]. Compared to the CHSMC and HNTSMC, the proposed controller demonstrated better performance when the IP suffered from more lumped disturbances and high initial errors of angular pendulum; (2) The disturbances of the IP were directly estimated by using the LESO, that is, the proposed controller compensated for these disturbances with more accuracy and efficiency; (3) The reaching law of the proposed controller compensated the estimation errors of the LESO results with respect to chattering reduction in the control input; (4) The overall stability of the control system was analyzed using the Lyapunov theorem. The proposed controller can extend to a general under-actuated system with input coupling.

The remainder of the article is organized as follows: Section 2 formulates the problem associated with the IP. Section 3 provides a comprehensive description of the proposed controller and system stability. Section 4 presents the results of numerical simulations for the IP. Finally, Section 5 draws some conclusions and outlines potential directions for future works.

## 2. Problem Formulation

The IP in Figure 1 is considered to be the most exemplary laboratory experiment to perform nonlinear control techniques. It comprises a cart moving along a guiding rail and an aluminium rod mounted on a cart, thereby allowing the rod to freely swing within a vertical plane. The cart is moved by a DC motor connected via a belt. The dynamic model of the IP (Figure 1) is derived using the Lagrange equation according to the following set of equations [47]:

$$\dot{x}_1(t) = x_2(t), \tag{1}$$

$$\dot{x}_2(t) = f_1(\mathbf{x}, t) + b_1(\mathbf{x}, t)u(t) + n_1(t), \tag{2}$$

$$\dot{x}_3(t) = x_4(t), \tag{3}$$

$$\dot{x}_4(t) = f_2(\mathbf{x}, t) + b_2(\mathbf{x}, t)u(t) + n_2(t), \tag{4}$$

where $f_1(\mathbf{x}, t), b_1(\mathbf{x}, t), f_2(\mathbf{x}, t), b_2(\mathbf{x}, t)$ are given as

$$f_1(\mathbf{x}, t) = \frac{m_t g sin(x_1(t)) - m_p L sin(x_1(t)) cos(x_1(t)) x_2(t)^2}{L(4/3m_t - m_p cos^2(x_1(t)))}, \tag{5}$$

$$b_1(\mathbf{x}, t) = \frac{cos(x_1(t))}{L(4/3m_t - m_p cos^2(x_1(t)))}, \tag{6}$$

$$f_2(\mathbf{x}, t) = \frac{-4/3 m_p L x(t)_2^2 sin(x_1(t)) + m_p g sin(x_1(t)) cos(x_1(t))}{4/3m_t - m_p cos^2(x_1(t))}, \tag{7}$$

$$b_2(\mathbf{x}, t) = \frac{4}{3(4/3m_t - m_p cos^2(x_1(t)))}, \tag{8}$$

where the system state vector $\mathbf{x} = [x_1, x_2, x_3, x_4]^T$, $x_1(t)$ is the angular position of the pendulum from the vertical axis; $x_2(t)$ is the angular velocity of the pendulum; $x_3(t)$ is the position of the cart; $x_4(t)$ is the velocity of the cart; $m_t$ is the total mass (in kg) of the cart mass $m_c$ and pendulum mass $m_p$; $L$ is the half-length of the pendulum (in m); $g$ is the acceleration of gravity (in m/s$^2$); $u(t)$ is the force applied to the cart (N); and $n_1(t)$ and $n_2(t)$ are lumped disturbances, which include the uncertainties and external disturbances of the system.

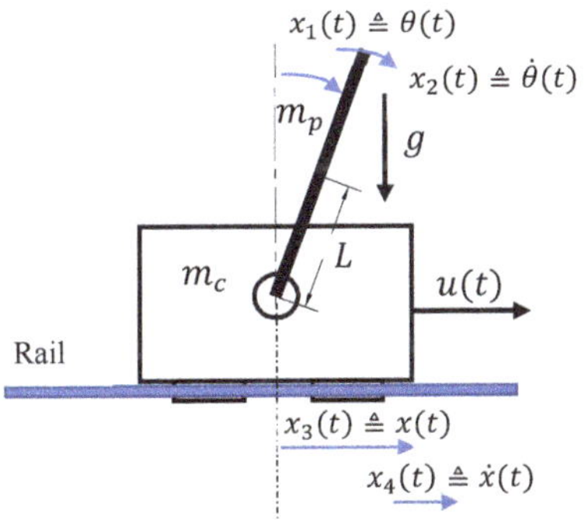


**Figure 1.** The mechanical model of the IP.

**Assumption 1.** *The unknown lumped disturbances of the system and its first time derivatives are bounded [48], with the specific bounds being unknown. This assumption represents all of the continuous and bounded disturbances, including but not limited to constant disturbances, harmonic disturbances, or neutral stable disturbances, thereby representing various disturbances in engineering applications.*

**Remark 1.** *When deriving a mathematical model, it is necessary to take into account the accuracy of the system model. However, a highly accurate model may comprise several terms, which can be cumbersome to determine. Therefore, in this work, the authors did not consider friction forces, including viscous, coulomb, and static (dry) friction forces, which can make the IP model more complex. Instead, these forces were assumed as parts of the lumped disturbance.*

**Remark 2.** *The actuator in the IP, consisting of a DC motor with a gearbox, is regarded as a subsystem of the IP model. While this dynamic can be neglected by comparing it with the dynamics of the IP, it is important to recognize that the actuator's dynamic can significantly affect the overall performance of the control system. This problem can be solved by implementing an internal control system, such as an open-loop proportional regulator or a PID controller, to control the DC motor [49].*

**Remark 3.** *Based on the system model (1)–(4), the same control input $u(t)$ appears in two dynamic equations. Therefore, the whole system is under-actuated (single input multioutput).*

**Definition 1.** *The detailed expressions of the notation $sig(e)^\alpha$ can be found in [50], which can be expressed as follows:*

$$sig(e)^\alpha = |e|^\alpha sgn(e), \tag{9}$$

*where $\alpha > 0 \forall e \in \mathbb{R}$. Note that the function $sig(e)^\alpha$ is smooth and monotonically increasing.*

The objective of this work is to propose a new hierarchical sliding mode controller that can navigate the cart to desired positions while balancing the pendulum on its upright position with minimal deflection in the presence of disturbances and uncertainties.

## 3. Control Strategy and Stability Analysis

The proposed controller (Figure 2) comprises a hierarchical structure and an LESO observer. The hierarchical structure uses the RNTSMC for the first level of the structure. The LESO is utilized to observe the lumped disturbance of each subsystem. Afterwards, the reaching law is applied to compensate for the approximation errors of the LESO.

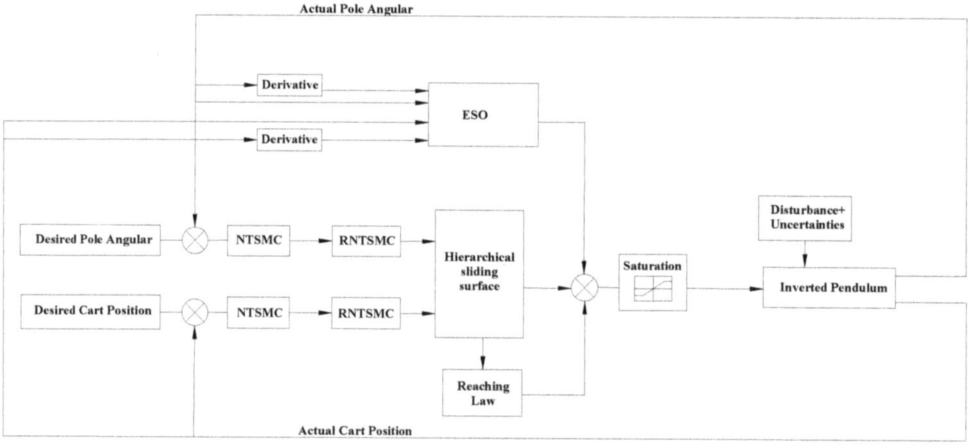


**Figure 2.** The structure of proposed control.

### 3.1. Design Hierarchical Recursive Nonsingular Terminal Sliding Mode Control

According to the dynamic model of the IP, the hierarchical structure in Figure 3 is employed to design controllers for the two subsystems [51]. The first subsystem is the pendulum, which has state variables $[x_1(t), x_2(t)]$, and the second is the cart, which includes state variables $[x_3(t), x_4(t)]$. The first layer of the hierarchical structure comprises the sliding mode surface of each subsystem: the subsystem sliding surface $\sigma_1(t)$ and the subsystem sliding surface $\sigma_2(t)$. The second layer of the hierarchical structure comprises sliding mode surfaces in the first layer. The total control law is synthesized from the hierarchical structure by employing the Lyapunov theory, thereby ensuring the stability of each sliding surface within the subsystem.

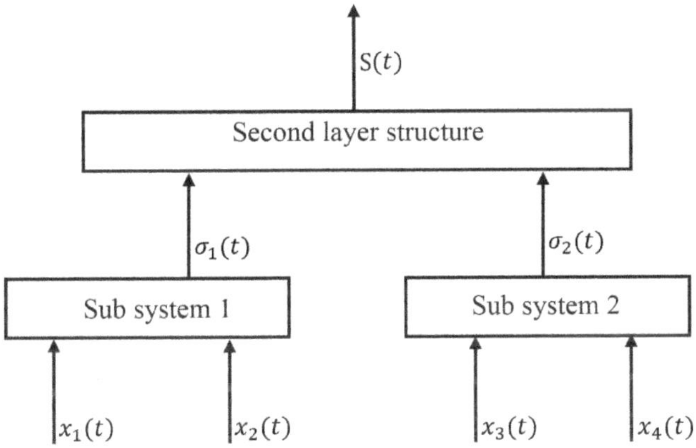


**Figure 3.** Hierarchical structure is given by [51].

The error of the angular pendulum is defined as follows:

$$e_1(t) \triangleq x_1(t) - x_{1d}(t), \tag{10}$$

where $x_1(t)$ is the actual angle of the pendulum, and $x_{1d}(t)$ is the desired angle of the pendulum. In addition, we define the error of the cart position as follows:

$$e_3(t) \triangleq x_3(t) - x_{3d}(t), \tag{11}$$

where $x_3(t)$ is the actual position of the cart, and $x_{3d}(t)$ is the desired position of the cart.

The NTSMC surface was presented by [24], who adapted for the first subsystem (1) and (2) as follows:

$$s_1(t) \triangleq \dot{e}_1(t) + k_1 e_1(t) + \gamma_1 sig(e_1(t))^{r_1}, \tag{12}$$

where $r_1 > 1; \gamma_1 > 0$; and $k_1 > 0$ are the control parameters to be designed. Then, the RNTSMC surface $\sigma_1(t)$ for the first subsystem is defined as follows:

$$\sigma_1(t) \triangleq s_1(t) + \gamma_2 \int_0^t sig(s_1(t))^{r_2}, \tag{13}$$

where the control parameters are $1 > r_2 > 0$ and $\gamma_2 > 0$ [52], and the variable $s_1(t)$ is defined in (12).

The first time derivative of (13) becomes the following:

$$\dot{\sigma}_1(t) = \dot{s}_1(t) + \gamma_2 sig(s_1(t))^{r_2}, \tag{14}$$

where the time derivative of $s_1(t)$ is defined as follows:

$$\dot{s}_1(t) = f_1(\mathbf{x}, t) + b_1(\mathbf{x}, t)u(t) + n_1(t) - \ddot{x}_{1d}(t) + k_1\dot{e}_1(t) + \gamma_1 r_1 |e_1(t)|^{r_1 - 1}\dot{e}_1(t). \tag{15}$$

The NTSMC surface for the second subsystem (3) and (4) is described as follows:

$$s_2(t) \triangleq \dot{e}_3(t) + k_2 e_3(t) + \gamma_3 sig(e_3(t))^{r_3}, \tag{16}$$

where the design parameters are set as $r_3 > 1; \gamma_3 > 0$; and $k_2 > 0$.

We introduce the RNTSMC surface $\sigma_2(t)$ for the second subsystem as follows:

$$\sigma_2(t) \triangleq s_2(t) + \gamma_4 \int_0^t sig(s_2(t))^{r_4}, \tag{17}$$

where $1 > r_4 > 0$ and $\gamma_4 > 0$ are the design parameters.

Differentiating (17) with respect to time yields the following:

$$\dot{\sigma}_2(t) = \dot{s}_2(t) + \gamma_4 sig(s_2(t))^{r_4}, \tag{18}$$

where

$$\dot{s}_2(t) = f_2(\mathbf{x},t) + b_2(\mathbf{x},t)u(t) + n_2(t) - \ddot{x}_{3d}(t) + k_2\dot{e}_3(t) + \gamma_3 r_2|e_3(t)|^{r_2-1}\dot{e}_3(t). \tag{19}$$

The second-level hierarchical structure based on the first level is defined as follows:

$$S(t) \triangleq \alpha\sigma_1(t) + \sigma_2(t), \tag{20}$$

where $\alpha$ is a positive or negative design constant, and $\sigma_1(t)$ and $\sigma_2(t)$ are defined, respectively, in (13) and (17).

Differentiating $S(t)$ with respect to time yields the following:

$$\dot{S}(t) = \alpha\dot{\sigma}_1(t) + \dot{\sigma}_2(t). \tag{21}$$

Based on (14) and (18), let $\dot{\sigma}_1(t) = 0$ and $\dot{\sigma}_2(t) = 0$; the equivalent control laws $u_{1eq}(t)$ and $u_{2eq}(t)$ can be obtained for the corresponding subsystems as follows:

$$\begin{aligned} u_{1eq}(t) = -1/b_1(\mathbf{x},t)(f_1(\mathbf{x},t) + n_1(t) - \ddot{x}_{1d}(t) + k_1\dot{e}_1(t) + \gamma_1 r_1|e_1(t)|^{r_1-1}\dot{e}_1(t) \\ +\gamma_2 sig(s_1(t))^{r_2}). \end{aligned} \tag{22}$$

$$\begin{aligned} u_{2eq}(t) = -1/b_2(\mathbf{x},t)(f_2(\mathbf{x},t) + n_2(t) - \ddot{x}_{3d}(t) + k_2\dot{e}_3(t) + \gamma_3 r_3|e_1(t)|^{r_3-1}\dot{e}_1(t) \\ +\gamma_4 sig(s_2(t))^{r_4}). \end{aligned} \tag{23}$$

A constant plus the proportion reaching law [53] $\dot{S}(t) = -\kappa_1 S(t) - \kappa_2 sign(S(t))$ are employed to design this controller, where $\kappa_1 > 0$ is the switching gain; $\kappa_2 > 0$ is the reaching control gain; and $sign(S(t))$ is the signum function defined as the form:

$$sign(S(t)) = \begin{cases} 1 & \text{if } S(t) > 0, \\ 0 & \text{if } S(t) = 0, \\ -1 & \text{if } S(t) < 0. \end{cases}$$

Based on [54,55], the total control law is determined as follows:

$$u(t) \triangleq \frac{\alpha b_1(\mathbf{x},t)u_{1eq}(t) + b_2(\mathbf{x},t)u_{2eq}(t) - \kappa_1 S(t) - \kappa_2 sign(S(t))}{\alpha b_1(\mathbf{x},t) + b_2(\mathbf{x},t)}. \tag{24}$$

$$u_{eq}(t) \triangleq \frac{\alpha b_1(\mathbf{x},t)u_{1eq}(t) + b_2(\mathbf{x},t)u_{2eq}(t)}{\alpha b_1(\mathbf{x},t) + b_2(\mathbf{x},t)}. \tag{25}$$

$$u_{sw}(t) \triangleq \frac{-\kappa_1 S(t) - \kappa_2 sign(S(t))}{\alpha b_1(\mathbf{x},t) + b_2(\mathbf{x},t)}. \tag{26}$$

The proposed control law $u(t)$ from (24) consists of $u_{eq}(t)$ and $u_{sw}(t)$, which are mentioned later.

To demonstrate the stability analysis of the second-level hierarchical structure, we choose a Lyapunov function candidate as follows:

$$V(t) \triangleq \frac{1}{2}S(t)^2. \tag{27}$$

The time derivative of the Lyapunov function is thus obtained from the following:

$$\dot{V}(t) = S(t)\dot{S}(t) = S(t)(\alpha\dot{\sigma_1}(t) + \dot{\sigma_2}(t)). \tag{28}$$

$$\begin{aligned} \dot{V}(t) = S(t)\dot{S}(t) = S(t)(\alpha(f_1(\mathbf{x},t) + b_1(\mathbf{x},t)u(t) + n_1(t) - \ddot{x}_{1d}(t) \\ +k_1\dot{e}_1(t) + \gamma_1 r_1|e_1(t)|^{r_1-1}\dot{e}_1(t) + \gamma_2 sig(s_1(t))^{r_2}) \\ +(f_2(\mathbf{x},t) + b_2(\mathbf{x},t)u(t) + n_2(t) - \ddot{x}_{3d}(t) \\ +k_2\dot{e}_3(t) + \gamma_3 r_2|e_3(t)|^{r_2-1}\dot{e}_3(t) + \gamma_4 sig(s_2(t))^{r_4})). \end{aligned} \tag{29}$$

We substitute (24) into (29) to yield the following:

$$\dot{V}(t) = S(t)(-\kappa_1 S(t) - \kappa_2 sign(S(t))). \tag{30}$$

$$\dot{V}(t) = -\kappa_1 S(t)^2 - \kappa_2|S(t)| < 0. \tag{31}$$

Therefore, $S(t)$ will converge to zero in finite time [56]. $\sigma_1(t)$ and $\sigma_2(t)$ will converge asymptotically to zero [4]. Then, the errors $e_1(t)$ and $e_3(t)$ converge to zero in finite time [24].

### 3.2. Design of Extended State Observer

The equivalent control $u_{1eq}(t)$ and $u_{2eq}(t)$ involves the knowledge of the lumped disturbances $n_1(t)$ and $n_2(t)$, which can be measured using intrinsic sensors. To estimate the knowledge of the lumped disturbances, the LESO [31] is employed for the IP system. The LESO, designed to estimate $n_1(t)$, is expressed as follows:

$$\varepsilon_1(t) \triangleq x_1(t) - \hat{x}_1(t), \tag{32}$$

$$\hat{\dot{x}}_1(t) \triangleq \hat{x}_2(t) + \frac{a_1\varepsilon_1}{\xi_1}, \tag{33}$$

$$\hat{\dot{x}}_2(t) \triangleq f_1(\mathbf{x},t) + b_1(\mathbf{x},t)u(t) + \hat{n}_1(t) + \frac{a_2\varepsilon_1(t)}{\xi_1^2}, \tag{34}$$

$$\hat{\dot{n}}_1(t) \triangleq \frac{a_3\varepsilon_1(t)}{\xi_1^3}, \tag{35}$$

where $\xi_1 > 0; a_1, a_2$, and $a_3$ are positive constants, and the polynomial $s^3 + a_1 s^2 + a_2 s + a_3$ is Hurwitz; $\hat{x}_1(t), \hat{x}_2(t)$, and $\hat{n}_1(t)$ are the observed values of $x_1(t), x_2(t)$, and $n_1(t)$, respectively.

**Proof.** The proof is detailed in Appendix A. □

**Remark 4.** *The selection of the LESO parameters has a significant effect on the observer's stability, as has been thoroughly described in [57]. The estimation states from (32)–(34) depend on the error between the measured and the estimated angular pendulum. However, when the position error $e_1(t)$ becomes small, it is necessary to select the high gains in (33)–(35) to achieve a reliable estimation of the lump disturbances. Therefore, to achieve an accurate estimation of the lump disturbances, it is advisable to select larger values for the parameters $\frac{a_1}{\xi_1}$, $\frac{a_2}{\xi_1}$, and $\frac{a_3}{\xi_1}$. In practical implementation, the initial observation errors are usually nonzero due to the difference between the initial value of the LESO and the plant. Since the system employs a small $\xi_1$, the initial peak values of the LESO* *can become relatively large. This can lead to a significant demand for control force, which, in turn, potentially damages the motor, degrades the overall control performance, and negatively affects the observer's convergence. To alleviate the peaking phenomenon, $\xi_1$ is selected as follows [58,59]:*

$$\frac{1}{\xi_1} = R = \begin{cases} 10t^3, 0 \le t \le 1, \\ 10, t > 1. \end{cases} \tag{36}$$

*The LESO, designed to estimate $n_2(t)$, is expressed as follows:*

$$\varepsilon_2(t) \triangleq x_3(t) - \hat{x}_3(t), \tag{37}$$

$$\dot{\hat{x}}_3(t) \triangleq \hat{x}_4(t) + \frac{a_4\varepsilon_2(t)}{\xi_2}, \tag{38}$$

$$\dot{\hat{x}}_4(t) \triangleq f_2(x,t) + b_2(x,t)u(t) + \hat{n}_2(t) + \frac{a_5\varepsilon_2(t)}{\xi_2^2}, \tag{39}$$

$$\dot{\hat{n}}_2(t) \triangleq \frac{a_6\varepsilon_2(t)}{\xi_2^3}, \tag{40}$$

*where $a_4, a_5$, and $a_6$ are positive constants, and the polynomial $s^3 + a_4 s^2 + a_5 s + a_6$ is Hurwitz; $\hat{x}_3, \hat{x}_4$, and $\hat{n}_2$ are the observed values of $x_1, x_2$, and $n_2$, respectively; $\xi_2$ is selected based on (36).*

**Proof.** The proof is similar to Appendix A. □

Based on (32) to (40), the lumped disturbances $n_1(t)$ and $n_2(t)$ are estimated by $\hat{n}_1(t)$ and $\hat{n}_2(t)$, respectively. We substitute $\hat{n}_1(t)$ and $\hat{n}_2(t)$ into (22) and (23), respectively; the equivalent controls $u_{1eq}(t)$ and $u_{2eq}(t)$ are then rewritten as follows:

$$\begin{aligned} u_{1eq}(t) = -1/b_1(\mathbf{x},t)(f_1(\mathbf{x},t) + \hat{n}_1(t) - \ddot{x}_{1d}(t) + k_1\dot{e}_1(t) + \gamma_1 r_1|e_1(t)|^{r_1-1}\dot{e}_1(t) \\ +\gamma_2 sig(s_1(t))^{r_2}). \end{aligned} \tag{41}$$

$$\begin{aligned} u_{2eq}(t) = -1/b_2(\mathbf{x},t)(f_2(\mathbf{x},t) + \hat{n}_2(t) - \ddot{x}_{3d}(t) + k_2\dot{e}_3(t) + \gamma_3 r_3|e_1(t)|^{r_3-1}\dot{e}_1(t) \\ +\gamma_4 sig(s_2(t))^{r_4}). \end{aligned} \tag{42}$$

To demonstrate the stability of the proposed controller, the Lyapunov is selected as follows:

$$V_2(t) \triangleq \frac{1}{2}S(t)^2. \tag{43}$$

Taking the time derivative of the Lyapunov function, one can obtain the following:

$$\dot{V}_2(t) = S(t)\dot{S}(t) = S(t)(\alpha\dot{\sigma}_1(t) + \dot{\sigma}_2(t)). \tag{44}$$

$$\begin{aligned} \dot{V}_2(t) = S(t)\dot{S}(t) = S(t)(\alpha(f_1(\mathbf{x},t) + b_1(\mathbf{x},t)u(t) + n_1(t) - \ddot{x}_{1d} \\ +k_1\dot{e}_1(t) + \gamma_1 r_1|e_1(t)|^{r_1-1}\dot{e}_1(t) + \gamma_2 sig(s_1(t))^{r_2}) \\ +(f_2(\mathbf{x},t) + b_2(\mathbf{x},t)u(t) + n_2(t) - \ddot{x}_{3d}(t) \\ +k_2\dot{e}_3(t) + \gamma_3 r_2|e_3(t)|^{r_2-1}\dot{e}_3(t) + \gamma_4 sig(s_2(t))^{r_4})). \end{aligned} \tag{45}$$

We denote the estimation errors of $n_1(t)$ and $n_2(t)$ as follows:

$$\tilde{n}_1(t) \triangleq \hat{n}_1(t) - n_1(t), \tag{46}$$

$$\tilde{n}_2(t) \triangleq \hat{n}_2(t) - n_2(t). \tag{47}$$

We substitute (24) with the new equivalent control laws (41) and (42) into (45), which yields the following:

$$\dot{V}_2(t) = S(t)(-\kappa_1 S(t) - \kappa_2 sign(S(t)) + \alpha\tilde{n}_1(t) + \tilde{n}_2(t)). \tag{48}$$

$$\dot{V}_2(t) = -\kappa_1 S(t)^2 - \kappa_2|S(t)| + S(t)(\alpha\tilde{n}_1(t) + \tilde{n}_2(t)). \tag{49}$$

$$\dot{V}_2(t) < -\kappa_1 S(t)^2 - \kappa_2|S(t)| + |S(t)|(|\alpha\tilde{n}_1(t)| + |\tilde{n}_2(t)|). \tag{50}$$

$\kappa_2$ is selected as follows:

$$\kappa_2 > |\alpha\tilde{n}_1(t)| + |\tilde{n}_2(t)| + \kappa_3, \tag{51}$$

where $\kappa_3$ is a small positive number. Substituting (51) into (50) yields the following:

$$\dot{V}_2(t) < -\kappa_1 S(t)^2 - \kappa_3|S(t)| < 0. \tag{52}$$

$$\dot{V}_2(t) < -2\kappa_1 V_2(t) - \kappa_3\sqrt{2}V_2(t)^{1/2}. \tag{53}$$

Therefore, $S(t)$ will converge to zero in finite time:

$$T \leq \frac{1}{\kappa_1} ln \frac{2\kappa_1 V_2^{1/2}(0) + \sqrt{2}\kappa_3}{\sqrt{2}\kappa_3}. \tag{54}$$

Then, $\sigma_1(t)$ and $\sigma_2(t)$ will converge asymptotically to zero [4], and the errors $e_1(t)$ and $e_3(t)$ converge to zero in finite time [24].

**Remark 5.** *The proposed control law $u(t)$ of (24) consists of two main components: $u_{eq}(t)$ and $u_{sw}(t)$. While $u_{eq}(t)$ is responsible for compensating the nominal parts, $u_{sw}(t)$ is designed to compensate for the impact of the estimation errors stemming from the LESO. The RNTSMC is designed for each subsystem, thus inheriting the merits of both the NTSMC and the high-order sliding mode control (HOSMC) [60] and ensuring finite time convergence and attenuated chattering.*

**Remark 6.** *The chattering of the proposed controller is significantly reduced by using the LESO, as it solely addresses the estimation errors of the lumped disturbances through the switching function (26) [61]. Despite its advantages, in practical scenarios, it is challenging to achieve a complete reduction of s to zero due to some factors such as noise, delay, and imperfection of the devices [62]. Consequently, the $sign(s)$ function in the control input is often substituted by the saturation to reduce the chattering effect. The sat function is given as follows [63]:*

$$sat(s) \triangleq \begin{cases} 1, s > \Delta, \\ ks, |s| \leq \Delta, k = \frac{1}{\Delta}, \\ -1, s < -\Delta. \end{cases} \tag{55}$$

**Remark 7.** *In practical application, the control input (24) cannot be applied directly to the DC motor. Instead, some experiments are conducted to determine the relationship between the pulse-width-modulated (PWM) input duty and the rotation speed of the motor. Subsequently, the relationship between the rotation speed and the force output of the motor is determined by using catalog information and a wheel radius. Afterward, the required PWM duty can be calculated and provided to the DC motor to achieve the desired control input (24).*

### 3.3. Parameters Selection

The selection of the control parameters is of great importance for practitioners in practical implementation. The control performance can be compromised by factors such as control saturation, measurement noise, and chattering in the control signal. When choosing parameters, it is essential to adhere to two fundamental rules: 1. Ensure that the conditions (12), (13), (16), (17), and (26) are satisfied in the designed controller. 2. Strive to keep the parameter values as small as possible to reduce the required force of the actuator.

Selecting the parameters of the sliding surfaces can be referred in Ref. [24]. In addition, the LESO parameters should align with the recommendation presented in Ref. [64]:

Selection of $k_1, \gamma_1$, and $r_1$: The parameters $k_1, \gamma_1$, and $r_1$ have a direct effect on the dynamic behaviours of the sliding surface (12). An increased value of these parameters leads to faster convergence of $e_1(t)$, but this comes at the expense of higher demand for the control input.

Selection of $\gamma_2$ and $r_2$: A larger $\gamma_2$ or $r_2$ will increase the convergent speed of the error $e_1(t)$ and reduce the steady-state tracking errors but at the cost of an increased control input requirement.

Selection of $\kappa_1$ and $\kappa_2$: These control parameters affect the robustness of the controller in (26). Higher values of these parameters improve the robustness but may lead to a less smooth control signal.

Selection of $a_1, a_2$, and $a_3$ : These parameters can be determined using Equation (A19), which solely requires the value of $\omega$. A higher value of $\omega$ improves the accuracy of the disturbances estimation. However, it is worth noticing that if the measurement signal contains noise, substantial observer errors may arise when dealing with large $\omega$ values.

**Remark 8.** *The HRNTSMC surface can be considered as a general form encompassing both the CHSMC surface and HNTSMC surface. When $k_1 = 0, r_1 = 1$, and $\gamma_2 = 0$, the HRNTSMC surface is the same as the CHSMC surface. When $k_1 = 0, 2 > r_1 > 1$, and $\gamma_2 = 0$, the HRNTSMC surface exhibits similar behavior to the HNTSMC surface. The control parameters for the simulation results were tuned as follows: Initially, all the control parameters of the proposed controller were set to zero values. Subsequently, the values of $\gamma_1$ and $\kappa_2$ were adjusted, and $r_1$ was fixed at one, which made the proposed controller similar to the CHSMC [65]. Following that, we gradually increased the control parameters $\gamma_1$ and $\kappa_2$ to reach the optimal performance of the CHSMC, with an emphasis on minimizing the values of $\kappa_2$ to reduce chattering in the control input and maximizing $\gamma_1$ to increase the convergence speed of the errors. Afterwards, we selected $r_1$ in the range (1–2) while adjusting $\gamma_1$ to improve the convergence errors of the proposed controller. Following most related works of the NTSMC, $r_1$ was fixed at 5/3. To enhance the reaching phase speed of the proposed control, the values of $\kappa_1$ were increased, while those of $\kappa_2$ were adjusted. It is important to note that the values of $\kappa_1$ should be large to increase the reaching speed, whereas the values of $\kappa_2$ should be kept small to reduce the chattering. At this point, the proposed controller shares the same structure with the HNTSMC [43], which gains a fast convergence speed. However, since the control input suffers from more chattering, and there is a need for improved convergence speed, the remaining control parameters should be tuned to obtain the merit properties for the proposed controller. To accomplish this, the value $\xi_1$ of the LESO was set according to (36), and the values of $\omega$ in (A19) were selected to determine $a_1, a_2$, and $a_3$ for the LESO. Adjusting the $\omega$ values is essential to achieve satisfactory disturbance estimations while simultaneously reducing the values of $\kappa_2$ as much as possible. Additionally, gradually adjusting $\gamma_2$, $r_2$, and $k_1$ from small to large values in the proposed controller contributes to an improvement in both the convergence speed and accuracy.*

**Remark 9.** *The versatility of the proposed controller allows for its widespread applicability in other under-actuated systems such as acrobot, pendubot, magnetic suspension, beam-and-ball, and the TORA system. However, the proposed controller has more parameters for tuning to gain the optimal performance, which could be the limitation of this approach. Therefore, future work endeavors are expected to employ an optimization algorithm to identify the most suitable control parameters. In this work, the LESO has been successfully employed to estimate the lumped disturbances, which were assumed to be constant or low-frequency disturbances. However, a limitation in performance will arise when dealing with high-frequency disturbances. Therefore, a generalized integrator extended state observer [66] will be taken into consideration to estimate both slow and rapid disturbances in future work.*

### 3.4. Detailed Step for Designing Proposed Controller

The design procedures are summarized as follows:
Step 1: Initialize the control parameters and system states.
Step 2: Design sliding surfaces for two subsystems (13) and (17).
Step 3: Design the observers to determine $\hat{n}_1$ as in (32)–(35) and $\hat{n}_2$ as in (37)–(40).
Step 4: Determine the equivalent control laws in (41) and (42).
Step 5: Obtain the control input for the motor (24).

## 4. Simulation Verification

This section presents several simulation results to demonstrate the effectiveness of the proposed controller with respect to tracking errors, energy consumption, and chattering reduction. A comparative analysis was conducted between the proposed controller HRNTSMC and two existing controllers: the HNTSMC in [43] and the CHSMC in [65]. The CHSMC, which is characterized by its simplicity and feasibility, has been widely designed for under-actuated systems. Meanwhile, the HNTSMC not only inherits the characteristics of the CHSMC, but also ensures the faster convergence rate of system states to zero in finite time. The control systems were simulated in the MATLAB SIMULINK environment using the ODE 45 solver with a fixed step size of 0.001 s.

### 4.1. Simulation Condition

In the IP system, the system parameters were chosen as g = 9.8 m/s$^2$; $m_t$ = 0.8 kg; $m_p$ = 0.188 kg; and $L$ = 0.34 m. In the first and second case, the initial conditions of the IP were set as follows: $[x_1\ x_2\ x_3\ x_4] = [\frac{-\pi}{15}\ 0\ 0\ 0]$; the initial conditions in the third case of the IP were set as $[x_1\ x_2\ x_3\ x_4] = [\frac{-\pi}{3}\ 0\ 0\ 0]$. The third case was designed to perform the advantage of the HRNTSMC over the other linear controllers, thus showcasing its ability to balance the IP even when starting from a large initial error of angular pendulum. The desired angle and position were obtained from [67,68] and were respectively defined as follows: $x_{1d} = 0$ rad and $x_{3d} = 0$ m.

The lumped disturbances of the pendulum and cart were assumed as follows [68]:

$$n_1(t) = 0.0873sin(t) + 0.5sin(x_1(t)); \tag{56}$$

$$n_2(t) = 0.0873sin(t) + 0.5sin(x_3(t)); \tag{57}$$

### 4.2. Simulation Results

It is worth noticing that the CHSMC, HNTSMC, and HRNTSMC use the same system parameters, initial conditions, and other common parameters [69]. The magnitude force was restricted to be less than 30 N [70]. The corresponding parameters of the different control systems were set with the same values [71,72]. The parameters of the controllers used in the first and second case are given in Table 1, while Table 2 shows the control parameters for the third case.

**Table 1.** Parameter settings of each controller for the first case and the second case.

| Controller | Tuning Parameters |
|---|---|
| CHSMC | $\gamma_1 = 3; \gamma_3 = \gamma_1; \alpha = -2; \kappa_2 = 0.5$ |
| HNTSMC | $\gamma_1 = 3; r_1 = 5/3; \gamma_3 = \gamma_1; r_3 = r_1; \alpha = -2; \kappa_1 = 2; \kappa_2 = 1$ |
| HRNTSMC | $k_1 = 2; \gamma_1 = 3; r_1 = 5/3; \gamma_2 = 2; r_2 = 0.7; k_2 = k_1; \gamma_3 = \gamma_1; r_3 = r_1;$<br>$\gamma_4 = \gamma_2; r_4 = r_2; \alpha = -2; \kappa_1 = 2; \kappa_2 = 0.01;$<br>$a_1 = 3; a_2 = 3; a_3 = 1; a_4 = a_1; a_5 = a_2; a_6 = a_3; \xi_2 = \xi_1$ |

**Table 2.** Parameter settings of each controller for the third case.

| Controller | Tuning Parameters |
|---|---|
| CHSMC | $\gamma_1 = 0.5; \gamma_3 = \gamma_1; \alpha = -2; \kappa_2 = 0.5$ |
| HNTSMC | $\gamma_1 = 0.1; r_1 = 5/3; \gamma_3 = \gamma_1; r_3 = r_1; \alpha = -2; \kappa_1 = 0.2; \kappa_2 = 1$ |
| HRNTSMC | $k_1 = 0.1; \gamma_1 = 0.1; r_1 = 5/3; \gamma_2 = 2; r_2 = 0.7; k_2 = k_1; \gamma_3 = \gamma_1; r_3 = r_1;$ $\gamma_4 = \gamma_2; r_4 = r_2; \alpha = -2; \kappa_1 = 0.2; \kappa_2 = 0.01;$ $a_1 = 3; a_2 = 3; a_3 = 1; a_4 = a_1; a_5 = a_2; a_6 = a_3; \xi_2 = \xi_1$ |

The sliding surfaces for each subsystem of the HNTSMC in [43] were defined as follows:

$$s_1(t) \triangleq e_1(t) + \gamma_1 sig(\dot{e}_1(t))^{r_1}. \tag{58}$$

$$s_2(t) \triangleq e_3(t) + \gamma_3 sig(\dot{e}_3(t))^{r_3}. \tag{59}$$

$$S(t) \triangleq \alpha s_1(t) + s_2(t). \tag{60}$$

$$\dot{S}(t) \triangleq -\kappa_1 S(t) - \kappa_2 sign(S(t)). \tag{61}$$

The sliding surfaces for each subsystem of the CHSMC [65] were defined as follows:

$$s_1(t) \triangleq e_1(t) + \gamma_1 \dot{e}_1(t). \tag{62}$$

$$s_2(t) \triangleq e_3(t) + \gamma_3 \dot{e}_3(t). \tag{63}$$

$$S(t) \triangleq \alpha s_1(t) + s_2(t). \tag{64}$$

$$\dot{S}(t) \triangleq -\kappa_2 sign(S(t)). \tag{65}$$

The CHSMC, HNTSMC, and HRNTSMC were compared in two different cases: (1) In the first case, the lumped disturbances $n_1(t)$ and $n_2(t)$ did not exist. (2) In the second case, the lumped disturbances $n_1(t)$ and $n_2(t)$ existed. (3) In the third case, besides the presence of the lumped disturbances, a substantial initial error of the angular pendulum was set, thereby emphasizing the advantages of the hierarchical sliding mode in its comparison with the other linear control methods.

Figures 4–6 show the simulation results of the three controllers in case 1. In this scenario, while the CHSMC failed to ensure the convergence of the tracking of the angular pendulum and the cart position error, the HNTSMC and HRNTSMC showcased their effectiveness in achieving this objective. In addition, the HRNTSMC demonstrated the fastest convergence speed and exhibited less oscillation near the equilibrium point, with $e_1(t) = 0$ and $e_3(t) = 0$. In the beginning, the CHSMC generated a higher cart position error $e_3(t)$ of 0.85 m, whereas the HNTSMC and HRNTSMC exhibited smaller errors, which measured 0.64 m and 0.53 m, respectively. After 40 s, both the HNTSMC and HRNTSMC performed satisfactory tracking of the angular pendulum and cart position with minimal errors. Specially, the maximum values of the steady-state tracking errors for the cart position were 0.35 m for the CHSMC, a significant improvement to 0.006 m for the HNTSMC, and an impressive reduction to just 0.001 m for the HRNTSMC. In terms of the angular pendulum errors of $e_1(t)$, the CHSMC, HNTSMC, and HRNTSMC generated the maximum steady-state errors of 0.179 rad, 0.005 rad, and 0.0005 rad, respectively. It is worth noticing that the steady-state performance outcomes of the HNTSMC and HRNTSMC significantly outperformed that of the CHSMC. Upon closer examination of the cart position and angular pendulum errors, the HRNTSMC's errors were smaller than those of the HNTSMC. Though the HNTSMC requires higher force control than the HRNTSMC (Figure 6) and experiences more chattering in the control input, the HNTSMC still converged lower than the HRNTSMC. The convergence time of tracking the cart position error under the HRNTSMC was about 25 s, while the HNTSMC demanded about 35 s, thereby demonstrating the faster convergence speed of the HRNTSMC. The HRNTSMC can effectively alleviate chattering in the control input, use lower energy, and achieve a fast convergence speed for tracking errors. The results strongly suggest that the HRNTSMC offers better tracking performance than the CHSMC and HNTSMC. For more comprehensive qualitative analysis, Table 3 provides the index values of the integrated absolute errors (IAEs), the energy of control input (ECI), and the absolute input chattering error (AICE). These indexes are defined from [73]:

$$IAE = \frac{1}{N}\sum_{k=1}^{N}|e(k)|, \tag{66}$$

$$ECI = \frac{1}{N}\sum_{k=1}^{N}|u(k)|, \tag{67}$$

$$AICE = \frac{1}{N}\sum_{k=1}^{N}|u(k+1) - u(k)|, \tag{68}$$

where $N$ is total number of samples, while $e(k)$ and $u(k)$ denote the position error and control input, respectively. The IAE index is responsible for qualifying tracking errors. A smaller IAE means fewer accumulated tracking errors, which in turn signifies better tracking performance. By contrast, a higher IAE implies a greater accumulation of tracking errors, thereby indicating a worse tracking capability. While the ECI is responsible for evaluating the required energy consumption, the AICE assesses the chattering in the control signal. Remarkably, the proposed controller gained the lowest index values for the IAE, ECI, and AECI. To be more specific, the IAE values of the tracking cart position in the CHSMC and HNTSMC were 0.2445 m and 0.0601 m, respectively, while only 0.00448 m was recorded value for the proposed control HRNTSMC. Similarly, when evaluating the IAE values for the tracking of the angular pendulum, the CHSMC recorded 0.1214 rad, the HNTSMC recorded 0.0381 rad, and the HRNTSMC recorded a mere 0.0217 rad. These results strongly prove the superiority of the proposed controller in terms of achieving the minimum IAE, which demonstrated fewer tracking errors compared to the CHSMC and HNTSMC. Additionally, the ECI of the proposed controller was 0.2156, thereby indicating its superior energy efficiency compared to the HNTSMC, with 0.5212, and the CHSMC, with 1.1972. In addition, the HRNTSMC significantly reduced the chattering in the control input, thus resulting in an ACEI value of 0.0605.

**Table 3.** Index values of control strategies in case 1.

| | IAE$x_1$ | IAE$x_3$ | ECI | AECI |
|---|---|---|---|---|
| CHSMC | 0.1214 | 0.2445 | 1.1972 | 0.3294 |
| HNTSMC | 0.0381 | 0.0601 | 0.5212 | 0.2814 |
| HRNTSMC | 0.0217 | 0.0448 | 0.2156 | 0.0605 |

Figures 7–11 depict the performance outcomes of the three controllers under the effects of lumped disturbances. In this context, due to the effects of the lumped disturbances, the tracking errors of the three controllers were higher compared to the first case. At the outset, the most significant error of the cart position $e_3(t)$ was recorded at 0.74 m for the CHSMC, 0.59 m for the HNTSMC, and 0.49 m for the HRNTSMC. Notably, the CHSMC produced the worst results among the three controllers despite the fact that it requires a higher control force. The three controllers performed different error convergences of the cart position and angular pendulum, as presented in Figures 7 and 8. Although the steady-state errors of the

HNTSMC and HRNTSMC were not precisely zero, they remained bounded and relatively small. The HRNTSMC stood out by providing the fastest tracking cart position and angular pendulum measurements compared to the other controllers, while still maintaining smooth force control. The effectiveness of the HRNTSMC can be attributed to its incorporation of the LESO, which can effectively estimate the lumped disturbances of the IP system with small observation errors, as are shown in Figures 10 and 11. These estimation errors of both the $n_1$ and $n_2$ came out to approximately 0.01 and 0.02, respectively. Table 4 illustrates the performance indexes of the three controllers, wherein the HRNTSMC clearly gained the smallest index values of the three controllers. As a result, this contributed to the exceptional performance of the proposed controller, which was characterized by superior tracking capability, lower energy consumption, and effective chattering reduction, even in the presence of lumped disturbances. Between the 50 s to 100 s interval, the CHSMC, HNTSMC, and HRNFTSMC controllers produced the IAE errors of 0.06 rad, 0.0023 rad, and 0.0023 rad, respectively, in tracking the angular pendulum, while they produced errors of 0.127 m, 0.0035 m, and 0.0061 m in cart position, respectively. Obviously, the HNTSMC slightly outperformed the HRNFTSMC due to the higher actuator force requirement (Figure 9) during the measurement of the steady-state errors. While the HNTSMC experienced high chattering, the HRNTSMC significantly reduced it by integrating the LESO to estimate the lumped disturbances. Table 2 provides the performance indexes of each controller, which clearly indicates that the HRNTSMC reached the lowest values of the IAE, ECI, and AECI when compared to the CHSMC and HNTSMC.

**Table 4.** Index values of control strategies in case 2.

| | IAE$x_1$ | IAE$x_3$ | ECI | AECI |
|---|---|---|---|---|
| CHSMC | 0.1303 | 0.2629 | 1.2877 | 0.3322 |
| HNTSMC | 0.0411 | 0.0651 | 0.5471 | 0.3117 |
| HRNTSMC | 0.0263 | 0.0555 | 0.2672 | 0.0629 |

Figures 12–14 demonstrate the performance outcomes of the three controllers under the condition of a significant initial error in the angular pendulum and the effects of lumped disturbances. In this scenario, due to the initial error of the angular pendulum, the biggest errors of the cart position were measured at 4.1 m for the CHSMC, 5 m for the HNTSMC, and 4.3 m for the HRNTSMC. Generally, despite requiring substantial control forces, the CHSMC failed to ensure the convergence errors of the angular pendulum and cart position. The tracking performance of the HNTSMC was drastically reduced, while the performance of the HRNTSMC was insignificantly affected by the initial error. Figures 12 and 13 present the different convergence errors of the three controllers. At the steady state, the HRNTSMC exhibited smaller errors in the angular pendulum compared to the HNTSMC and CHSMC, while the tracking performance of the cart position in the HRNTSMC outperformed the other controllers. Figure 14 depicts the control forces of the three controller, thus indicating that the HRNTSMC required large control input at the initial stages to balance the IP. However, the chattering problem of the HRNTSMC was significantly reduced compared to the other controllers, which became apparent after 40 s. Table 5 provides the performance indexes of the three controllers, thus showing that the HRNTSMC effectively demonstrated the smallest index values of the IAE and AECI. Despite its advantages, the ECI index value of the HRNTSMC was slightly higher than the HNTSMC, which is attributed to the large control force at the initial stage from 20 s. As a result, the proposed controller demonstrates effective performance of the IP in the case of a large initial error of the angular pendulum.

**Table 5.** Index values of control strategies in case 3.

| | IAE$x_1$ | IAE$x_3$ | ECI | AECI |
|---|---|---|---|---|
| CHSMC | 0.6058 | 1.2271 | 9.2318 | 4.0046 |
| HNTSMC | 0.0807 | 0.8674 | 0.9029 | 0.6959 |
| HRNTSMC | 0.0746 | 0.1816 | 1.0154 | 0.4238 |

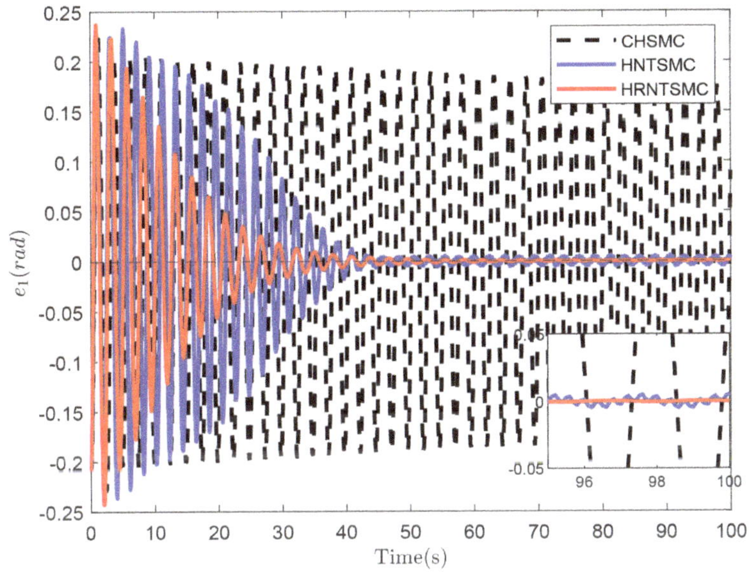


**Figure 4.** The error convergence of angular pendulum in case 1.

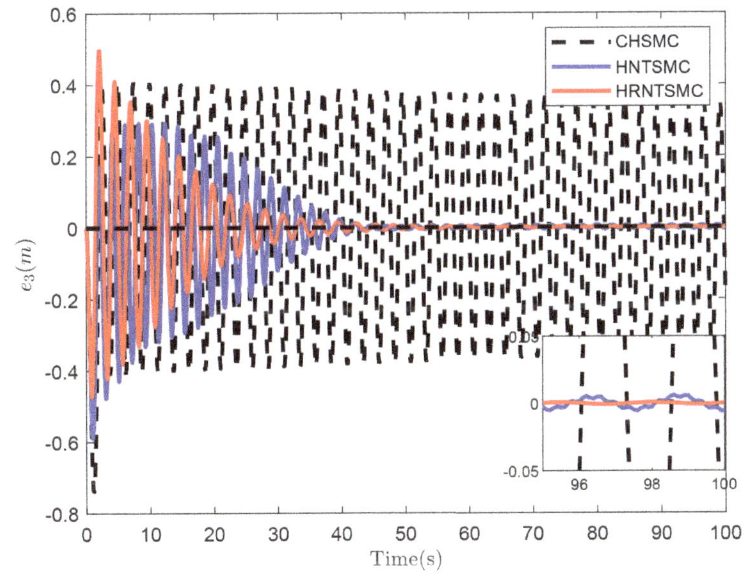


**Figure 5.** The error convergence of cart position in case 1.

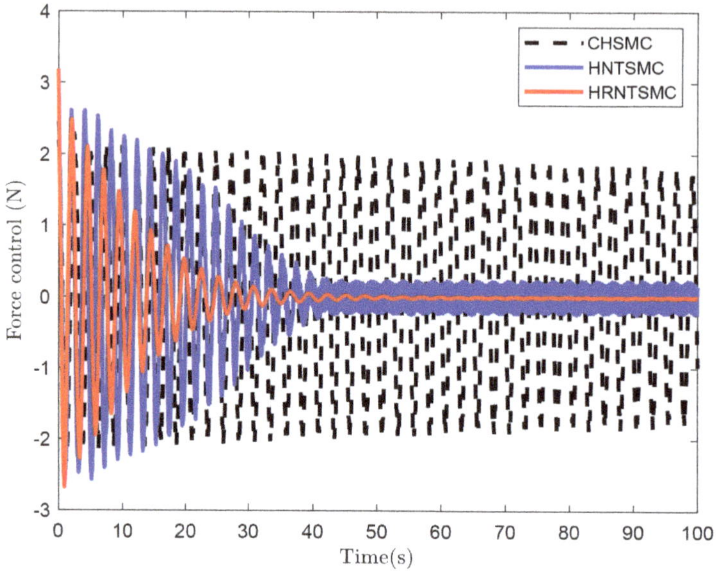


**Figure 6.** Force control in case 1.

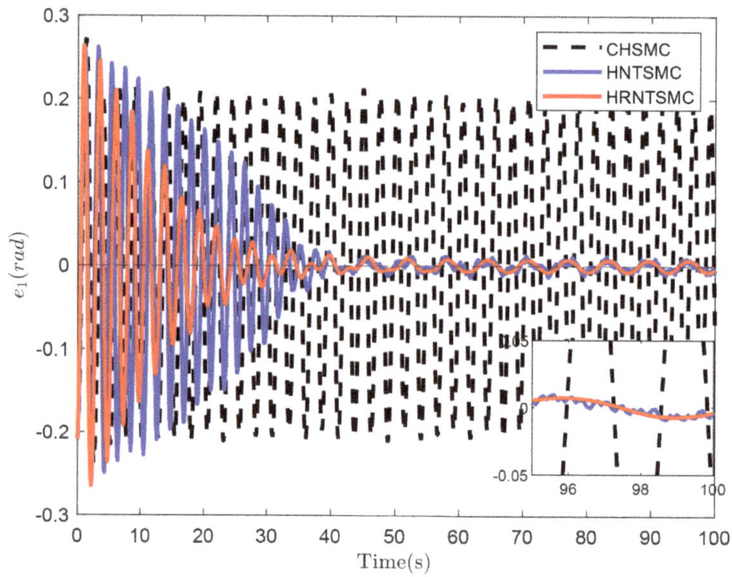


**Figure 7.** The error convergence of angular pendulum in case 2.

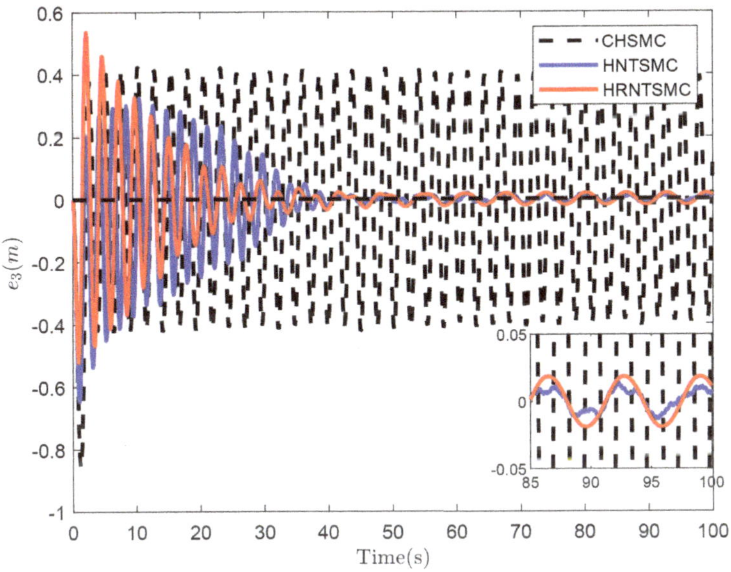


**Figure 8.** The error convergence of cart position in case 2.

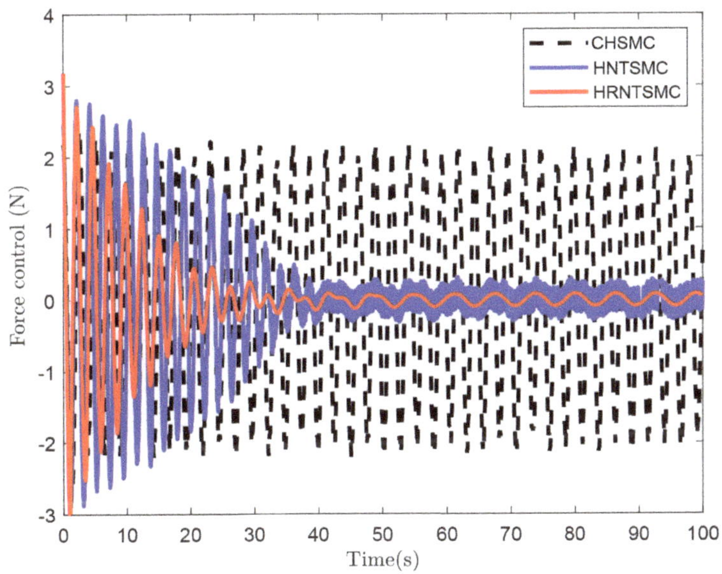


**Figure 9.** Force control in case 2.

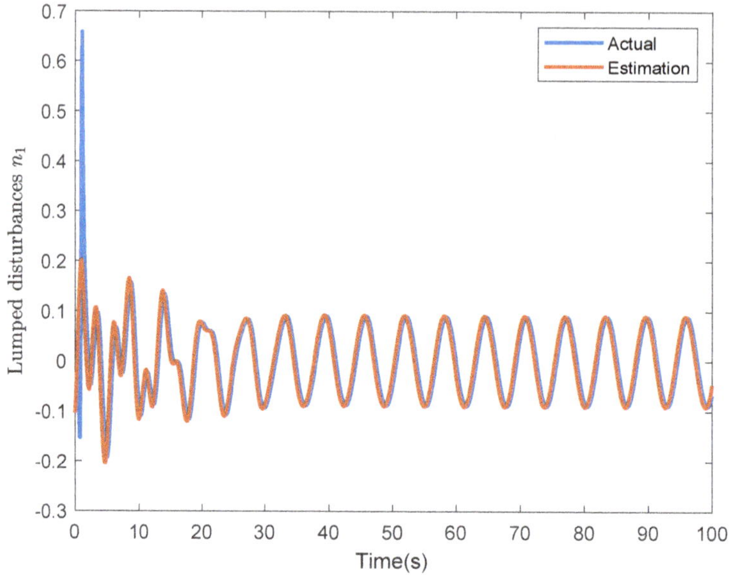


**Figure 10.** The estimated values of lumped disturbances $n_1$.

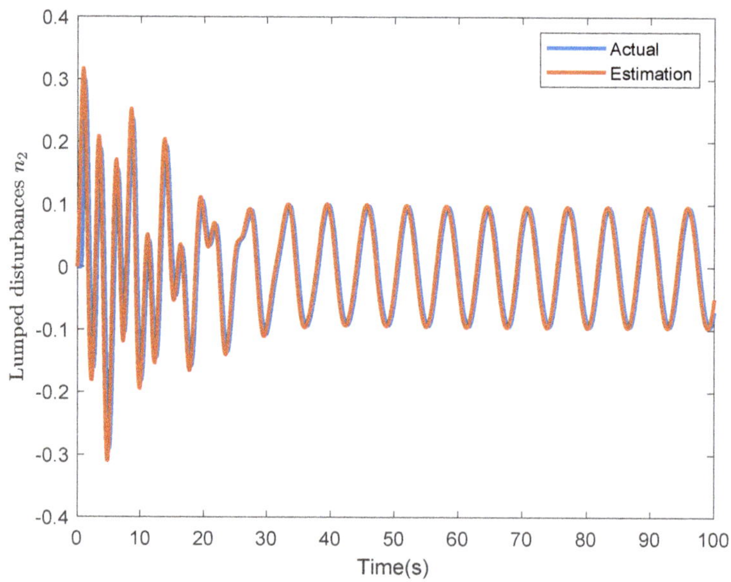


**Figure 11.** The estimated values of lumped disturbances $n_2$.

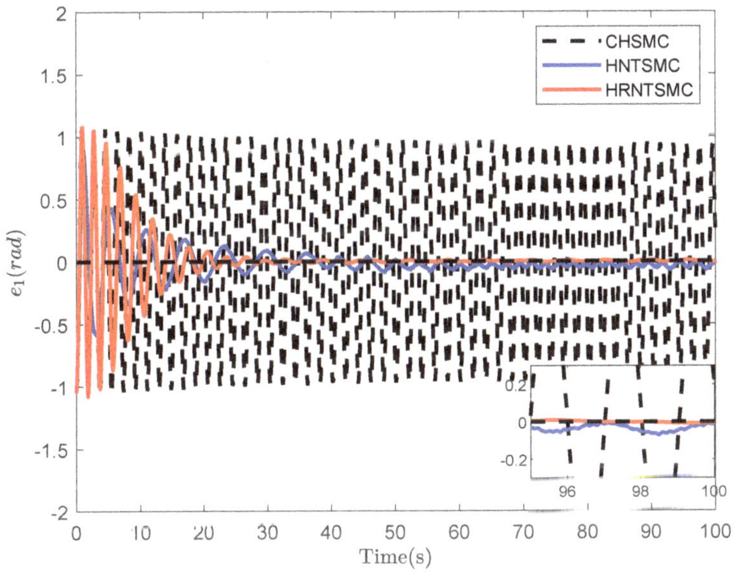


**Figure 12.** The error convergence of angular pendulum in case 3.

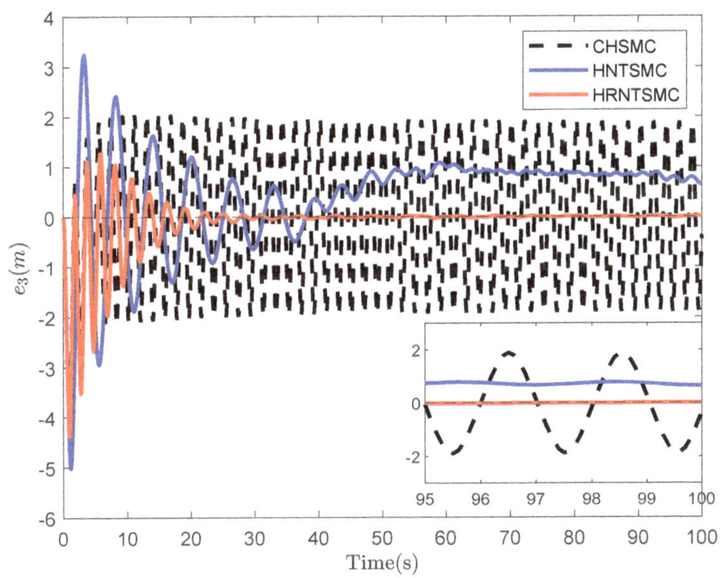


**Figure 13.** The error convergence of cart position in case 3.

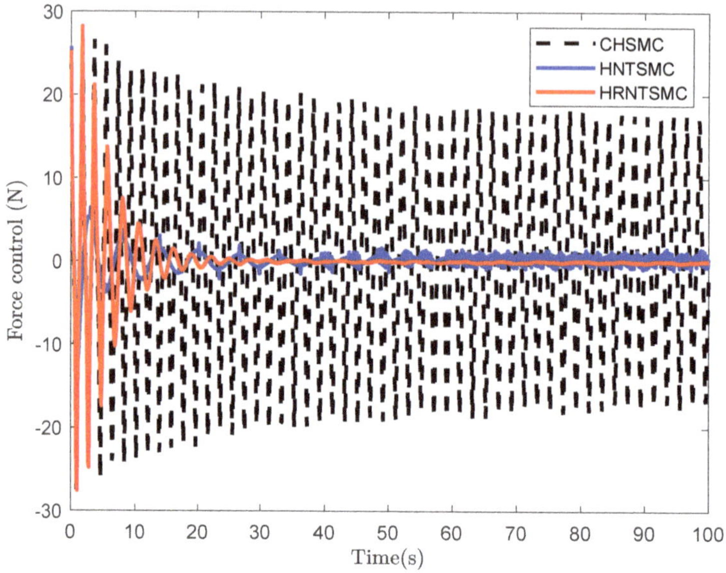


**Figure 14.** Force control in case 3.

## 5. Conclusions

This work proposes the HRNTSMC as an effective approach to increase convergence speed and reduce control input chattering in the IP system. Throughout this study, the simulation results confirm the applicability and performance of the proposed controller. Despite outperforming other compared controllers, it is acknowledged that the proposed controller contains more parameters for tuning, which could be determined using optimization methods in future work. However, it is essential to note that the present work is limited to a simulation-based investigation, which is aimed at showing the feasibility of this approach and establishing a theoretical foundation. Hence, to verify the practical implementation of the proposed controller in real-world scenarios, the next phase should involve experimental work.

**Author Contributions:** Conceptualization, H.D.L.; Methodology, H.D.L.; Software, H.D.L.; Validation, H.D.L.; Formal analysis, H.D.L.; Investigation, H.D.L.; Writing—original draft, H.D.L.; Writing—review and editing, T.N.; Visualization, H.D.L. and T.N.; Supervision, T.N.; Funding acquisition, T.N. All authors have read and agreed to the published version of the manuscript.

**Funding:** This work was supported by the Research Program of the DAAD (No. 91799091), and the APC was funded by journal Actuators (ISSN 2076-0825).

**Data Availability Statement:** The authors are willing to share data with interested researchers upon request.

**Acknowledgments:** This work was supported by the Research Program of the DAAD.

**Conflicts of Interest:** The authors declare that they have no known competing financial interests or personal relationships that could have appeared to influence the work reported in this paper.

## Appendix A. Stability of the LESO

We define the scaled estimation error vector of the LESO as follows:

$$\mathbf{v}(t) \triangleq [v_1(t), v_2(t), v_3(t)]^T \triangleq [\frac{x_1(t) - \hat{x}_1(t)}{\xi_1^2}, \frac{x_2(t) - \hat{x}_2(t)}{\xi_1}, n_1(t) - \hat{n}_1(t)]. \tag{A1}$$

By combining (1), (2), and (32)–(34), we obtain the following:

$$\xi_1 \dot{v}_1(t) = \frac{\dot{x}_1(t) - \dot{\hat{x}}_1(t)}{\xi_1(t)} = -a_1 v_1(t) + v_2(t). \tag{A2}$$

$$\xi_1 \dot{v}_2(t) = \dot{x}_2(t) - \dot{\hat{x}}_2(t) = -a_2 v_1(t) + v_3(t). \tag{A3}$$

$$\xi_1 \dot{v}_3(t) = \dot{n}_1(t) - \dot{\hat{n}}(t) = -a_3 v_1(t) + \xi_1 \dot{n}_1(t). \tag{A4}$$

The estimation error dynamics equation can be rewritten as follows:

$$\xi_1 \dot{\mathbf{v}}(t) = \mathbf{A}\mathbf{v}(t) + \xi_1 \mathbf{B} \dot{n}_1(t), \tag{A5}$$

where

$$\mathbf{A} = \begin{bmatrix} -a_1 & 1 & 0 \\ -a_2 & 0 & 1 \\ -a_3 & 0 & 0 \end{bmatrix}; \tag{A6}$$

$$\mathbf{B} = \begin{bmatrix} 0 \\ 0 \\ 1 \end{bmatrix}; \tag{A7}$$

The characteristic equation of matrix $A$ is defined as follows:

$$|\lambda \mathbf{I} - \mathbf{A}| = \begin{bmatrix} \lambda + a_1 & -1 & 0 \\ a_2 & \lambda & -1 \\ a_3 & 0 & \lambda \end{bmatrix} = 0. \tag{A8}$$

Then,

$$(\lambda + a_1)\lambda^2 + a_2 \lambda + a_3 = 0. \tag{A9}$$

And,

$$\lambda^3 + a_1 \lambda^2 + a_2 \lambda + a_3 = 0. \tag{A10}$$

By choosing suitable values for $a_1, a_2$, and $a_3$, we can make $\mathbf{A}$ a Hurwitz matrix, which has the real part of its eigenvalues being negative. For any given three-dimensional symmetric positive definite matrix $\mathbf{Q}$, there exists a symmetric positive definite matrix $\mathbf{P}$, which satisfies the following condition:

$$\mathbf{A}^{\mathrm{T}}\mathbf{P} + \mathbf{P}\mathbf{A} + \mathbf{Q} = 0. \tag{A11}$$

The Lyapunov function of the LESO is defined as follows:

$$V_e(t) \triangleq \xi_1 \mathbf{v}(t)^T \mathbf{P} \mathbf{v}(t). \tag{A12}$$

The time derivative of $V_e(t)$ is defined as follows:

$$\dot{V}_e(t) = \xi_1 \dot{\mathbf{v}}(t)^T \mathbf{P} \mathbf{v}(t) + \xi_1 \mathbf{v}(t)^T \mathbf{P} \dot{\mathbf{v}}(t). \tag{A13}$$

$$\dot{V}_e(t) = (\mathbf{A}\mathbf{v}(t) + \xi_1 \mathbf{B} \dot{n}_1)^T \mathbf{P} \mathbf{v}(t) + \xi_1 \mathbf{v}(t)^T \mathbf{P} (\mathbf{A}\mathbf{v}(t) + \xi_1 \mathbf{B} \dot{n}_1(t)). \tag{A14}$$

$$\dot{V}_e(t) = \mathbf{v}(t)^T (\mathbf{A}^T \mathbf{P} + \mathbf{P}\mathbf{A}) \mathbf{v}(t) + 2\xi_1 \mathbf{v}(t)^T \mathbf{P}\mathbf{B} \dot{n}_1(t). \tag{A15}$$

$$\dot{V}_e(t) \leq -\mathbf{v}^T \mathbf{Q} \mathbf{v} + 2\xi_1 ||\mathbf{P}\mathbf{B}|| \cdot |\mathbf{v}(t)| \cdot |\dot{n}_1(t)|. \tag{A16}$$

We assume that $|\dot{n}_1(t)| \leq L$; then:

$$\dot{V}_e(t) \leq -\lambda_{min}(\mathbf{Q})||\mathbf{v}(t)||^2 + 2\xi_1||\mathbf{PB}|| \cdot |\mathbf{v}(t)| \cdot |L|, \tag{A17}$$

where $\lambda_{min}(\mathbf{Q})$ is the minimum eigenvalue of $\mathbf{Q}$, where $\dot{V}_e(t) \leq 0$; the observer error convergence is defined as follows:

$$||\mathbf{v}(t)|| \leq \frac{2\xi_1||\mathbf{PB}|| \cdot |L|}{\lambda_{min}(\mathbf{Q})}. \tag{A18}$$

To simplify the complexity of the LESO, matrix $\mathbf{Q}$ is selected as the identity matrix $\mathbf{I}$ [74], and the polynomial (A10) is selected based on [75]:

$$\lambda^3 + a_1\lambda^2 + a_2\lambda + a_3 = (s + \omega)^3, \tag{A19}$$

where $\omega$ presents the bandwidth of the LESO. It can be seen that the observer error depends on the bound of the derivative disturbance $|L|$ and $||\mathbf{PB}||$. The value of $||\mathbf{PB}||$ is influenced by the eigenvalues of the matrix $\mathbf{A}$, as $\mathbf{P}$ is determined by (A11), with $\mathbf{Q}$ being the identity matrix. To enhance the accuracy of the disturbances estimation, selecting a large $\omega$ value results in a small value of $||\mathbf{PB}||$. However, in a practical system, $\omega$ is limited by the bandwidth of the controller and the sampling frequency of the system [57]. When the high-frequency disturbances exist in the IP system, the LESO requires maximum $\omega$ values to achieve the fast convergence of estimation, which can introduce more noise and diminish the observer performance. Addressing this challenge will be a focus of the future work.

## References


1. Irfan, S.; Mehmood, A.; Razzaq, M.T.; Iqbal, J. Advanced sliding mode control techniques for inverted pendulum: Modelling and simulation. *Eng. Sci. Technol. Int. J.* **2018**, *21*, 753–759. [CrossRef]
2. Kim, S.; Kwon, S. Nonlinear optimal control design for underactuated two-wheeled inverted pendulum mobile platform. *IEEE/ASME Trans. Mechatronics* **2017**, *22*, 2803–2808. [CrossRef]
3. Sentis, L. Compliant control of whole-body multi-contact behaviors in humanoid robots. In *Motion Planning for Humanoid Robots*; Springer: Berlin/Heidelberg, Germany, 2010; pp. 29–66.
4. Wang, W.; Yi, J.; Zhao, D.; Liu, D. Design of a stable sliding-mode controller for a class of second-order underactuated systems. *IEE Proc. -Control Theory Appl.* **2004**, *151*, 683–690. [CrossRef]
5. Horibe, T.; Sakamoto, N. Optimal swing up and stabilization control for inverted pendulum via stable manifold method. *IEEE Trans. Control. Syst. Technol.* **2017**, *26*, 708–715. [CrossRef]
6. Ismail, N. Fuzzy Logic Controller Design for Inverted Pendulum System. Ph.D. Thesis, Universiti Tun Hussein Malaysia, Johor, Malaysia, 2013.
7. Bennett, S. Development of the PID controller. *IEEE Control Syst. Mag.* **1993**, *13*, 58–62.
8. Ghosh, A.; Krishnan, T.; Subudhi, B. Robust proportional–integral–derivative compensation of an inverted cart–pendulum system: An experimental study. *IET Control Theory Appl.* **2012**, *6*, 1145–1152. [CrossRef]
9. Batista, J.G.; Souza, D.A.; dos Reis, L.L.; Filgueiras, L.V.; Ramos, K.M.; Junior, A.B.; Correia, W.B. Performance comparison between the PID and LQR controllers applied to a robotic manipulator joint. In Proceedings of the IECON 2019-45th Annual Conference of the IEEE Industrial Electronics Society, Lisbon, Portugal, 14–17 October 2019; Volume 1, pp. 479–484.
10. Wang, H.; Dong, H.; He, L.; Shi, Y.; Zhang, Y. Design and simulation of LQR controller with the linear inverted pendulum. In Proceedings of the 2010 International Conference on Electrical and Control Engineering, Wuhan, China, 25–27 June 2010; pp. 699–702.
11. Camacho, O.; Smith, C.A. Sliding mode control: An approach to regulate nonlinear chemical processes. *ISA Trans.* **2000**, *39*, 205–218. [CrossRef]
12. Huang, C.H.; Wang, W.J.; Chiu, C.H. Design and implementation of fuzzy control on a two-wheel inverted pendulum. *IEEE Trans. Ind. Electron.* **2010**, *58*, 2988–3001. [CrossRef]
13. Yang, C.; Li, Z.; Cui, R.; Xu, B. Neural network-based motion control of an underactuated wheeled inverted pendulum model. *IEEE Trans. Neural Netw. Learn. Syst.* **2014**, *25*, 2004–2016. [CrossRef]
14. Shtessel, Y.; Edwards, C.; Fridman, L.; Levant, A. *Sliding Mode Control and Observation*; Springer: Berlin/Heidelberg, Germany, 2014; Volume 10.
15. Krafes, S.; Chalh, Z.; Saka, A. Linear, nonlinear and intelligent controllers for the inverted pendulum problem. In Proceedings of the 2016 International Conference on Electrical and Information Technologies (ICEIT), Tangiers, Morocco, 4–7 May 2016; pp. 136–141.

16. Utkin, V.I. Sliding mode control in discrete-time and difference systems. In *Variable Structure and Lyapunov Control*; Springer: Berlin/Heidelberg, Germany, 2005; pp. 87–107.
17. Utkin, V.I. *Sliding Modes in Control and Optimization*; Springer Science & Business Media: Berlin/Heidelberg, Germany, 2013.
18. Basin, M.V.; Yu, P.; Shtessel, Y.B. Hypersonic missile adaptive sliding mode control using finite-and fixed-time observers. *IEEE Trans. Ind. Electron.* **2017**, *65*, 930–941. [CrossRef]
19. Wang, G.; Xu, Q. Design and precision position/force control of a piezo-driven microinjection system. *IEEE/ASME Trans. Mechatronics* **2017**, *22*, 1744–1754. [CrossRef]
20. Liu, Y.; Zhang, Y.; Xu, Q. Design and control of a novel compliant constant-force gripper based on buckled fixed-guided beams. *IEEE/ASME Trans. Mechatronics* **2016**, *22*, 476–486. [CrossRef]
21. Zhihong, M.; Paplinski, A.P.; Wu, H.R. A robust MIMO terminal sliding mode control scheme for rigid robotic manipulators. *IEEE Trans. Autom. control* **1994**, *39*, 2464–2469. [CrossRef]
22. Yu, X.; Zhihong, M. Fast terminal sliding-mode control design for nonlinear dynamical systems. *IEEE Trans. Circuits Syst. I Fundam. Theory Appl.* **2002**, *49*, 261–264.
23. Feng, Y.; Yu, X.; Man, Z. Non-singular terminal sliding mode control of rigid manipulators. *Automatica* **2002**, *38*, 2159–2167. [CrossRef]
24. Shao, K.; Zheng, J.; Huang, K.; Wang, H.; Man, Z.; Fu, M. Finite-time control of a linear motor positioner using adaptive recursive terminal sliding mode. *IEEE Trans. Ind. Electron.* **2019**, *67*, 6659–6668. [CrossRef]
25. Utkin, V.; Lee, H. Chattering problem in sliding mode control systems. In Proceedings of the International Workshop on Variable Structure Systems, 2006. VSS'06, Alghero, Sardinia, 5–7 June 2006; pp. 346–350.
26. Asad, M.; Bhatti, A.I.; Iqbal, S. A novel reaching law for smooth sliding mode control using inverse hyperbolic function. In Proceedings of the 2012 International Conference on Emerging Technologies, Islamabad, Pakistan, 8–9 October 2012; pp. 1–6.
27. Le, H.D.; Nestorović, T. Adaptive Proportional Integral Derivative Nonsingular Dual Terminal Sliding Mode Control for Robotic Manipulators. *Dynamics* **2023**, *3*, 656–677. [CrossRef]
28. Kim, K.S.; Rew, K.H.; Kim, S. Disturbance observer for estimating higher order disturbances in time series expansion. *IEEE Trans. Autom. control* **2010**, *55*, 1905–1911.
29. Zhang, J.; Liu, X.; Xia, Y.; Zuo, Z.; Wang, Y. Disturbance observer-based integral sliding-mode control for systems with mismatched disturbances. *IEEE Trans. Ind. Electron.* **2016**, *63*, 7040–7048. [CrossRef]
30. Mobayen, S.; Tchier, F. Nonsingular fast terminal sliding-mode stabilizer for a class of uncertain nonlinear systems based on disturbance observer. *Sci. Iran.* **2017**, *24*, 1410–1418. [CrossRef]
31. Chen, W.H.; Yang, J.; Guo, L.; Li, S. Disturbance-observer-based control and related methods—An overview. *IEEE Trans. Ind. Electron.* **2015**, *63*, 1083–1095. [CrossRef]
32. Guo, B.Z.; Zhao, Z.L. Extended state observer for nonlinear systems with uncertainty. *IFAC Proc. Vol.* **2011**, *44*, 1855–1860. [CrossRef]
33. Liu, Y.C.; Laghrouche, S.; Depernet, D.; N'Diaye, A.; Djerdir, A.; Cirrincione, M. Super-twisting sliding-mode observer-based model reference adaptive speed control for PMSM drives. *J. Frankl. Inst.* **2023**, *360*, 985–1004. [CrossRef]
34. Zhang, J.; Gao, W.; Guo, Q. Extended State Observer-Based Sliding Mode Control Design of Two-DOF Lower Limb Exoskeleton. *Actuators* **2023**, *12*, 402. [CrossRef]
35. Chen, H.; Sun, N. Nonlinear control of underactuated systems subject to both actuated and unactuated state constraints with experimental verification. *IEEE Trans. Ind. Electron.* **2019**, *67*, 7702–7714. [CrossRef]
36. Utkin, V.; Guldner, J.; Shi, J. *Sliding Mode Control in Electro-Mechanical Systems*; CRC Press: Boca Raton, FL, USA, 2017.
37. Idrees, M.; Muhammad, S.; Ullah, S. Robust hierarchical sliding mode control with state-dependent switching gain for stabilization of rotary inverted pendulum. *Kybernetika* **2019**, *55*, 455–471. [CrossRef]
38. Zhao, H.; Zong, G.; Zhao, X.; Wang, H.; Xu, N.; Zhao, N. Hierarchical Sliding-Mode Surface-Based Adaptive Critic Tracking Control for Nonlinear Multiplayer Zero-Sum Games Via Generalized Fuzzy Hyperbolic Models. *IEEE Trans. Fuzzy Syst.* **2023**, *31*, 4010–4023. [CrossRef]
39. Hwang, C.L.; Chiang, C.C.; Yeh, Y.W. Adaptive fuzzy hierarchical sliding-mode control for the trajectory tracking of uncertain underactuated nonlinear dynamic systems. *IEEE Trans. Fuzzy Syst.* **2013**, *22*, 286–299. [CrossRef]
40. Hwang, C.L.; Wu, H.M. Trajectory tracking of a mobile robot with frictions and uncertainties using hierarchical sliding-mode under-actuated control. *IET Control Theory Appl.* **2013**, *7*, 952–965. [CrossRef]
41. Nafa, F.; Labiod, S.; Chekireb, H. A structured sliding mode controller for a class of underactuated mechanical systems. In Proceedings of the International Workshop on Systems, Signal Processing and their Applications, WOSSPA, Tipaza, Algeria, 9–11 May 2011; pp. 243–246.
42. Yue, S.; Niu, B.; Wang, H.; Zhang, L.; Ahmad, A.M. Hierarchical sliding mode-based adaptive fuzzy control for uncertain switched under-actuated nonlinear systems with input saturation and dead-zone. *Robot. Intell. Autom.* **2023**, *43*, 523–536. [CrossRef]
43. Min-xiu, Y.; Yuan-wei, J. Terminal sliding mode decomposed control for a class of nonlinear systems. In Proceedings of the 2008 Chinese Control and Decision Conference, Yantai, China, 2–4 July 2008; pp. 4988–4991.
44. Bayramoglu, H.; Komurcugil, H. Nonsingular decoupled terminal sliding-mode control for a class of fourth-order nonlinear systems. *Commun. Nonlinear Sci. Numer. Simul.* **2013**, *18*, 2527–2539. [CrossRef]

45. Edwards, C.; Shtessel, Y.B. Adaptive continuous higher order sliding mode control. *Automatica* **2016**, *65*, 183–190. [CrossRef]
46. Wang, H.; Li, Z.; Jin, X.; Huang, Y.; Kong, H.; Yu, M.; Ping, Z.; Sun, Z. Adaptive integral terminal sliding mode control for automobile electronic throttle via an uncertainty observer and experimental validation. *IEEE Trans. Veh. Technol.* **2018**, *67*, 8129–8143. [CrossRef]
47. Liu, W.; Chen, S.; Huang, H. Actor-Critic learning hierarchical sliding mode control for a class of underactuated systems. In Proceedings of the 2019 Chinese Automation Congress (CAC), Hangzhou, China, 22–24 November 2019; pp. 1–6.
48. Li, S.; Yang, J.; Chen, W.H.; Chen, X. Generalized extended state observer based control for systems with mismatched uncertainties. *IEEE Trans. Ind. Electron.* **2011**, *59*, 4792–4802. [CrossRef]
49. Czyżniewski, M.; Łangowski, R.; Klassa, D.; Matwiszyn, M. A case study of robust sliding mode control applied to inverted pendulum on a cart. In Proceedings of the 2021 25th International Conference on Methods and Models in Automation and Robotics (MMAR), Miedzyzdroje, Poland, 23–26 August 2021; pp. 156–161.
50. Haimo, V.T. Finite time controllers. *SIAM J. Control Optim.* **1986**, *24*, 760–770. [CrossRef]
51. Qian, D.; Yi, J.; Zhao, D.; Hao, Y. Hierarchical sliding mode control for series double inverted pendulums system. In Proceedings of the 2006 IEEE/RSJ International Conference on Intelligent Robots and Systems, Beijing, China, 9–15 October 2006; pp. 4977–4982.
52. Chiu, C.S. Derivative and integral terminal sliding mode control for a class of MIMO nonlinear systems. *Automatica* **2012**, *48*, 316–326. [CrossRef]
53. Bartoszewicz, A. A new reaching law for sliding mode control of continuous time systems with constraints. *Trans. Inst. Meas. Control* **2015**, *37*, 515–521. [CrossRef]
54. Yue, M.; Liu, B. Adaptive control of an underactuated spherical robot with a dynamic stable equilibrium point using hierarchical sliding mode approach. *Int. J. Adapt. Control Signal Process.* **2014**, *28*, 523–535. [CrossRef]
55. Qian, D.; Yi, J.; Zhao, D. Multiple layers sliding mode control for a class of under-actuated systems. In Proceedings of the Multiconference on Computational Engineering in Systems Applications, Beijing, China, 4–6 October 2006; Volume 1, pp. 530–535.
56. Wang, H.; Shi, L.; Man, Z.; Zheng, J.; Li, S.; Yu, M.; Jiang, C.; Kong, H.; Cao, Z. Continuous fast nonsingular terminal sliding mode control of automotive electronic throttle systems using finite-time exact observer. *IEEE Trans. Ind. Electron.* **2018**, *65*, 7160–7172. [CrossRef]
57. Gao, Z. Scaling and bandwidth-parameterization based controller tuning. In Proceedings of the ACC, Denver, CO, USA, 4–6 June 2003; pp. 4989–4996.
58. Sussmann, H.; Kokotovic, P. The peaking phenomenon and the global stabilization of nonlinear systems. *IEEE Trans. Autom. Control* **1991**, *36*, 424–440. [CrossRef]
59. Huang, Y.; Han, J. Analysis and design for the second order nonlinear continuous extended states observer. *Chin. Sci. Bull.* **2000**, *45*, 1938–1944. [CrossRef]
60. Wang, J.; Wu, Y.; Dong, X. Recursive terminal sliding mode control for hypersonic flight vehicle with sliding mode disturbance observer. *Nonlinear Dyn.* **2015**, *81*, 1489–1510. [CrossRef]
61. Shi, S.L.; Li, J.X.; Fang, Y.M. Extended-state-observer-based chattering free sliding mode control for nonlinear systems with mismatched disturbance. *IEEE Access* **2018**, *6*, 22952–22957. [CrossRef]
62. González, I.; Salazar, S.; Lozano, R. Chattering-free sliding mode altitude control for a quad-rotor aircraft: Real-time application. *J. Intell. Robot. Syst.* **2014**, *73*, 137–155. [CrossRef]
63. Liu, Q.; Lv, Z.; Wu, Y. Design of sliding mode controller based on high-gain observer of inverted pendulum on a cart. In Proceedings of the 2019 Chinese Control Conference (CCC), Guangzhou, China, 27–30 July 2019; pp. 846–851.
64. Cui, M.; Liu, W.; Liu, H.; Jiang, H.; Wang, Z. Extended state observer-based adaptive sliding mode control of differential-driving mobile robot with uncertainties. *Nonlinear Dyn.* **2016**, *83*, 667–683. [CrossRef]
65. Muhammad, S.; Idrees, M. Comparative study of hierarchical sliding mode control and decoupled sliding mode control. In Proceedings of the 2017 12th IEEE Conference on Industrial Electronics and Applications (ICIEA), Siem Reap, Cambodia, 18–20 June 2017; pp. 818–824.
66. Guo, B.; Bacha, S.; Alamir, M.; Hably, A.; Boudinet, C. Generalized integrator-extended state observer with applications to grid-connected converters in the presence of disturbances. *IEEE Trans. Control Syst. Technol.* **2020**, *29*, 744–755. [CrossRef]
67. Mahmoodabadi, M.J.; Khoobroo Haghbayan, H. An optimal adaptive hybrid controller for a fourth-order under-actuated nonlinear inverted pendulum system. *Trans. Inst. Meas. Control* **2020**, *42*, 285–294. [CrossRef]
68. Haghighi, D.A.; Mobayen, S. Design of an adaptive super-twisting decoupled terminal sliding mode control scheme for a class of fourth-order systems. *ISA Trans.* **2018**, *75*, 216–225. [CrossRef] [PubMed]
69. Boukattaya, M.; Gassara, H.; Damak, T. A global time-varying sliding-mode control for the tracking problem of uncertain dynamical systems. *ISA Trans.* **2020**, *97*, 155–170. [CrossRef] [PubMed]
70. Mei, K.; Ding, S. Second-order sliding mode controller design subject to an upper-triangular structure. *IEEE Trans. Syst. Man Cybern. Syst.* **2018**, *51*, 497–507. [CrossRef]
71. Qiao, L.; Zhang, W. Trajectory tracking control of AUVs via adaptive fast nonsingular integral terminal sliding mode control. *IEEE Trans. Ind. Inform.* **2019**, *16*, 1248–1258. [CrossRef]
72. Yao, Q. Adaptive finite-time sliding mode control design for finite-time fault-tolerant trajectory tracking of marine vehicles with input saturation. *J. Frankl. Inst.* **2020**, *357*, 13593–13619. [CrossRef]

73. Sai, H.; Xu, Z.; He, S.; Zhang, E.; Zhu, L. Adaptive nonsingular fixed-time sliding mode control for uncertain robotic manipulators under actuator saturation. *ISA Trans.* **2022**, *123*, 46–60. [CrossRef] [PubMed]
74. Li, H.; Yang, S.; Kong, L.; Wen, T. High-precision angular speed tracking control of gimbal system with harmonic reducer. *IEEE Trans. Ind. Electron.* **2021**, *69*, 8168–8177. [CrossRef]
75. Lau, J.Y.; Liang, W.; Tan, K.K. Motion control for piezoelectric-actuator-based surgical device using neural network and extended state observer. *IEEE Trans. Ind. Electron.* **2019**, *67*, 402–412. [CrossRef]

actuators

MDPI

*Article*

# An Improved Analytical Model of a Thrust Stand with a Flexure Hinge Structure Considering Stiffness Drift and Rotation Center Offset

**Xingyu Chen, Liye Zhao *, Jiawen Xu and Zhikang Liu**

School of Instrument Science and Engineering, Southeast University, Nanjing 210096, China; 230218947@seu.edu.cn (X.C.); jiawen.xu@seu.edu.cn (J.X.); 220213617@seu.edu.cn (Z.L.)
* Correspondence: liyezhao@seu.edu.cn

**Abstract:** Micro-newton thrust stands are widely used in thruster ground calibration procedures for a variety of space missions. The conventional analytical model does not consider the gravity-induced extension effect and systematic error in displacement for thrust stands consisting of hanging pendulums based on flexure hinge structures. This paper proposes an improved analytical model of a hanging pendulum for thrust measurement, where an elliptical notched flexure hinge is the key component. A parametric model of the bending stiffness of the flexure hinge is developed. Equally, both the bending stiffness shift under the gravity-induced extension effect and the systematic error in displacement due to the assumed rotational center offset of the hinge are investigated. The presented stiffness equations for elliptical notched hinges can be degenerated into stiffness equations for circular notched as well as leaf-type hinges. The improved model aims to evaluate and highlight the influence of the two considered factors for use in thrust stand parameter design and thrust analysis. A finite element modeling solution is proposed to validate the proposed analytical model. The results show that the proposed model can quantify the hinge bending stiffness shift, which also demonstrates that even a small bending stiffness shift may introduce great uncertainty into the thrust analysis.

**Keywords:** micro-newton thrust stand; analytical model; flexure hinge; bending stiffness; uncertainty

## 1. Introduction

With the development of space investigations, the control of satellites requires a rising degree of accuracy. These high-precision controls include pointing adjustment in satellite attitude control, absolute positioning in drag-free control, and relative positioning in multi-satellite coordinated flight. The realizations of these desired controls are often accompanied by the need for high-precision, high-resolution thrusters [1,2]. In recent years, space science missions that have relied on micro-newton thrusters include Gaia, Darwin, Aspics, Microscope, LISA Pathfinder (LPF), Taiji, and so on. Among them, the space gravitational wave detection mission represented by LPF was extremely demanding on the thruster [3–5], which is required to achieve a wide dynamic range of 0.1 μN–100 μN thrust. This means the noise is required to be less than $0.1\mu\text{N}/\sqrt{\text{Hz}}$ in the frequency range of 1 mHz–1 Hz [6,7].

To characterize the high-performance thrusters on the ground, many different types of high-sensitivity thrust stands have been developed [8–10]. Typically, these thrust stands are based on different types of pendulum structures. The typical pendulums used are of three types: the hanging pendulum [11,12], the torsional pendulum [13,14], and the inverted pendulum [15,16]. Various measurement systems have been derived from the three prototypes: for example, a double pendulum system based on the hanging pendulum structure [17] and a null-displacement control system based on the torsional pendulum structure [18]. Due to the inherent instability of the inverted pendulums, it is difficult to meet the requirements of a wide range of thrust measurements and large weight loads.

For torsional pendulums that rotate around fibers, a symmetrical configuration in the horizontal direction may be difficult to achieve. Accordingly, a hanging pendulum with the simplest solution is widely adopted. A quasi-static thrust analytical model, which indicates its measurement principles with generalized Hooke's law, is given in all thrust stand design schemes, as well as in those based on hanging pendulums [19–21]. James E. Polk et al. give details of the thrust measurement principle of a hanging pendulum, which also represents a standard [22].

The selection and parameter setting of the "rotating shaft component" is an essential part of the design of a hanging pendulum. This is because it both carries the loads and provides the rotation function. In compliant mechanisms, flexure hinges have been applied by many designers due to their excellent mechanical properties without wear and mechanical friction. Then, flexure hinges are manufactured in different shapes according to different application scenarios. Among them, notched hinges are the most advisable because they have a more stable axis of rotation, for example, elliptical, V-shaped, parabolic, circular, cycloidal [23–26], etc. Stefano et al. use flexure hinges with symmetrical semicircular notches to connect the "tilt plate" to a rigid block [27]. Xu et al. choose a beryllium copper strip hinge as the rotating mechanism with a length of only 1 mm to avoid uncertainty in the position of the rotating axis [28]. Nevertheless, almost all thrust stand designers utilize only an equivalent spring with constant $K$ to characterize the flexure hinge in their analytical models, and few have investigated the mechanical properties of the hinge as operating conditions change. Indeed, the equivalent spring stiffness will be affected by the axial force due to the extension. This effect may bring greater uncertainty to the thrust analysis, especially in the case of heavy loads. Therefore, it is necessary to improve the "standard" analysis model of the thrust measurement principle.

To improve the theoretical system of thrust stand modeling, this study develops a new analytical model for a thrust stand consisting of a hanging pendulum based on an elliptical notched flexure hinge. In the new model, parametric modeling of the bending stiffness of the hinge is emphasized as it is the basis for the subsequent analysis. On the one hand, the gravity-induced extension effect on the hinge which leads to a bending stiffness shift will be considered. On the other hand, during pendulum deflection, a major source of errors in displacement measurements is the offset of the assumed rotational center. It can be misinterpreted as an action from the thruster and also needs to be considered. One benefit of the presented analytical model for thrust measurement is such that the notch boundary and dimensional characteristics of the flexure hinge are parametrically characterized, thus facilitating the quantitative calculation of the bending stiffness of the "rotating mechanism" and the traceability of the uncertainty budget. Another benefit is that the established relationship between the axial tensile force and bending stiffness of the hinge, as well as the relationship between the offset of the assumed rotational center and thrust quantification, will provide a great reference value and high convenience to the optimal design of the thrust stand.

The remainder of the paper is organized as follows: Section 2 presents the conceptual illustration. The conventional measurement model of the hanging pendulum is introduced in Section 3. In Section 4, the improved analytical model of the thrust measurement is derived. The finite element modeling strategies and comparison results of the simulation experiments are presented and discussed to validate the proposed analytical model in Section 5. In Section 6, a summary of this research and future work is given.

## 2. Conceptual Illustration

In this section, a conceptual illustration of improvements to the analytical model of the hanging pendulum thrust stand with a flexure hinge structure is presented.

Figure 1a,b present the conventional schematic model and the improved one, respectively. The red line indicates the bent flexure hinge and the thin black line indicates the pendulum arm that is equivalent to a rigid body. In the past, the extension effect of the flexure hinge and displacement error were omitted. The hinge during pendulum deflection is not a purely ideal bend in the conventional analytical model (see Figure 1a). It suffers due to two points: first of all, the point $o$, the assumed rotational center of the hinge, would drift to point $o'$ (the offset is $\Delta u$, as shown in Figure 1b). This offset of the assumed rotational center would lead to errors in the measured displacement. Secondly, the hinge would be extended due to the axial component of gravity $G_{axial}$. Such an extension can be intuitively expressed as the lengthening of the hinge (the green part in Figure 1b) and further reflected in changes in other geometric parameters such as the hinge thickness. This, in turn, brings about a change in the line stiffness $K_{line}$ of the pendulum. Therefore, it is necessary to improve the conventional model by considering the load effects and rotational center offset. In particular, the conventional analytical model $F = K_{line}u$ is revised into the form of $F = K'_{line}(u - \Delta u)$, where $K'_{line}$ is the linear stiffness of the pendulum after taking into account the extension effect, and $\Delta u$ is the measurement displacement error considering the offset of the assumed rotation center.

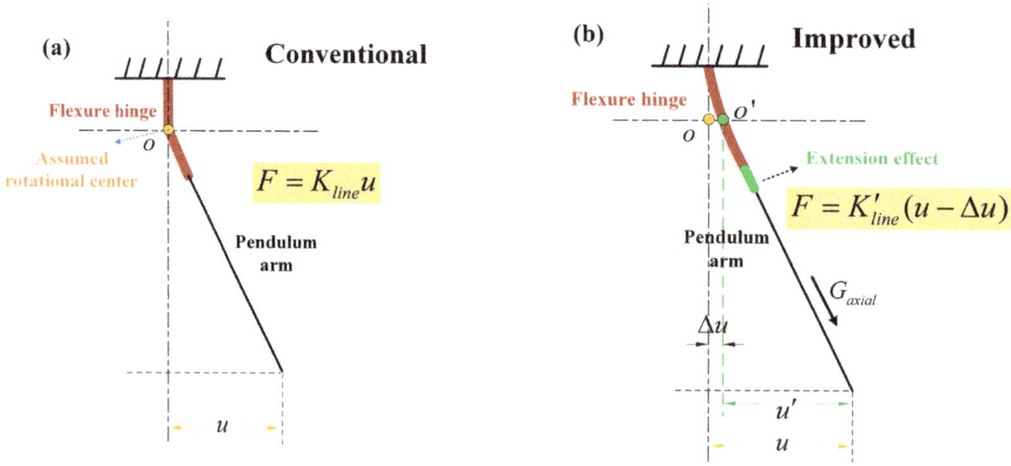


**Figure 1.** Comparison before and after analytical model improvement. (**a**) The conventional analytical model and pendulum deflection schematic; (**b**) the new analytical model and pendulum deflection schematic.

It should be noted that the bending stiffness of a flexure hinge is determined by the geometrical boundary of a given notch. Equally, the change in the geometrical boundary under the influence of the loads is deduced by bearing an axial tensile force on the hinge. Therefore, "modified bending stiffness" is equal to "modified boundary" plus "bending stiffness". The general methodology for deriving the improved analytical model is to integrate the modified flexure hinge bending stiffness and the measured displacement error. Based on the force analysis while the pendulum is working in the equilibrium state, the derivation and mathematical equations of the analytical model are given in detail in the following sections.

## 3. Conventional Analytical Model of the Thrust Stand

In this section, we briefly present the designed thrust stand and use it to give a conventional thrust analytical model.

Like most of the micro-newton thrust balances that are being or have been developed [29,30], we have also developed a thrust stand consisting of a compound pendulum, as shown in Figure 2. Figure 2a shows the 3D structural schema of the thrust stand. In the pendulum, there are two elliptical notched flexure hinges located on both sides of the center axis of the compound pendulum. Each flexure hinge can be thought of as a small-length frictionless pivot, which provides a rotating function and is located in the middle of the hinge. Figure 2b displays a physical diagram of the thrust stand.

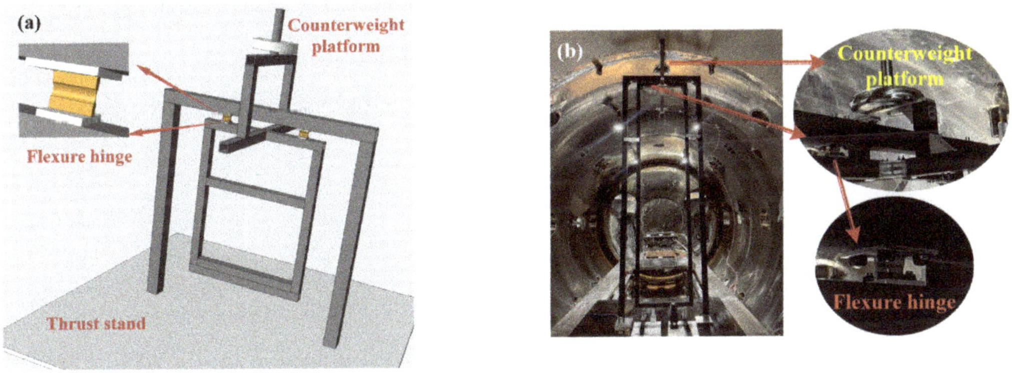


**Figure 2.** The designed thrust stand consisting of the compound pendulum (partial). (**a**) 3D rendering diagram; (**b**) physical diagram.

Without loss of generality, the pendulum is assumed to operate in an open-loop condition. As a common practice, the micro-newton thrust adopts the quasi-static force measurement method. Figure 3 shows the schematic principle. Table 1 lists the parameters, the corresponding symbols, and the units. Equally, CW denotes the counterweight, while $C$ denotes the center of mass of the entire pendulum. The measurement principle is based on the equilibrium of the pendulum torque, given as:

$$Fl_{Th.} = K_p\theta + g\sin\theta(m_{Th.}l_{Th.} + m_{ro.}l_{ro.} - m_{cout.}l_{cout.}) \tag{1}$$

where $K_p$ is the bending stiffness of the flexure hinge. The left side of Equation (1) is the torque provided by the thrust, and the right side of the equation represents the torques associated with the flexure hinge itself and the gravity tangential component, respectively. When the deflection angle $\theta$ is small, for example, less than $5^\circ$, it can be considered that $\sin\theta \approx \theta$ [22]. Simplifying Equation (1) yields

$$Fl_{Th.} \approx K_{ang}\theta = (K_p + K_{G_{tangent}})\theta \tag{2}$$

where $K_{ang}$ is the torsional stiffness of the pendulum. $K_{G_{tangent}}$ is the equivalent spring stiffness supplied by the tangential component of gravity $G_{tangent}$, and $K_{G_{tangent}} = g(m_{Th.}l_{Th.} + m_{ro.}l_{ro.} - m_{cout.}l_{cout.})$.

**Table 1.** Description of parameters adopted in the model.

| Parameter | Symbol | Unit |
|---|---|---|
| Applied thrust | $F$ | N |
| Distance from thrust point to torsion center | $l_{Th.}$ | m |
| Distance from pendulum arm centroid to torsion center | $l_{ro.}$ | m |
| Distance from counterweight to torsion center | $l_{cout.}$ | m |
| Distance from measurement point to torsion center | $l_{sen.}$ | m |
| Thruster mass | $m_{Th.}$ | kg |
| Pendulum arm mass | $m_{ro.}$ | kg |
| Counterweight mass | $m_{cout.}$ | kg |
| The gravitational acceleration | $g$ | $ms^{-2}$ |
| Gravity of the whole pendulum | $G$ | N |
| Tangential component of the gravity of the whole pendulum | $G_{tangent}$ | N |
| Deflection angle of the pendulum | $\theta$ | rad |
| Horizontal displacement at measuring point | $u$ | m |

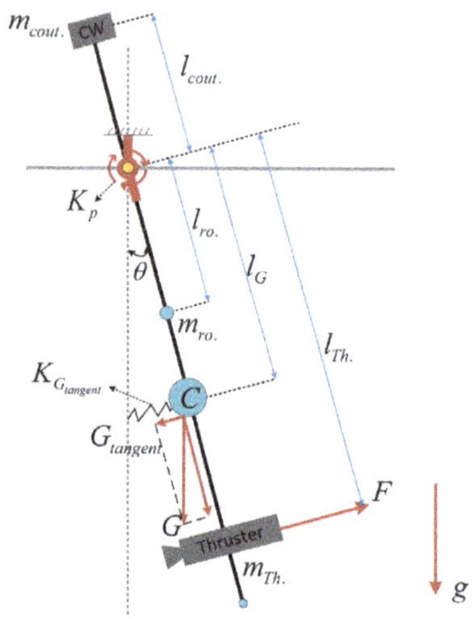


**Figure 3.** Schematic diagram of a moment of the compound pendulum swing.

Rather than directly measuring the angle $\theta$, we characterize the magnitude of the thrust by measuring the horizontal displacement of a point on the pendulum arm at a distance $l_{sen.}$ from the assumed rotational axis in real applications. Therefore, we have

$$F = \frac{K_{ang}}{l_{Th.}} * \frac{u}{l_{sen.}} = K_{line} * u = \frac{K_p + K_{G_{tangent}}}{l_{Th.}l_{sen.}} * u \tag{3}$$

where $u$ is the horizontal displacement of the measurement point. $K_{line}$ is the linear stiffness of the pendulum, satisfying $K_{line} = K_{ang}/[l_{Th.}l_{sen.}]$. From Equation (3), it can be found that all symbols except the bending stiffness $K_p$ of the flexure hinge are characterized by the most refined parameters. The bending stiffness of the hinge, as the key component of the compound pendulum deflection, requires a more explicit parametric equation. Furthermore, if the mass parameter $m_*$ (the subscript "$*$" denotes "$Th.$", "$ro.$", and "$cout.$") will affect the stiffness $K_p$, or if the non-ideal deformation of the hinge will affect the displacement measurement value $u$, then continuing to use Equation (3) to calculate and analyze will result in deviations. Hence, an improvement needs to be made to the conventional model.

## 4. The Improved Analytical Model of the Thrust Stand

In the previous section, the thrust measurement principle of the compound pendulum under quasi-static conditions is given in detail. Nevertheless, the effect of gravitational effects on the thrust measurement system for each component of the pendulum is only partially considered—providing the torsional return stiffness of the pendulum. The component of gravity in the normal direction of rotation of the pendulum will exert an axial tensile force on the flexure hinge, which will introduce a shifting of the original $K_p$. Moreover, an offset in the assumed center of rotation during pendulum deflection can also lead to a false displacement measurement $u$. Therefore, an improved analytical model of the pendulum based on the establishment of a comprehensive theory of flexure hinge bending is presented.

### *4.1. Hinge Bending Deflection Modeling*

The flexure hinge used in the thrust stand is a symmetrical elliptical notched structure, whose physical style is shown in Figure 2, and the structural schematic is given in Figure 4. Based on the Euler–Bernoulli beam theory, it is assumed that the small deflection satisfies the material elastic deformation theory. Moreover, the hinge can only rotate in the XY plane.

Here, we define the $X$-axis as the axial direction of the hinge. The deformations caused by shear and torsion in the other directions are ignored.

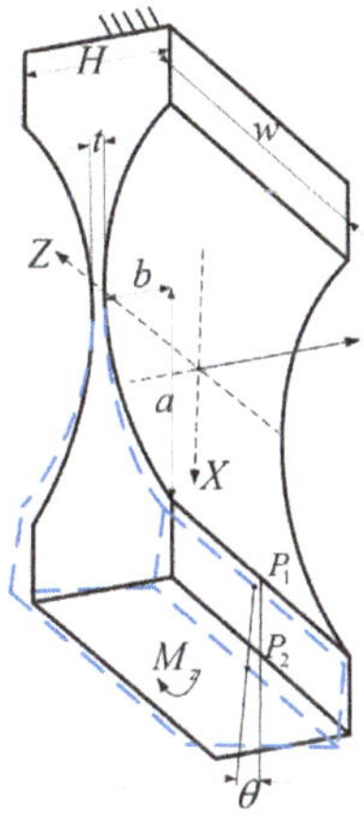


**Figure 4.** Bending diagram of a flexure hinge with elliptical notched boundary.

In Figure 4, the blue dashed part indicates the small deflection around the $Z$-axis (the assumed rotational center) under the bending moment $M_Z$. $P_1$ and $P_2$ are located at the midpoints of the upper and lower sides of the deformed end, respectively. The line connecting the two can be used to indicate the deflection angle $\theta$. $w$ is the width of the hinge and $H$ is the height of the hinge. $a$ and $b$ are the long and short axes of the elliptical notched portion, respectively. $t$ is the thickness of the thinnest part in the middle of the hinge, where $t = H - 2b$. The elliptical notched structure represents an intermediate solution between the semicircular structure and the leaf structure. When the long and short axes are equal ($a = b$), the hinge of the elliptical notch evolves into a semicircular flexure hinge. When $b$ approximates to 0, the elliptical notched hinge can be equivalent to a hinge of the leaf structure. Therefore, the analysis of this elliptical notched hinge is of general significance.

There are many methodologies to calculate $\theta$ under the bending moment $M_Z$ [31–35]. The commonly used one is to obtain a concise formula for calculating the deflection angle by introducing the elliptic centrifugal angle as an integral variable. The schematic of the bending deflection modeling is shown in Figure 5.

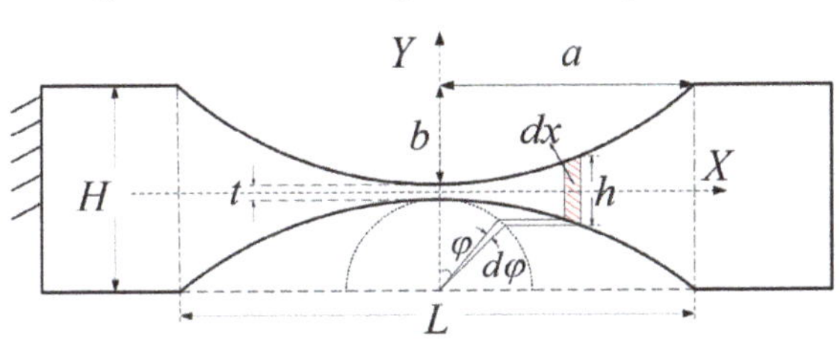


**Figure 5.** Schematic diagram of bending deflection modeling based on elliptic centrifugal angle.

In Figure 5, $\varphi$ is defined as the centrifugal angle of the ellipse, so an arbitrary point $q$ on the ellipse is given [36]:

$$\begin{cases} x_q = a \sin \varphi \\ y_q = b \cos \varphi \\ h(x) = h(\varphi) = 2b + t - 2b \cos \varphi \end{cases} \quad (4)$$

where $\varphi$ has a range of $[-\frac{\pi}{2}, \frac{\pi}{2}]$.

It is assumed that the deformation of the hinge is mainly concentrated in the elliptical part $L$ where $L = 2a$ and the deformation of other parts is neglected. The bending of the continuous beam of the variable section is regarded as the result of the accumulation of many tiny bending deformations. Each tiny section is regarded as a rectangular section beam of length $dx$ (the red shaded part in Figure 5). In general, the size of the hinge is much smaller than that of the pendulum arm, and the bending moment variation can be considered small enough to treat $M_Z$ as a constant.

According to the theory of a small deformation of the cantilever beam, we have the following differential equation of the deflection line:

$$EI(x)\frac{d^2y}{dx^2} = M_Z \tag{5}$$

where $E$ is the elastic modulus of the material, and $I(x)$ is the moment of inertia of the section about the central axis with $I(x) = (1/12)w[h(x)]^3$.

Combining Equations (4) and (5), the hinge deflection angle can be obtained,

$$\theta = \frac{dy}{dx} = \int_{-a}^{a} \frac{12M_Z}{Ew[h(x)]^3}dx = \frac{12M_Z a}{Ewt^3}\int_{-\pi/2}^{\pi/2} \frac{\cos\varphi}{\Phi(\varphi)^3}d\varphi \tag{6}$$

where $\Phi(\varphi) = h(\varphi)/t = 2\sigma + 1 - 2\sigma\cos\varphi$ and $\sigma = b/t$. The expression of the integral term is obtained using the MATHEMATICA:

$$\int_{-\pi/2}^{\pi/2} \frac{\cos\varphi}{\Phi(\varphi)^3}d\varphi = \frac{12\sigma^2 + 8\sigma + 2}{(4\sigma+1)^2(2\sigma+1)} + \frac{12\sigma(2\sigma+1)\arctan\sqrt{4\sigma+1}}{(4\sigma+1)^{5/2}} \tag{7}$$

In summary, the bending stiffness of the flexure hinge in the direction of thrust measurement can be parametrically modelled as

$$K_p = \frac{M_Z}{\theta} = \frac{Ewt^3}{12a\gamma_1} \tag{8}$$

where the coefficient $\gamma_1$ is

$$\gamma_1 = \frac{12\sigma^2 + 8\sigma + 2}{(4\sigma+1)^2(2\sigma+1)} + \frac{12\sigma(2\sigma+1)\arctan\sqrt{4\sigma+1}}{(4\sigma+1)^{5/2}} \tag{9}$$

The above parametric modeling successfully establishes the relationship between hinge bending stiffness and its geometric dimensions and refines the characterization of $K_p$. Its modeling method also provides the basis for other stiffness modeling research to follow.

### 4.2. Gravity-Induced Extension Effect

As a high-precision compliant mechanism, a flexure hinge often exhibits an increase in length and a decrease in thickness in the axial extension effect under a whole pendulum load (including the counterweight, thruster, and pendulum arm, excluding the hinge itself). These will change the original bending stiffness of the hinge. The bending stiffness shift introduced by the axial extension effect is studied below.

The pendulum, considering the extension effect, is shown in Figure 6. In Figure 6, $G_{axial}$ is the axial component of the gravity of the whole pendulum load, perpendicular to the swing direction. The restoring force generated by the tangential component of gravity $G_{tangent}$ has been considered in the conventional analytical model, and it is modeled as an equivalent spring $K_{G_{tangent}}$. While the angle $\theta$ is very small, there is $G \approx G_{tangent}$. If the load on the hinge is large, the axial extension effect due to $G_{axial}$ is more pronounced, which is manifested as a dimensional change. For this reason, we propose characterizing the gravity-induced extension effect by modeling an additional equivalent spring $K_{G_{axial}}$

(see Figure 6). The zero potential energy point of this spring is in the free vertical state of the pendulum.

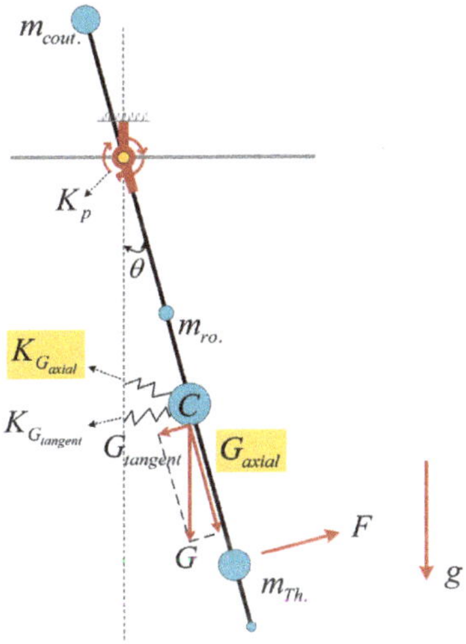


**Figure 6.** Schematic diagram of a moment of pendulum swing under extension effect.

The equivalent spring $K_{G_{axial}}$ is attached to $K_p$ so that it is shifted and

$$K'_p = K_p + K_{G_{axial}} \tag{10}$$

During the preparation phase of the thrust measurement, the loads are mounted, and the extension effect of the static axial force $T$ on the hinge is shown in Figure 7. $\Delta L$ is the deformation in the axial direction, and the yellow point $o$ and the green point $o'$ are the centers before and after the deformation, respectively. $t'$ is the thickness after deformation at the center in the middle part, with $t' = t - \Delta t$ where $\Delta t$ is the variation due to the extension effect. In the case of a large cross-sectional aspect ratio, the variation in the hinge width $w$ is negligible.

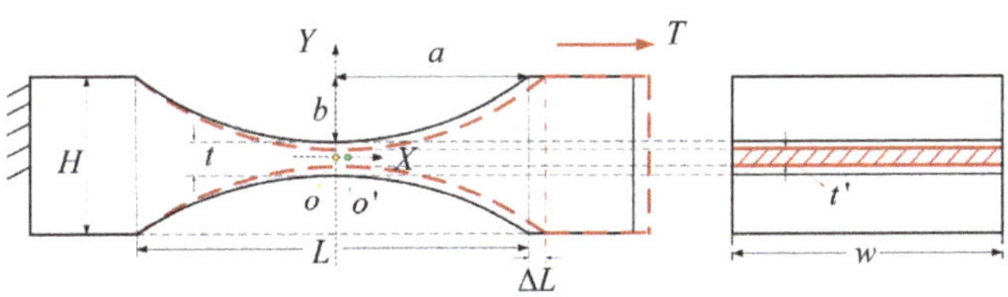


**Figure 7.** Schematic diagram of the geometric dimensional changes under axial tension of the flexure hinge.

Applying Castigliano's second theorem, it follows that

$$\Delta L = \frac{\partial U}{\partial T} \tag{11}$$

where $U$ is the deformation energy, which is given by

$$U = \int \frac{T^2}{2EA(x)} dx \tag{12}$$

where $A(x)$ is the cross-sectional area of the hinge, and $A(x) = w * h(x)$. Combining Equations (11) and (12) gives

$$\Delta L = \int_{-L/2}^{L/2} \frac{T}{Ewh(x)} dx = \frac{T}{Ew} \int_{-L/2}^{L/2} \frac{1}{h(x)} dx \tag{13}$$

Using the "elliptic centrifugal angle" integration methodology (see Section 4.1 for details), the length elongation is

$$\Delta L = \frac{aT}{Ewt} \int_{-\pi/2}^{\pi/2} \frac{\cos\varphi}{\Phi(\varphi)} d\varphi = \frac{aT}{Ewt} \gamma_2 \tag{14}$$

where $\gamma_2$ satisfies the equation obtained from the MATHEMATICA:

$$\gamma_2 = \frac{-\pi\sqrt{4\delta + 1} + 4(2\delta + 1)\arctan\sqrt{4\delta + 1}}{2\delta\sqrt{4\delta + 1}} \tag{15}$$

where $\delta = b/t$.

Furthermore, from the theory of Poisson's ratio ($v$) of material, the minimum thickness change in the hinge can be described by

$$\Delta t = t * \varepsilon_{lateral} = tv\varepsilon_{lateral} = tv\frac{T}{Ewt} = \frac{Tv}{Ew} \tag{16}$$

Notably, there exists $G_{axial} \approx G$ at tiny deflection angles, so that

$$T \approx G \tag{17}$$

Referring to Equations (8) and (9), the improved bending stiffness of the hinge under the tensile pre-stress of the whole pendulum load can be parameterized as

$$K'_p = K_p + K_{G_{axial}} = \frac{Ew(t - \Delta t)^3}{12(a + \frac{\Delta L}{2})\gamma_1[t - \Delta t]} \tag{18}$$

where $\gamma_1[t - \Delta t]$ denotes the value of $\gamma_1$ after the minimum thickness has been reduced by $\Delta t$.

It is worthwhile to report that $K_{G_{tangent}}$ can be represented explicitly, while $K_{G_{axial}}$ originates from the additional effect of the hinge's own geometrical boundary change, which cannot be expressed separately due to the complexity of the calculation.

### 4.3. Rotational Center Offset Effect

In this section, we proceed to model the offset of the assumed rotational center. An accurate calculation of the thrust based on an analytical model is not only related to the stiffness coefficient $K_{line}$ (see Equation (3)), but also depends on the reliability of the displacement $u$. That is, the displacement $u$ has to be generated by the thrust rather than other unknown factors. However, the hinge is subject to a non-ideal situation during bending deformation—the assumed rotational center would have an offset.

Figure 8a shows that the assumed rotational center will be offset from the position of point $o$ (the yellow dot) to the position of $o'$ (the green dot) during the bending deformation of the hinge. Figure 8b exhibits the uncertainty of the displacement measurement due to the assumed rotational center offset effect $\Delta u$. Here, the red part is the bent hinge, the black line is the pendulum arm, $u$ is the measured displacement, and $u'$ is the displacement generated by the actual thrust. Note that the interference of other environmental noises such as seismic vibration and thermal effects on the measurement is not included in this study.

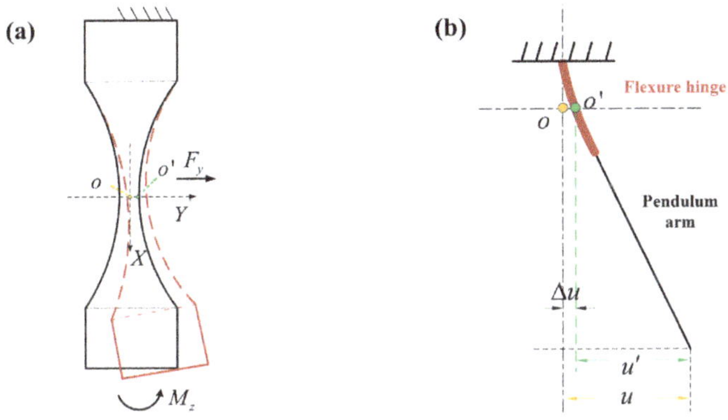


**Figure 8.** Flexure hinge bending deformation. (**a**) Axial section diagram; (**b**) measurement displacement schematic.

In order to characterize the offset effect at point $o$, a zero-value force $F_y$ is assumed to be applied at point $o$ along the Y-axis. According to Castigliano's second theorem, the following equations exist:

$$\begin{cases} \Delta u = \frac{\partial U}{\partial F_y} \\ U = \int \frac{[-M_Z - F_y(-x)]^2}{2EI(x)} dx \end{cases} \tag{19}$$

The symbols used in Equation (19) and the subsequent formulas in this section refer to the previous two subsections. Thus, $\Delta u$ can be derived from the above formula:

$$\Delta u = \frac{12M_Z}{Ew} \int_{-a}^{0} \frac{-x}{[h(x)]^3} dx = \frac{12M_Z a^2}{Ewt^3} \int_{-\pi/2}^{0} \frac{-\sin\varphi\cos\varphi}{\Phi(\varphi)^3} d\varphi = \frac{12M_Z a^2}{Ewt^3}\gamma_3 \tag{20}$$

The coefficient $\gamma_3$ is expressed as

$$\gamma_3 = \frac{1}{2+4\sigma} \tag{21}$$

where $\sigma = b/t$. As such, we define the offset line stiffness of the assumed rotational center as

$$K_d = \frac{M_Z}{\Delta u} = \frac{Ewt^3}{12a^2\gamma_3} \tag{22}$$

*4.4. The Improved and Corrected Analytical Model*

In this section, we formulate the improved model of the pendulum considering the extension effect and assumed rotational center offset. During the measurement operation of the thrust stand, the flexure hinge exhibits the characteristics first of elongation and thinning, and then bending. The bending line stiffness ($K_{line}$) of the pendulum changes accordingly. Moreover, as the hinge bends, the assumed center of rotation is offset, resulting in an error in the displacement measurement ($u$).

Compared to the conventional analytical model represented by Equation (3), the improved analytical model is given as

$$F = K'_{line} * u' = K'_{line}(u - \Delta u) = K'_{line}u - K'_{line}\frac{12l_{Th.}Fa^2}{Ewt^3}\gamma_3 = K'_{line}u - K'_{line}\frac{12l_{Th.}a^2\gamma_3}{Ewt^3}F \tag{23}$$

where $K'_{line}$ is the modified linear stiffness of the pendulum, and $K'_{line} = \frac{K'_p + K_{G_{tangent}}}{l_{Th.} l_{sen.}}$. $\Delta u = \frac{12 l_{Th.} F a^2}{E w t^3}$ is obtained from the offset stiffness using Equation (22) with $M_Z = l_{Th.} F$. Simplifying the resulting expression yields

$$F = \frac{K'_{line}}{(1 + K'_{line} \frac{12 l_{Th.} a^2 \gamma_3}{E w t^3})} * u \tag{24}$$

Finally, the improved and corrected thrust analytical model is given as

$$F = \frac{K'_{line}}{(1 + K'_{line} W)} * u \tag{25}$$

where

$$\begin{cases} K'_{line} = \frac{K'_p + K_{G_{tangent}}}{l_{Th.} l_{sen.}} \\ W = \frac{12 l_{Th.} (a + \frac{\Delta L}{2})^2 \gamma_3 [t - \Delta t]}{E w (t - \Delta t)^3} \\ K'_p = \frac{E w (t - \Delta t)^3}{12 (a + \frac{\Delta L}{2}) \gamma_1 [t - \Delta t]} \\ K_{G_{tangent}} = g(m_{Th.} l_{Th.} + m_{ro.} l_{ro.} - m_{cout.} l_{cout.}) \\ \gamma_1 [t - \Delta t] = \frac{12\sigma^2 + 8\sigma + 2}{(4\sigma + 1)^2 (2\sigma + 1)} + \frac{12\sigma (2\sigma + 1) \arctan\sqrt{4\sigma + 1}}{(4\sigma + 1)^{5/2}} \\ \gamma_3 [t - \Delta t] = \frac{1}{2 + 4\sigma} \\ \sigma = \frac{b + \Delta t / 2}{t - \Delta t} \\ \Delta t = \frac{G v}{E w} \\ \Delta L = \frac{a G}{E w t} \gamma_2 \\ G = g(m_{Th.} + m_{ro.} + m_{cout.}) \\ \gamma_2 = \frac{-\pi\sqrt{4\delta + 1} + 4(2\delta + 1)\arctan\sqrt{4\delta + 1}}{2\delta\sqrt{4\delta + 1}} \\ \delta = \frac{b}{t} \end{cases}$$

Here, the conventional Equation (3) is revised as Equation (25). In particular, the bending stiffness shift in the flexure hinge under an extension effect and the offset of the assumed center of rotation after bending deformation are all taken into account in the new model. In addition, the parametric equations for the bending and tensile deformations of the hinge are derived in detail. Since the elongation and thinning behavior of the hinge occurs before the bending behavior, the initial conditions for the respective modeling are different, so different symbols ($\delta$ and $\sigma$) denote the intermediate variables involved.

## 5. Case Study and Discussion

In this section, case studies and discussions are presented. Firstly, we propose a finite element modeling solution for an elliptical notched hinge. Then, the validity and generalizability of the proposed parametric analytical model are verified using the finite element analysis (FEA) method. Finally, the necessity of the proposed improvements is demonstrated by analyzing the parametric design of multiple sets of thrust stands.

### *5.1. The Improved and Corrected Analytical Model*

The FEA technique is a powerful tool for research in the field of structural mechanics. The well-modeled finite element model (FEM) has very high accuracy in the problem domain of this study. In this research, the FEM results are adopted as a benchmark to verify the validity of the proposed analytical model for thrust analysis. Reasonably, we are interested in the elastic deformation of the flexure hinge. Hence, the bending and tensile deformations of the pendulum arm are neglected by assuming the pendulum arm to be rigid. Moreover, the proposed analytical model points out that the mechanical properties of hinges not previously considered are the core link of the research. Therefore, the focus of finite element modeling is the flexure hinge.

COMSOL Multiphysics 6.0 as a very mature FEA simulation software will be applied. In the FEA, the accuracy of the solution is related to the mesh size. In principle, a finer custom meshing brings higher accuracy. However, it is difficult to be adopted in real applications due to the limitation of time and computing resources. Nevertheless, unstructured meshing algorithms represented by tetrahedral elements are prone to low-quality meshes with distorted tetrahedral shapes when meshing structures with large aspect ratios such as flexure hinges. Consequently, a solution for meshing is given to make the FEM results convincing.

In the finite element modeling, as shown in Figure 9a, the left side of the hinge is the fixed end, and the right side is the free end. A shear force $F$ is applied at the middle position of the free end to generate torque with respect to the assumed center of rotation. Based on Saint-Venant's principle, the length of the non-hinged part is designed to be $L_c \geq 3H$ to avoid the effect of fixed restraint and load on the hinge part with inhomogeneous stress. For the building of the hinge mesh, we divide it into three parts: the fine region A, the transition region B, and the coarse region C, which are described as follows,

(1) Region A is the thinnest region of the hinge, has a length about 1/3 of the entire one, and has the characteristics of a large aspect ratio. The stresses and strains in both bending and tensile deformation are large. It is a key concern in force analysis. Therefore, a controlled structured hexahedral mesh is used instead of an unstructured one for the meshing (see Figure 9b). A three-level hexahedral mesh is established in the thickness direction, and 30 and 100 elements are divided in the axial and width directions, respectively, using "mapped" and "swept" techniques to accomplish the above operations. Accordingly, region A is equivalent to a large number of healthy micro-cantilevers.

(2) Region B contains the part of the hinge root with larger curvature, the hexahedral element is no longer applicable, and the physical field-controlled tetrahedral element is used to build the mesh. Furthermore, in order to avoid a poor-quality mesh in the narrow region of the hinge root, a virtual mesh technique is applied to its root to supplement a circular arc-shaped region (see Figure 9a); this region is only used to distinguish the difference between the meshes, and does not have an actual physical partitioning function (i.e., the machining of the hinge shown in Figure 9a is shaped in one piece).

(3) Region C is the part outside the hinge, which is not the focus of attention, so it is subjected to a coarser free-division tetrahedral mesh.

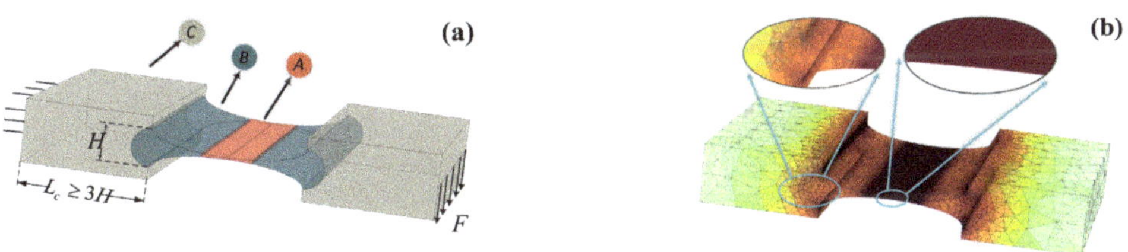


**Figure 9.** Meshing of the finite element model. (**a**) Mesh area division; (**b**) mesh division results.

The above meshing scheme for the finite element modeling of the flexure hinge can provide a feasible FEA solution strategy not only for thrust stands but also for flexure hinge structures in other devices.

### 5.2. Results and Discussion

In this section, we present comparisons between the parametric equations and FEA in order to evaluate the accuracy and generalizability of the proposed parametric equations to describe the mechanical behavior of the hinge. Beryllium copper material is used in the flexure hinge due to its advantages of high mechanical strength and high temperature stability. Table 2 gives the initial geometric parameters and mechanical parameters of the hinge.

**Table 2.** Initialization of the flexure hinge parameters.

| Parameter | Symbol | Value |
|---|---|---|
| Length | $L$ | 12 mm |
| Width | $W$ | 20 mm |
| Height | $H$ | 3 mm |
| Minimum thickness | $t$ | 0.1 mm |
| Elliptic long axis | $a$ | 6 mm |
| Elliptic short axis | $b$ | 1.45 mm |
| Young's modulus | $E$ | 110 Gpa |
| Poisson's ratio | $v$ | 0.3 |
| Density | $\rho$ | 8750 kg/m$^3$ |

The primary concern is the reliability of the equations describing the assumed rotational center offset of the hinge in the proposed model. It is noticeable that the length change in the hinge under axial tension can also be interpreted as an axial offset of the assumed rotational center (see Figure 7), except that this offset does not lead to uncertainty in the displacement measurement. The linear stiffness $K_{T\_line}$ and $K_{d\_line}$ obtained from the analytical model are $K_{T\_line} = Ewt/a\gamma_2$ and $K_{d\_line} = Ewt^3/[12a^2(a + L_c)\gamma_3]$ (refer to Equation (14) and Equation (22)), respectively, which are compared with the FEM results to verify the stiffness equations under a small deformation. Figure 10 shows a schematic of the access of FEM results. Five points are selected at equal intervals at the front (fixed end), middle, and rear (free end) of the hinge section, distinguished by black, green, and red in turn (see Figure 10). The mean values of their respective displacements are calculated to represent the displacements at the front, middle, and rear of the hinge. Thus, the bending and tensile deformations caused by the black points are subtracted from the total deformation obtained from the red points or green points to obtain the pure deformation of the flexure hinge itself.

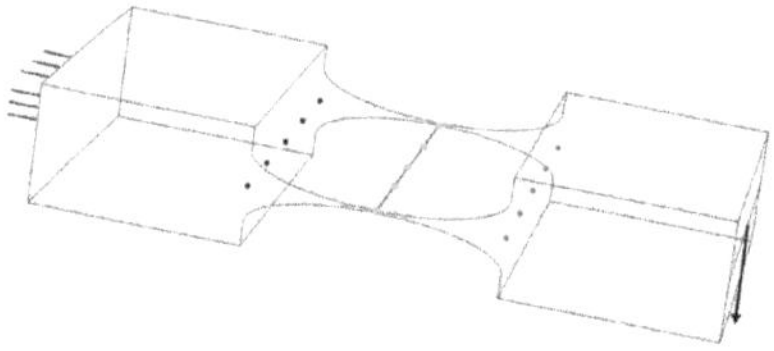

**Figure 10.** A schematic of the access of FEM results.

In real applications, the minimum thickness of the flexure hinge frequently becomes the point of penetration for customizing the mechanical properties of the hinge. Therefore, the variation range of the minimum thickness $t$ is set to be 0.1~0.4 mm, and the hinge height $H$ is kept constant. As formula $H = 2b + t$ needs to be satisfied, the short axis $b$ matches 1.3~1.45 mm. The other parameters are listed in Table 2. The comparisons within this parameter range are given in Figure 11.

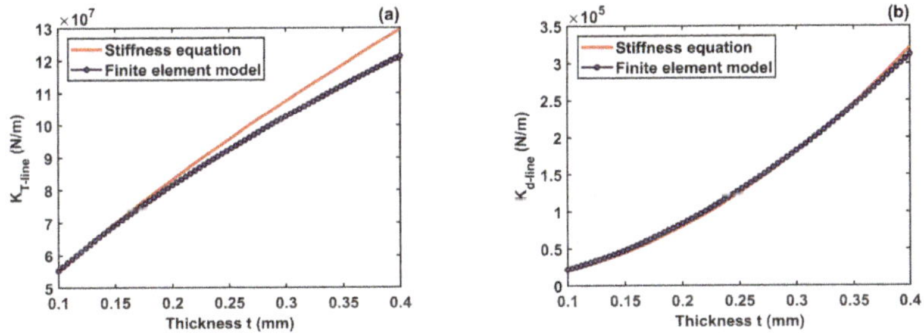


**Figure 11.** Comparison of stiffness equation and FEM. (**a**) Tension line stiffness $K_{T\_line}$; (**b**) offset line stiffness $K_{d\_line}$.

Figure 11a,b present the comparative results of tension line stiffness and offset line stiffness under the stiffness equation and FEM, respectively. It can be obtained that the result curves calculated by the stiffness equations agree well with the result curves of FEM, and they also have the same trend of change. In addition, the hinge tension line stiffness is much greater than the offset line stiffness by about two orders of magnitude over a range of variations in parameter $t$. This is because the function of the hinge is to provide bending rather than axial elongation. The comparative results seen for an incomplete fit mainly come from the assumptions made in those theoretical derivations. It can be substantiated that the equations (Equations (14) and (22)) describing the hinge rotational center offset in the proposed analytical model are valid when the thickness $t$ is within the allowed variation range.

The accuracy of the hinge bending stiffness equation is central to the persuasiveness of the proposed model. With reference to $K_p$ (see Equation (8)), we define the bending line stiffness of the hinge as $K_{p\_line} = Ewt^3/[12a(a+L_c)a\gamma_1]$. It is important to point out that the actual displacement produced by the force $F$ in the FEM should be the displacement of the free end of the hinge minus the offset of the middle of the hinge due to the offset of the assumed rotational center. The comparison is plotted in Figure 12. The results of the equation calculation and the FEM results also have small discrepancies and the same trends, which verifies the stiffness equation (Equation (8)). Comparing Figure 11b, the offset line stiffness is only about 10 times higher than the bending line stiffness for the hinge itself. Even if the thrust is applied at the end of the pendulum arm rather than at the end of the hinge in practice, this multiplier is magnified again. However, the system still has significant measurement uncertainty due to the offset of the assumed rotational center. The subsequent analysis will give a detailed confirmation of this point.

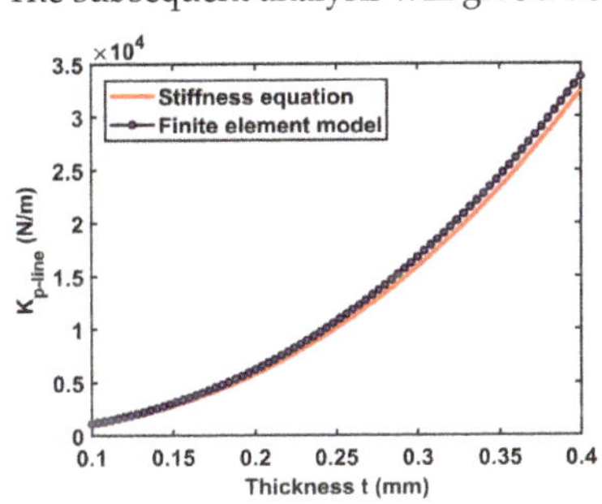


**Figure 12.** Comparison of stiffness equation for bending line stiffness $K_{p\_line}$ with FEM.

In order to further investigate the influence of dimensional changes on the bending stiffness under a gravity-induced extension effect in the hinge proposed in the analytical model (refer to Equation (18)), an axial tensile force is first applied to the hinge in the FEM. Then, the changing dimensional parameters are recorded and modified before applying force $F$, which provides torque. It is significant to emphasize that the axial tensile force needs to be removed after the correction of the hinge dimensional parameters; otherwise, an additional stiffening effect will be introduced. Meanwhile, the offset of the assumed rotational center is also considered in the FEA. Based on the thrust stand we designed, the whole pendulum load needs to reach about 10 kg. Thus, we choose $T \approx G \rightarrow 100\mathrm{N}$. The results of the bending line stiffness variation obtained using both approaches are shown in Figure 13. Such results show that the overall effect of the dimensional change is a decrease in the bending line stiffness. It can be observed that the stiffness equation successfully predicts the variation in the bending line stiffness under the axial tensile force, and it coincides very well with the FEM fitting (the green curve in Figure 13). Among them, the best prediction is achieved when the thickness $t$ lies in the range of 0.1 mm to 0.2 mm.

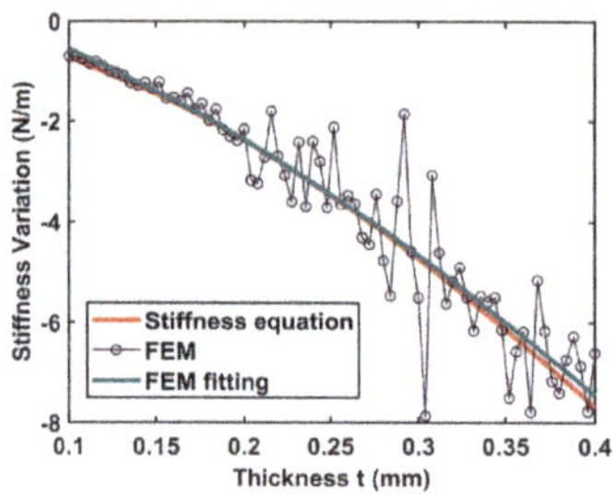


**Figure 13.** Bending line stiffness variation of the flexure hinge under gravity-induced extension effect.

In addition, combined with Equation (10), it can be seen that the variation in stiffness is equivalent to $K_{G_{axial}}$ in the formula, and the difference between the two is a conversion multiple of linear stiffness and torsional stiffness. In other words, the presented analytical model can accurately give the numerical solution of the bending stiffness shift in the flexure hinge under a certain pendulum load, indicating that it makes sense.

To demonstrate the generalizability of the improved analytical model for different notch boundary sizes and dimensions of the hinge, two additional sets of studies (see Table 3) are carried out as follows. Figure 14 shows the comparisons for Case I. It is clear from Figure 14a–c those stiffness parametric equations give a precise description of the corresponding stiffness even under the condition of a wide range of variation in the elliptic long axis parameter a. Since there are several outlier points in the FEM result, the first fitting (the green curve) is not convincing, as shown in Figure 14d. The fitting result of the post-processing FEM (the yellow curve) is highly consistent with that of the stiffness equation. The comparisons under Case II are plotted in Figure 15. The results demonstrate the predictive performance of these parametric equations. Notably, the bending line stiffness variation under an axial extension effect is independent of the hinge width.

**Table 3.** Two additional sets of the flexure hinge geometric parameters traversed.

| | $t$ | $b$ | $a$ | $H$ | $L$ | $W$ |
|---|---|---|---|---|---|---|
| | | | [mm] | | | |
| Case I | 0.1 | 1.45 | 1.45~6 | 3 | $2 * a$ | 20 |
| Case II | 0.1 | 1.45 | 6 | 3 | 12 | 10~20 |

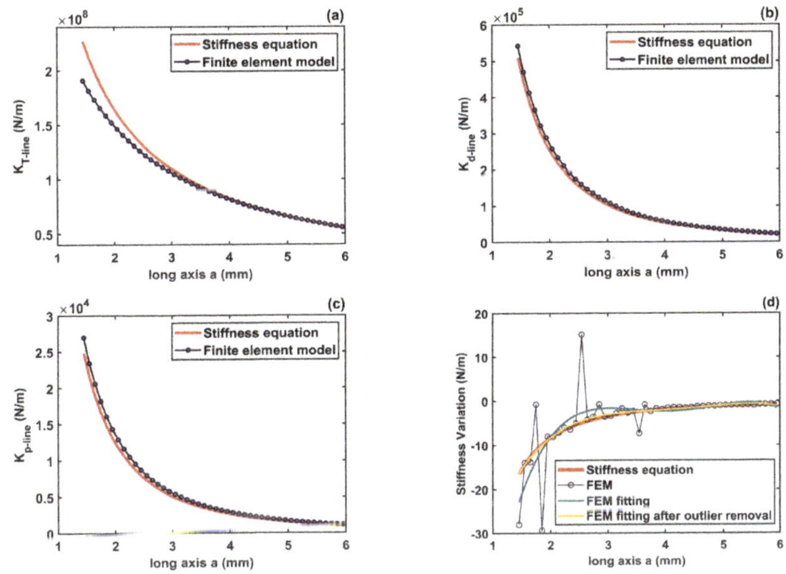


**Figure 14.** Comparisons of the stiffness equations and FEM under Case I. (**a**) Tension line stiffness $K_{T\_line}$; (**b**) offset line stiffness $K_{d\_line}$; (**c**) bending line stiffness $K_{p\_line}$; (**d**) bending line stiffness variation under axial tensile force.

Furthermore, Figure 16 demonstrates the comparisons under three typical flexure hinges. The three hinges maintain a consistent minimum thickness of $t = 0.1$ mm, and their key boundary parameters are provided by the table in the figure. The results again verify the improved analytical model. In the meantime, it can be summarized that the elliptical notched hinge has both the advantage of the circular notched hinge, which has a clear center of rotation, and the advantage of the leaf-type hinge, which has less bending line stiffness. This is because with a very large bending line stiffness, the displacement from a micro-thrust may be not perceptible.

In summary, these case studies on the flexure hinge reflect, on the one hand, the validity and generalizability of the proposed new quasi-static thrust analytical model. On the other hand, the variation laws of those stiffness properties of the hinge with the changes in its geometrical parameters are shown in those comparative analyses. Therefore, based on the parametric equation of the bending stiffness, the appropriate $K_p$ can be easily found for the thrust stand with the selection of its independent variables. In the next section, we demonstrate the necessity of improvements to the conventional model analytically.

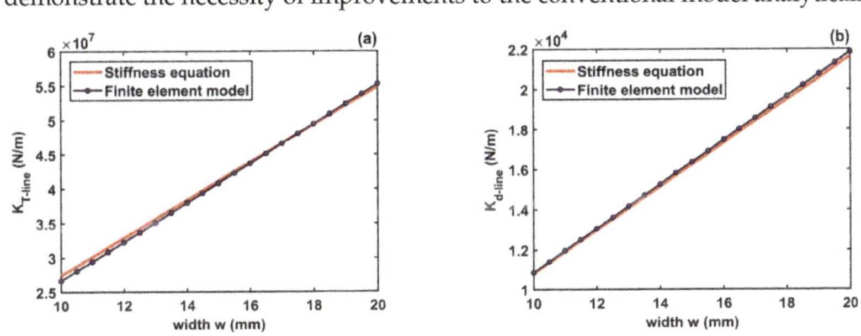


**Figure 15.** *Cont.*

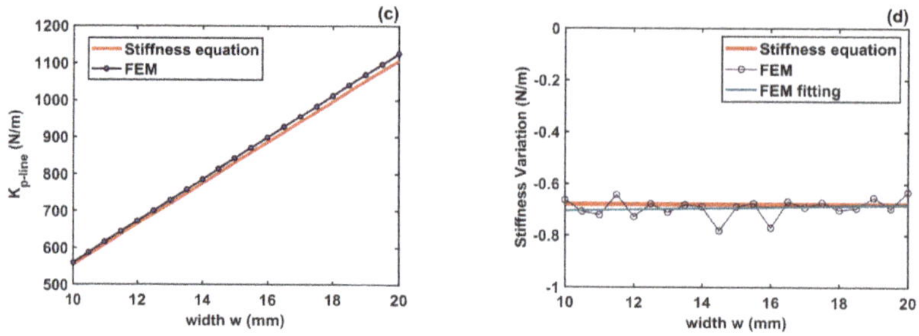


**Figure 15.** Comparisons of the stiffness equations and FEM under Case II. (**a**) Tension line stiffness $K_{T\_line}$, (**b**) offset line stiffness $K_{d\_line}$, (**c**) bending line stiffness $K_{p\_line}$, (**d**) bending line stiffness variation under axial tensile force.

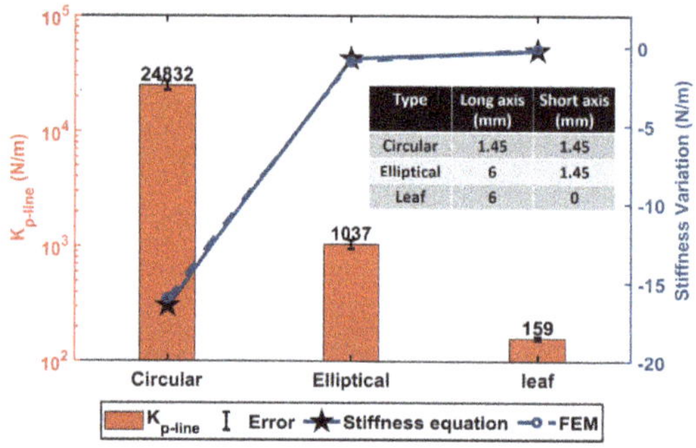


**Figure 16.** A bending line stiffness comparison of three typical structures of flexure hinges.

## 5.3. Discussion for Thrust Measurement

In the previous sections, we verified the accuracy of the improved analytical model in characterizing the neglected parts in conventional thrust analytical models, including the shift in the bending stiffness $K_p$ and the offset of the assumed rotational center during the actual operation. In this section, we illustrate that the presence of these non-ideal factors can introduce large uncertainties into the theoretical analysis of thrust measurements, which is worthy of attention by designers of thrust stands.

On the basis of what is studied in Figures 12 and 13 (the parameter $b$ changes with $t$ ($t \in (0.1, 0.4)$mm, and $H$ remains constant), attention is paid to the impact of the change in the hinge bending stiffness on the thrust analysis. We rewrite Equation (24) as follows:

$$F_{revise} = \frac{K'_p + K_{G_{tangent}}}{l_{Th.}l_{sen.}} * u' = \left(\frac{K_p + K_{G_{tangent}}}{l_{Th.}l_{sen.}} + \frac{K_{G_{axial}}}{l_{Th.}l_{sen.}}\right) * u'$$

where $F_{revise}$ is the revised thrust calculated by the improved analytical model. And

$$\begin{cases} K'_p = K_p + K_{G_{axial}} \\ K_{G_{tangent}} = g(m_{Th.}l_{Th.} + m_{ro.}l_{ro.} - m_{cout.}l_{cout.}) \end{cases}$$

Figures 12 and 13 show that the equivalent spring stiffness $K_{G_{axial}}$ introduced by the gravitational force component $G_{axial}$ is a negative number and its magnitude is small with respect to the original stiffness $K_p$ (about 0.02% to 0.06% of the total). Thus, if the pendulum is designed so that the pendulum load gravity component $G_{tangent}$ provides the main torsional return stiffness, then the effect of the shift in $K_p$ will be very slight. On the other hand, a problem with this is that the heavy load requirements of the pendulum often make the equivalent stiffness $K_{G_{tangent}}$ too large, making displacement measurements much more difficult. Therefore, the designer will adjust the counterweight to make $K_{G_{tangent}}$ as small as possible, or even have a negative equivalent stiffness $K_{G_{tangent}}$ to weaken the stiffness effect of $K_p$, so as to obtain a lower total stiffness coefficient $K_{line}$. This is exactly what our team is trying to undertake. In such a case, the effect of the shift in $K_p$ on the thrust will be magnified, and needs to be seriously considered.

To quantify the above analyses, several case studies are carried out, with the parameters for the compound pendulum given in Table 4. The six designs are divided into three groups: A, B, and C. The minimum thickness of the hinge is $t = 0.1$mm in both groups A and B, while $t$ is 0.3 mm in group C. The other geometric parameters of the hinge are provided in Table 2. The difference between groups A and B lies in the different thrusters and counterweight masses; the total stiffness coefficient $K_{line}$ is adjusted by changing the height of the counterweight bar within each group. The position of the thruster is fixed in different groups, and the length and mass of the pendulum arm vary with the height of the counterweight bar. The acceleration of gravity is chosen as $g = 9.8\text{ms}^{-2}$. The distance from the measurement point to the assumed rotational center is $l_{sen.} = 0.5$m, and the true value of the displacement at the measurement point is assumed to be $u' = 100$um. The results of the stiffness and thrust obtained with these three groups of parameters are reported in Table 5, where $F_{abs\_error}$ is the absolute error of thrust and $F_{re\_error}$ is the relative error.

**Table 4.** Designs of the compound pendulum parameters.

| | $t$ | $m_{Th.}$ | $l_{Th.}$ | $m_{ro.}$ | $l_{ro.}$ | $m_{cout.}$ | $l_{cout.}$ |
|---|---|---|---|---|---|---|---|
| | [mm] | [kg] | [m] | [kg] | [m] | [kg] | [m] |
| A1 | 0.1 | 3 | 0.5 | 0.4529 | 0.1377 | 7 | 0.2246 |
| A2 | 0.1 | 3 | 0.5 | 0.4500 | 0.1400 | 7 | 0.2200 |
| B1 | 0.1 | 2 | 0.5 | 0.4528 | 0.1378 | 4 | 0.2245 |
| B2 | 0.1 | 2 | 0.5 | 0.4519 | 0.1385 | 4 | 0.2230 |
| C1 | 0.3 | 3 | 0.5 | 0.4656 | 0.1275 | 7 | 0.2450 |
| C2 | 0.3 | 3 | 0.5 | 0.4438 | 0.1450 | 7 | 0.2100 |

**Table 5.** Calculation results of stiffness and thrust under different parameter designs.

| | $K_{G_{axial}}$ | $K_p$ | $K_{G_{tangent}}$ | $F_{revise}$ | $F_{abs\_error}$ | $F_{re\_error}$ |
|---|---|---|---|---|---|---|
| | | [N m rad$^{-1}$] | | | [N] | |
| A1 | $-4.8976 \times 10^{-5}$ | 0.0996 | −0.0964 | $1.2511 \times 10^{-6}$ | $1.9590 \times 10^{-8}$ | **1.57%** |
| A2 | $-4.8976 \times 10^{-5}$ | 0.0996 | 0.2254 | $1.2998 \times 10^{-4}$ | $1.9590 \times 10^{-8}$ | 0.015% |
| B1 | $-2.9388 \times 10^{-5}$ | 0.0996 | −0.0894 | $4.0579 \times 10^{-6}$ | $1.1755 \times 10^{-8}$ | **0.29%** |
| B2 | $-2.9388 \times 10^{-5}$ | 0.0996 | 0.0155 | $4.6040 \times 10^{-5}$ | $1.1755 \times 10^{-8}$ | 0.026% |
| C1 | $-2.9064 \times 10^{-4}$ | 1.5292 | −1.5252 | $1.4831 \times 10^{-6}$ | $1.1626 \times 10^{-7}$ | **7.84%** |
| C2 | $-2.9064 \times 10^{-4}$ | 1.5292 | 0.9246 | $9.8139 \times 10^{-4}$ | $1.1626 \times 10^{-7}$ | 0.022% |

From these results, it can be observed that the original bending stiffness $K_p$ is determined by its own dimensions (here $t$). In addition, the shift ($K_{G_{axial}}$) in $K_p$ is determined by both the pendulum load gravity and the hinge dimensions. A negative value of the equivalent stiffness $K_{G_{tangent}}$ may cancel most of the hinge's own stiffness $K_p$, making the effect of the stiffness shift ($K_{G_{axial}}$) amplified, leading to large uncertainties in the thrust calculations, such as 1.57% in A1, 0.29% in B1, and 7.84% in C1. However, the high-precision metrology standard generally needs to reach an error level of 0.05% or even higher, and only the designs in A2, B2, and C2 can reach this error level.

Table 6 shows the effect of the offset of the assumed rotational center on the theoretical calculation of thrust under several sets of different parameters. The geometric parameters of the hinge, except for the minimum thickness $t$, are shown in Table 2, and the structural parameters of the compound pendulum are shown in group A2 in Table 4. Again, $g = 9.8\text{ms}^{-2}$ and $l_{sen.} = 0.5$m are set. The thrust calculation error due to the bending stiffness shift is not considered at this point. From the relative error results (see Table 6), it can be obtained that the offset of the assumed rotational center introduces non-negligible uncertainty into the thrust calculation. And this relative error depends mainly on the geometric parameters of the hinge, independent of the measured displacement.

**Table 6.** Calculation results of thrust in case of offset of the assumed rotational center.

| | $t$ | $u$ | $F_{revise}$ | $F_{abs\_error}$ | $F_{re\_error}$ |
|---|---|---|---|---|---|
| | [mm] | [um] | [N] | | |
| Case 1 | 0.1 | 1 | $1.2974 \times 10^{-6}$ | $2.7602 \times 10^{-9}$ | 0.21% |
| Case 2 | 0.1 | 20 | $2.5948 \times 10^{-5}$ | $5.5203 \times 10^{-8}$ | 0.21% |
| Case 3 | 0.1 | 100 | $1.2974 \times 10^{-4}$ | $2.7602 \times 10^{-7}$ | 0.21% |
| Case 4 | 0.2 | 100 | $3.1343 \times 10^{-4}$ | $4.0241 \times 10^{-7}$ | 0.13% |
| Case 5 | 0.3 | 100 | $7.0094 \times 10^{-4}$ | $8.9445 \times 10^{-7}$ | 0.13% |
| Case 6 | 0.4 | 100 | 0.0013 | $1.8243 \times 10^{-6}$ | 0.14% |

To sum up, it is necessary to always pay attention to the influence of $K_{G_{axial}}$ when the designer of the compound pendulum tries to customize the stiffness coefficient $K_{line}$ by changing the weight or height of the counterweight. The shift ($K_{G_{axial}}$) of the hinge bending stiffness can be regarded as the systematic error of the thrust measurement device together with the offset of the assumed rotational center. The systematic error of the equipment needs to be taken seriously in the design or measurement process. Ultimately, an improved analytical model that takes these factors into account is given in Equation (25).

## 6. Conclusions

This study investigated a new analytical model of thrust measurement for a thrust stand with a flexure hinge structure, in an attempt to compensate for the uncertainty that is not considered in conventional thrust analytical models. It is pointed out that the uncertainty comes from two neglected aspects: first, the pendulum load during calibration and the measurement of thrust will produce an axial extension effect on the flexure hinge, which will change the stiffness coefficient that has been considered a constant; second, the assumed center of rotation will be offset when the pendulum is deflected by the thrust, which will introduce the displacement measurement error. Hinge models under different parameter designs are analyzed using finite element methods and theoretical models, and the results prove the validity and generalizability of the proposed analytical model.

The significance of this work is to point out the systematic errors that exist in the thrust stand itself, which should be paid attention to as high-precision metrology equipment. The improved model will guide the design of the structural parameters of the thrust stand composed of the compound pendulum, so that it can minimize the uncertainty introduced by the variation of the system's own characteristics.

Meanwhile, it is worth emphasizing that the accuracy of the parametric modeling of the flexure hinge is the basis for maintaining consistency between theory and practice. Compared to other stiffness modeling methods for elliptical notched flexure hinges such as inverse conformal mapping and empirical equations from FEA, etc., the modeling method based on the Euler beam linear integral and Castigliano's second theorem used in this study is more applicable and available. Within a wide range of hinge geometric parameters, the proposed stiffness equation achieves results that are consistent with simulation practices. Additionally, further improvements to the hinge stiffness equation will enhance our work.

**Author Contributions:** Conceptualization, X.C. and Z.L.; methodology, X.C. and Z.L.; software, X.C.; investigation, X.C.; resources, L.Z.; data curation, L.Z. and J.X.; writing—original draft preparation, X.C.; supervision, J.X.; project administration, L.Z.; funding acquisition, L.Z. All authors have read and agreed to the published version of the manuscript.

**Funding:** This work is supported by the National Key R&D Program of China (Grant No. 2021YFC2202703), the Postgraduate Research and Practice Innovation Program of Jiangsu Province (Grant No. KYCX23_0231), and the National Natural Science Foundation of China (Grant Nos. 51905094 and 52275093).

**Data Availability Statement:** Data will be made available on request.

**Acknowledgments:** We would like to thank the above funders for their technical and financial support.

**Conflicts of Interest:** The authors declare no conflict of interest.

## References

1. Li, H.; Wu, Z.; Sun, G. A model for macro-performances applied to low power coaxial pulsed plasma thrusters. *Acta Astronaut.* **2020**, *170*, 154–162. [CrossRef]
2. Xia, Q.; Wang, N.; Wu, X. The influence of external magnetic field on the plume of vacuum arc thruster. *Acta Astronaut.* **2019**, *164*, 69–76. [CrossRef]
3. Anderson, G.; Anderson, J.; Anderson, M. Experimental results from the ST7 mission on LISA Pathfinder. *Phys. Rev. D* **2018**, *98*, 102005. [CrossRef]
4. Armano, M.; Audley, H.; Auger, G. The LISA pathfinder mission. In *Journal of Physics: Conference Series, Proceedings of the 10th International LISA Symposium, Gainesville, FL, USA, 18–23 May 2014*; IOP Publishing: Bristol, UK, 2015; Volume 610, p. 610. [CrossRef]
5. Armano, M.; Audley, H.; Baird, J. LISA Pathfinder platform stability and drag-free performance. *Phys. Rev. D* **2019**, *99*, 082001. [CrossRef]
6. Tajmar, M. Overview of indium LMIS for the NASA-MMS mission and its suitability for an In-FEEP thruster on LISA. In Proceeding of the International Electric Propulsion Conference, IEPC-2011, Wiesbaden, Germany, 11–15 September 2011.
7. Danzmann, K.; Prince, T.A.; Binetruy, P. LISA: Unveiling a hidden Universe. *Assess. Study Rep. ESA/SRE* **2011**, *3*. [CrossRef]
8. Anselmo, M.R.; Marques, R.I. Torsional thrust balance for electric propulsion application with electrostatic calibration device. *Meas. Sci. Technol.* **2019**, *30*, 055903. [CrossRef]
9. Lun, J.; Law, C. Direct thrust measurement stand with improved operation and force calibration technique for performance testing of pulsed micro-thrusters. *Meas. Sci. Technol.* **2014**, *25*, 095009. [CrossRef]
10. Frollani, D.; Coletti, M.; Gabriel, S.B. A thrust balance for low power hollow cathode thrusters. *Meas. Sci. Technol.* **2014**, *25*, 065902. [CrossRef]
11. Orieux, S.; Rossi, C.; Esteve, D. Thrust stand for ground tests of solid propellant microthrusters. *Rev. Sci. Instrum.* **2002**, *73*, 2694–2698. [CrossRef]
12. Grubišić, A.N.; Gabriel, S.B. Development of an indirect counterbalanced pendulum optical-lever thrust balance for micro-to millinewton thrust measurement. *Meas. Sci. Technol.* **2010**, *21*, 105101. [CrossRef]
13. Acosta-Zamora, A.; Flores, J.R.; Choudhuri, A. Torsional thrust balance measurement system development for testing reaction control thrusters. *Measurement* **2013**, *46*, 3414–3428. [CrossRef]
14. Lam, J.K.; Koay, S.C.; Lim, C.H. A voice coil based electromagnetic system for calibration of a sub-micronewton torsional thrust stand. *Measurement* **2019**, *131*, 597–604. [CrossRef]
15. Xu, K.G.; Walker, M.L.R. High-power, null-type, inverted pendulum thrust stand. *Rev. Sci. Instrum.* **2009**, *80*, 055103. [CrossRef] [PubMed]
16. Knoll, A.; Lamprou, D.; Lappas, V. Thrust balance characterization of a 200 W quad confinement thruster for high thrust regimes. *IEEE Trans. Plasma Sci.* **2014**, *43*, 185–189. [CrossRef]
17. Hey, F.G.; Keller, A.; Braxmaier, C. Development of a highly precise micronewton thrust balance. *IEEE Trans. Plasma Sci.* **2014**, *43*, 234–239. [CrossRef]
18. Zhang, H.; Duan, B.; Wu, L. Development of a steady-state microthrust measurement stand for microspacecrafts. *Measurement* **2021**, *178*, 109357. [CrossRef]
19. Trezzolani, F.; Magarotto, M.; Manente, M. Development of a counterbalanced pendulum thrust stand for electric propulsion. *Measurement* **2018**, *122*, 494–501. [CrossRef]
20. Wachs, B.N.; Jorns, B.A. Sub-millinewton thrust stand and wireless power coupler for microwave-powered small satellite thrusters. *Rev. Sci. Instrum.* **2022**, *93*, 083507. [CrossRef]
21. Wang, A.; Wu, H.; Tang, H. Development and testing of a new thrust stand for micro-thrust measurement in vacuum conditions. *Vacuum* **2013**, *91*, 35–40. [CrossRef]
22. Polk, J.E.; Pancotti, A.; Haag, T. Recommended practice for thrust measurement in electric propulsion testing. *J. Propul. Power* **2017**, *33*, 539–555. [CrossRef]
23. Wei, H.; Tian, Y.; Zhao, Y. Two-axis flexure hinges with variable elliptical transverse cross-sections. *Mech. Mach. Theory* **2023**, *181*, 105183. [CrossRef]
24. Xu, N.; Dai, M.; Zhou, X. Analysis and design of symmetric notch flexure hinges. *Adv. Mech. Eng.* **2017**, *9*, 1687814017734513. [CrossRef]
25. Ma, W.; Wang, R.; Zhou, X. The performance comparison of typical notched flexure hinges. *Proc. Inst. Mech. Eng. Part C J. Mech. Eng.* **2020**, *234*, 1859–1867. [CrossRef]
26. Liu, M.; Zhang, X.; Fatikow, S. Design and analysis of a multi-notched flexure hinge for compliant mechanisms. *Precis. Eng.* **2017**, *48*, 292–304. [CrossRef]

27. Cesare, S.; Musso, F.; D'Angelo, F. Nanobalance: The European balance for micro-propulsion. In Proceedings of the 31st International Electric Propulsion Conference, Ann Arbor, MI, USA, 20–24 September 2009; p. 2009-0182.
28. Xu, H.; Gao, Y.; Mao, Q.B. A compound pendulum for thrust measurement of micro-Newton thruster. *Rev. Sci. Instrum.* **2022**, *93*, 064501. [CrossRef]
29. Rocca, S.; Menon, C.; Nicolini, D. FEEP micro-thrust balance characterization and testing. *Meas. Sci. Technol.* **2006**, *17*, 711. [CrossRef]
30. Luna, J.P.; Edwards, C.H.; Del Amo, J.G. Development status of the ESA micro-Newton thrust balance. In Proceedings of the 32nd International Electric Propulsion Conference, Wiesbaden, Germany, 11–15 September 2011.
31. Paros, J.M. How to design flexure hinges. *Mach. Des.* **1965**, *37*, 151–156.
32. Liu, M.; Zhang, X.; Fatikow, S. Design and analysis of a high-accuracy flexure hinge. *Rev. Sci. Instrum.* **2016**, *87*, 055106. [CrossRef]
33. Linß, S.; Gräser, P.; Räder, T. Influence of geometric scaling on the elasto-kinematic properties of flexure hinges and compliant mechanisms. *Mech. Mach. Theory* **2018**, *125*, 220–239. [CrossRef]
34. Zhu, Z.W.; Zhou, X.Q.; Wang, R.Q. A simple compliance modeling method for flexure hinges. *Sci. China Technol. Sci.* **2015**, *58*, 56–63. [CrossRef]
35. Li, T.M.; Zhang, J.L.; Jiang, Y. Derivation of empirical compliance equations for circular flexure hinge considering the effect of stress concentration. *Int. J. Precis.* **2015**, *16*, 1735–1743. [CrossRef]
36. Chen, G.; Liu, X.; Du, Y. Elliptical-arc-fillet flexure hinges: Toward a generalized model for commonly used flexure hinges. *J. Mech. Des.* **2011**, *133*, 081002. [CrossRef]

*actuators*

MDPI

*Article*

# Calibration Optimization of Kinematics and Dynamics for Delta Robot Driven by Integrated Joints in Machining Task

**Zhenhua Jiang [1,*], Yu Wang [1], Dongdong Liu [1] and Tao Sun [2]**

[1] Institute of Artificial Intelligence, Donghua University, Shanghai 201620, China; jason.wang@mail.dhu.edu.cn (Y.W.); 2221062@mail.dhu.edu.cn (D.L.)

[2] Shanghai Institute of Spacecraft Equipment, Shanghai 200240, China; suntao_kb@163.com

* Correspondence: zhenhuajiang@hotmail.com

**Abstract:** For the application of Delta robots with a 3-R(RPaR) configuration in machining tasks, this paper constructed a 54-parameter kinematic error model and a simplified dynamic model incorporating an integrated joint's position error and friction, respectively. Utilizing Singular Value Decomposition (SVD) of the Linear Model Coefficient Matrix (LMCM) and the coefficient chart, a criterion for identifiability of error components is established. For good identification results, the optimal measurement surface with Fourier series form is obtained using a combination of the Hook–Jeeves Direct Search Algorithm (DSA) and Inner Point Method (IPM). The friction coefficients and other dynamic parameters are obtained through fitting the integrated joint torque-angle pairs measured along specific trajectories using nonlinear least squares regression. The validation of the calibration process is conducted through simulations and experiments. The calibration results provide a foundation for the precise control of integrated joints and the high-precision motion of robots.

**Keywords:** integrated joint's friction; integrated joint's position error; kinematic and dynamic calibration; measurement surface optimization; delta robot; machining task

## 1. Introduction

In comparison to serial robots, parallel robots offer significant advantages, including enhanced stiffness and improved repeatability. Delta robots, as a highly successful configuration of parallel robots, have found extensive applications in various fields such as Pick-and-Place [1], 3D printing [2], medical fields [3], and machining [4]. For machining applications, Delta robots demand exceptional motion precision. Precision in robotic systems is typically achieved through two approaches: precise analysis during the design stage and error compensation following assembly. Error calibration and compensation, facilitated using algorithms and software, substantially enhance the precision of deployed robots [5,6].

Kinematic modeling serves as the cornerstone of robot error calibration. Various methods are employed, including the Denavit–Hartenberg (DH) method [7], exponential product [8], and vector loop method [9]. The vector loop method formulates closed-loop motion equations through position vectors at the link ends or joints within kinematic chains and is highly suitable for parallel robots. Vischer et al. [10] utilized the vector loop method to construct two kinematic models, "Model 54" and "Model 24", for the 3-R(2-SS) configuration. "Model 54" considers geometric errors in all mechanical components except for the spherical joints; if the spatial parallelogram is assumed to remain perfect, "Model 54" can be simplified to "Model 24". In "Model 24", the direction of the Delta robot's moving platform is assumed to be constant. Li et al. [11], also focusing on the 3-R(2-SS) configuration, demonstrated a significant influence of parallel mechanism errors on motion precision, especially in the Z-direction, with an increase in the tilt angle of the moving platform. Shen et al. [12], utilizing the variational method, studied the effect of error components (including the parallelogram mechanism) on the motion precision of Delta-RU robots, significantly enhancing the accuracy of the end effector through compensation. While scholars have recently introduced data-driven approaches for kinematic calibration, the model-based method continues to be predominant. This preference stems from its clear physical significance and the extensive measurement data demanded by the former [13].

Kinematic parameter identification is typically based on redundant equations formed using kinematic models and multiple measurement points. Considering noises in measurement data, the final parameter solution is obtained through nonlinear least squares solutions, with the Levenberg–Marquardt algorithm being the most commonly used algorithm [14]. Other probabilistic methods, such as maximum likelihood estimation and Kalman filtering, are less commonly used. Different parameter error components have varying degrees of influence on the total pose error, typically characterized using sensitivity analysis [15,16]. A local parameter sensitivity can be defined as the ratio of the sensitivity coefficient of a single error component to the combined one of all errors at a measurement point. A global sensitivity can be defined as the integration of local sensitivity over the robot workspace. Sensitivity analysis can identify the dominant error components of the robot. In addition to the calibration for link dimensions and assembly errors, scholars have also investigated the calibration of joint clearance. Kim et al. [17] discussed the relationship between the end effector pose error of the Stewart platform and the tolerance limits of its joints. Tannous et al. [18] conducted a sensitivity analysis on joint clearance using interval analysis methods, taking three manipulators as examples.

The problem of identifying dynamic parameters of robots involves several aspects, including specific identification parameters, dynamic models, excitation trajectories, and solution algorithms. Jan et al. [19] proposed a statistical framework for the dynamic parameter identification of multi-joint serial robots. The excitation trajectories were defined as finite Fourier series on each joint, with the optimal criterion being the uncertainty on the estimated parameters or a lower bound for it. The parameters identified included mass, inertia parameters, and friction parameters. In another study, the coefficients of the Fourier series were optimized to minimize the sensitivity of the identification to measurement disturbances [20]. Zafer et al. [21] used the least squares (LS) method and particle swarm optimization (PSO) technique to estimate distinct inertia parameters of the Staubli RX-60 robot, showing that the PSO method had a smaller identification error. Grotjahn et al. [22] investigated the identification of friction and rigid-body dynamics in parallel kinematic structures, achieving their separation via point-to-point motions. Han et al. [23] considered measurement errors in the process of optimizing excitation trajectories, integrating weighted least squares (WLS) and nonlinear error models, and iteratively reweighted least squares to unify the modeling process of friction and rigid-body dynamics. Dong et al. [24] first identified the friction parameters in the nonlinear Tustin friction model using the LS method during forward and reverse rotations, then estimated the remaining dynamic parameters of the Stäubli TX-90 robot using the symbiotic organisms search (SOS) algorithm. Song et al. [25] implemented the load parameter identification of parallel robots based on the Extended Kalman Filter. Ohno et al. [26] designed target trajectories for the detection of joint clearances based on actuation torque fluctuation. Diaz-Rodriguez et al. [27] developed a simplified dynamic model for a 3-DOF prismatic-revolute-spherical parallel manipulator to improve trajectory tracking precision and reduce the real-time computational burden. Abed Azad et al. [28] obtained the base inertial parameters through singular value decomposition, constructing an optimized path containing several harmonics within the desired bandwidth, achieving offline–online identification of the base inertial parameters.

Measurement processes can be categorized into two types: with or without external measurement devices. Without external measurement device support, the end effector of the robot is virtually constrained to a point, line, or surface [29,30], commonly applied in serial robots. The calibration process of parallel robots often employs external measurement devices, including laser trackers, laser interferometers, coordinate measurement machines (CMMs), vision-based measurement systems, ball-bar systems, etc. [10,31–33]. The selection of measurement points directly affects the quality of error parameter identification, leading to an optimization problem, typically aiming for an observation metric as the optimization objective along with constraints. Specific solving methods include local search methods represented by gradient descent and global search algorithms represented by meta-heuristic algorithms. Daney et al. [34] proposed a constrained optimization method for iteratively selecting measurement poses to minimize indicators based on the Jacobi matrix, utilizing the tabu search metaheuristic algorithm to avoid local optima.

This study tackles the relatively unexplored challenge of kinematic error modeling and dynamic calibration for the 3-R(RPaR) configuration Delta robot integrated into a self-built five-axis blade grinding device. It specifically investigates how geometric errors affect the poses of parallelogram mechanism while considering the constraints imposed by the robot's DoF. Departing from conventional measurement point selection methods, this research reframes the issue as an optimization problem aimed at shaping the measurement workpiece surface, under the assumption that pose errors can be inferred from surface machining errors. Given the complexity of existing parameter sensitivity approaches, the paper seeks to establish criteria for parameter identifiability through the singular value decomposition (SVD) of the LMCM. For machining tasks with smooth and uniform motion, a simplified dynamic model with friction coefficients and other dynamic parameters is established and identified. The objective of this study is to devise a Delta robot calibration methodology tailored for machining environments.

The structure of the subsequent sections is as follows: Section 2 provides an introduction to the blade grinding device and details the kinematic and dynamic models of the Delta robot. Section 3 examines the influence of geometric errors on the parallel mechanism and describes the development of the kinematic error model tailored for the 3-R(RPaR) configuration. In Section 4, the identifiability criteria for error components are proposed, along with the definition of periodic error measurement surfaces and the optimization models under constraints. Section 5 delves into solution algorithms for the optimization problem and validates the kinematic calibration process through simulation. Section 6 validates the dynamic calibration method through experiments. The paper concludes with a summary of the findings and final remarks.

## 2. Kinematic and Dynamic Modeling

The blade grinding equipment, depicted in Figure 1, includes an optical platform, an aluminum alloy frame, a grinding head unit with a C and B axis, and a Delta robot unit suspended at the top of the frame. The Delta robot's static platform is secured to the frame's top, while a six-axis force sensor is mounted on its moving platform and connected to the workpiece. The C axis rotates vertically on the optical platform, and the B axis is perpendicular to its rotation axis. A spindle, mounted perpendicular to the B axis, holds an abrasive flap wheel via an extension rod. In this paper, only the Delta robot unit of the equipment is addressed, with the angular positions of the C and B axis held fixed during the calibration process.

The Delta robot is composed of a stationary platform (base), a moving platform, and three interconnected kinematic chains. This study focuses on the 3-R(RPaR) configuration, illustrated in Figure 2. Each chain connects the base to an active arm via an active revolute joint denoted as $A_i$ ($i$ = 1, 2, 3). The active arm is linked to a parallelogram mechanism, consisting of four revolute joints ($\boldsymbol{D}_i$, $\boldsymbol{E}_i$, $\boldsymbol{F}_i$, $\boldsymbol{G}_i$) and four links, through another passive revolute joint labeled as $\boldsymbol{B}_i$. The moving platform connects to the parallel mechanism through a passive revolute joint denoted as $\boldsymbol{C}_i$. The three chains are evenly distributed around the stationary platform, spaced 120° apart, with the $Z_0$ axis defined as the central axis. The origin $\boldsymbol{S}$ of the stationary coordinate system $\{\boldsymbol{W}_0\}$ is at the intersection of the plane containing $\boldsymbol{A}_1\boldsymbol{A}_2\boldsymbol{A}_3$ with the $\boldsymbol{X}_0$ axis, and $\boldsymbol{Y}_0 = \boldsymbol{Z}_0 \times \boldsymbol{X}_0$. Coordinate systems $\{\boldsymbol{W}_{i,1}\}$ and $\{\boldsymbol{W}_{i,2}\}$ are established at $\boldsymbol{A}_i$, fixed to the stationary platform and the active arm, respectively. $\boldsymbol{Z}_{i,1}$ is oriented along the rotation axis of joint $\boldsymbol{A}_i$, while $\boldsymbol{Y}_{i,1}$ is parallel to $\boldsymbol{Z}_0$ but in the opposite direction. $\boldsymbol{Z}_{i,2}$ is aligned with $\boldsymbol{Z}_{i,1}$, and $\boldsymbol{X}_{i,2}$ is oriented along the vector $\boldsymbol{A}_i\boldsymbol{B}_i$. The angle between $X_{i,2}$ and $X_{i,1}$ is denoted as $\theta_i$, representing the rotation angle of the active arm. The unit vector of $D_iE_i$ is denoted as $Z_{i,3}$, and the unit vector of $C_iG_i$ is denoted as $Z_{i,4}$. $Z_{i,4}$ also serves as the Z-axis in $\{W_{i,4}\}$, fixed to point $C_i$ and the moving platform. The direction of $Y_{i,4}$ is consistent with $Z_0$. The lengths $SA_i$ and $TC_i$ represent the radius $r_1$ of the stationary platform and the radius $r_2$ of the moving platform, respectively. $l_p$ and $l_n$ denote the lengths of active arm $A_iB_i$ and the vector $B_iC_i$, respectively, defining the four basic parameters of the Delta robot. $\Phi_1 = 0°$, $\Phi_2 = 120°$, and $\Phi_3 = 240°$ indicate the angles between the three active arm rotation planes and $Z_0X_0$ plane for chains 1, 2, and 3, respectively. Under ideal conditions, the moving platform only undergoes translation, maintaining a fixed orientation. A local coordinate system $\{W_5\}$ is established with $T$ as the origin, aligned with $\{W_0\}$. $\{W_7\}$, $\{W_8\}$ and $\{W_9\}$ in Figure 2 are, respectively, associated with the C axis, B axis, and the grinding tool, which are not involved in the subsequent sections.

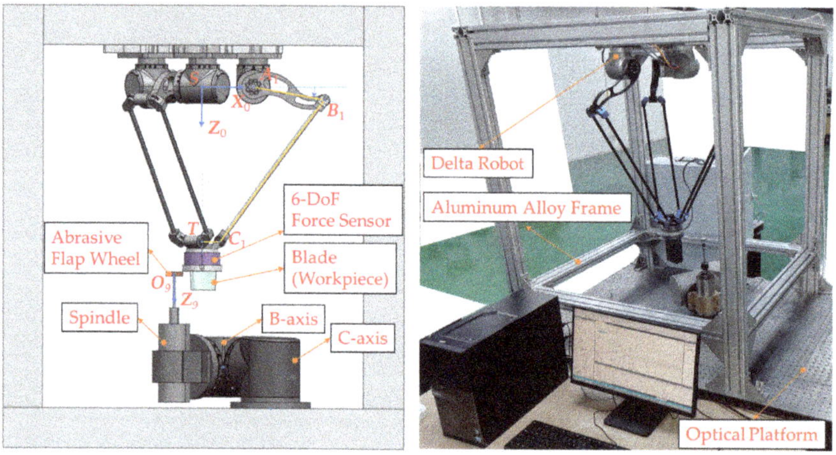


**Figure 1.** Five-axis blade grinding equipment based on the Delta robot.

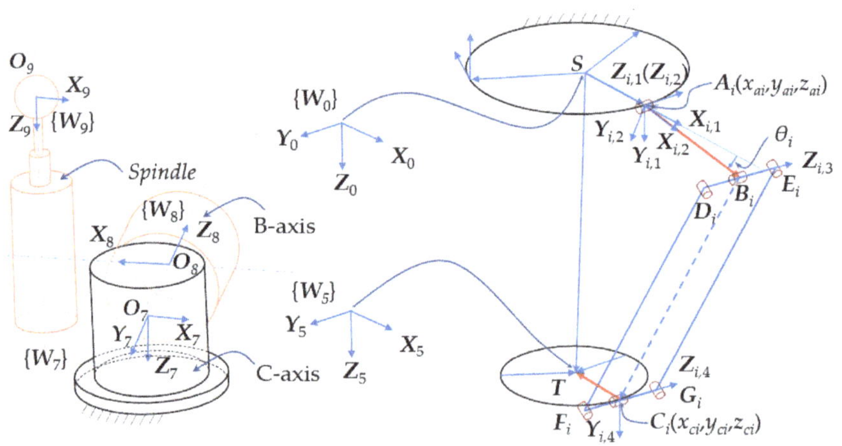


**Figure 2.** Coordinate systems of the blade grinding equipment.

$T$ is the center of $C_1$, $C_2$ and $C_3$ and is denoted as ${}^0T\ (x_t, y_t, z_t)$ and ${}^{i,1}T({}^{i,1}x_t, {}^{i,1}y_t, {}^{i,1}z_t)$ in $\{W_0\}$ and $\{W_{i,1}\}$, respectively. In $\{W_{i,1}\}$,

$$B_iC_i = TC_i - (SA_i + A_iB_i), \tag{1}$$

where $\boldsymbol{SA}_i = [-r_1, 0, 0]^T$, $\boldsymbol{A}_i\boldsymbol{B}_i = [l_p\cos\theta_i, l_p\sin\theta_i, 0]^T$, $\boldsymbol{TC}_i = \left[{}^{i,1}x_t, {}^{i,1}y_t, {}^{i,1}z_t\right]^T + [r_2, 0, 0]^T$. Substituted into Equation (1),

$$\boldsymbol{B}_i\boldsymbol{C}_i = \left[{}^{i,1}x_t + r_2, {}^{i,1}y_t, {}^{i,1}z_t\right]^T - [l_p\cos\theta_i, l_p\sin\theta_i, 0]^T. \tag{2}$$

$\boldsymbol{B}_i\boldsymbol{C}_i$ always satisfies:

$$\|\boldsymbol{B}_i\boldsymbol{C}_i\| = l_n \text{ or } \boldsymbol{B}_i\boldsymbol{C}_i \cdot \boldsymbol{B}_i\boldsymbol{C}_i = l_n^2. \tag{3}$$

The relationship of $\boldsymbol{T}$ coordinates in $\{\boldsymbol{W}_0\}$ and $\{\boldsymbol{W}_{i,1}\}$ can be represented as the following:

$$\left[{}^{i,1}x_t, {}^{i,1}y_t, {}^{i,1}z_t\right]^T = Rot(\mathbf{X}, -90^\circ)Rot(\mathbf{Z}, -\Phi_i)[x_t, y_t, z_t]^T - [r_1, 0, 0]^T. \tag{4}$$

where *Rot(axis, angle)* represents the rotation matrix around *axis* by *angle*. The fundamental kinematic equations of the Delta robot are formed by substituting Equation (3) for $i$ = 1,2,3, respectively. Given $\theta_1$, $\theta_2$, and $\theta_3$, if the goal is to determine $x_t$, $y_t$, $z_t$, it constitutes a forward kinematics problem. Conversely, it represents an inverse kinematics problem. By substituting Equation (2) into Equation (3),

$$\left(l_p\cos\theta_i - {}^{i,1}x_t - r_2\right)^2 + \left(l_p\sin\theta_i - {}^{i,1}y_t\right)^2 + \left({}^{i,1}z_t\right)^2 = l_n^2. \tag{5}$$

Beginning with the inverse kinematics, Equation (5) yields the following:

$$\cos\psi_i\cos\theta_i + \sin\psi_i\sin\theta_i = \cos(\theta_i + \psi_i) = M_i/2l_pN_i. \tag{6}$$

where $\psi_i = \text{atan2}\left({}^{i,1}y_t, {}^{i,1}x_t + r_2\right)$, $M_i = l_p^2 - l_n^2 + \left({}^{i,1}x_t + r_2\right)^2 + {}^{i,1}y_t^2 + {}^{i,1}z_t^2$, $N_i = \sqrt{\left({}^{i,1}x_t + r_2\right)^2 + {}^{i,1}y_t^2}$. Solving Equation (6),

$$\theta_i = \text{acos}(M_i/2l_pN_i) + \psi_i, \text{ or } \theta_i = -\text{acos}(M_i/2l_pN_i) + \psi_i. \tag{7}$$

Proceeding to obtain the forward solution, similarly derived from Equation (5),

$$x_t^2 + y_t^2 + z_t^2 + a_ix_t + b_iy_t + c_iz_t + d_i = 0. \tag{8}$$

where $a_i = -2l_p\cos\theta_i\cos\Phi_i - 2\Delta r\cos\Phi_i$, $b_i = -2l_p\cos\theta_i\sin\Phi_i - 2\Delta r\sin\Phi_i$, $c_i = -2l_p\sin\theta_i$, $d_i = l_p^2 + \Delta r^2 + 2\Delta r l_p\cos\theta_i - l_n^2$, and $\Delta r = r_1 - r_2$. The solutions for $x_t$, $y_t$ and $z_t$ are obtained through the resolution of quadratic equations, the details of which are omitted here for brevity. In both forward and inverse solutions, the selection among multiple solutions is necessary. The viable solution must adhere to inequality (9).

$$\left({}^0(\boldsymbol{A}_i\boldsymbol{B}_i) \times {}^0(\boldsymbol{B}_i\boldsymbol{C}_i)\right) \cdot {}^0\mathbf{Z}_{i,3} > 0 \tag{9}$$

where the left superscript "0" denotes the vector value in $\{\boldsymbol{W}_0\}$, and ${}^0\mathbf{Z}_{i,3} = Rot(\mathbf{Z}, f_i)Rot(\mathbf{X}, 90^\circ)[0,0,1]^T$, ${}^0\boldsymbol{A}_i = Rot(\mathbf{Z}, f_i)Rot(\mathbf{X}, 90^\circ)[r_1, 0, 0]^T$, ${}^0\boldsymbol{B}_i = Rot(\mathbf{Z}, f_i)Rot(\mathbf{X}, 90^\circ)[l_p\cos\theta_i + r_1, l_p\sin\theta_i, 0]^T$, ${}^0\boldsymbol{C}_i = {}^0\boldsymbol{T}_i + Rot(\mathbf{Z}, f_i)Rot(\mathbf{X}, 90^\circ)[r_2, 0, 0]^T$.

In accordance with the principle of virtual work, the virtual work carried out using all non-inertial forces should equal the virtual work carried out using inertial forces. This principle is applied to driving integrated joints:

$$\left(\boldsymbol{\Gamma} + \boldsymbol{\Gamma}_c + \boldsymbol{J}^T\boldsymbol{F}_G + \boldsymbol{\Gamma}_{CM}\right)\delta\boldsymbol{\Theta} = \left(\boldsymbol{I}\ddot{\boldsymbol{\Theta}} + \boldsymbol{J}^T\boldsymbol{F}_A\right)\delta\boldsymbol{\Theta}, \tag{10}$$

where $\boldsymbol{\Gamma}$ represents the motor torque, $\boldsymbol{\Gamma}_c$ represents frictional torque, $\boldsymbol{I}$ denotes the moment of inertia of the integrated joints (including active arms), $\boldsymbol{J}$ signifies the Jacobian matrix, $\boldsymbol{F}_A$

stands for the inertial force generated by the moving platform mass ($m_e$), $\boldsymbol{F}_G$ represents the gravitational force acting on $m_e$, $\boldsymbol{\Gamma}_{CM}$ accounts for the gravitational moment produced by the unbalanced mass of the integrated joints (including active arms), $\delta\boldsymbol{\Theta}$ represents virtual displacement, and $\ddot{\boldsymbol{\Theta}}$ denotes joint acceleration. Here, it is assumed that the mass of the parallelogram linkage segment is negligible, or it can be separately accounted for at either end of integrated joints or the moving platform.

## 3. Kinematic Error Modeling for the 3-R(RPaR) Configuration

Geometric errors of the Delta robot unit result in pose deviations of the workpiece mounted on the moving platform during grinding operations. Under error conditions, the relationships between coordinate systems are adjusted. Variables labeled with the right superscript "0" represent their nominal values.

### 3.1. Transfer Relationships between Coordinate Systems or Points

The orientation and origin of $\{\boldsymbol{W}_{i,1}\}$ in $\{\boldsymbol{W}_0\}$ are expressed as Equations (11) and (12), respectively.

$$ {}^{0}_{i,1}\boldsymbol{R} = Rot(\boldsymbol{Z}, \varphi_i)Rot(\boldsymbol{X}, \gamma_i) \tag{11} $$

$$ {}^{0}\boldsymbol{A}_i = [x_{ai}, y_{ai}, z_{ai}]^T \tag{12} $$

where the nominal values of $\varphi_i, \gamma_i, x_{ai}, y_{ai}, z_{ai}$ are denoted as $\varphi_i^0, \gamma_i^0, x_{ai}^0, y_{ai}^0, z_{ai}^0$, respectively, and $\varphi_i^0 = \Phi_i, \gamma_i^0 = 90°, x_{ai}^0 = r_1\cos\Phi_i, y_{ai}^0 = r_1\sin\Phi_i, z_{ai}^0 = 0$.

The origin of $\{\boldsymbol{W}_{i,2}\}$ is the same as that of $\{\boldsymbol{W}_{i,1}\}$ and the relative orientation is the following:

$$ {}^{i,1}_{i,2}\boldsymbol{R} = Rot(\boldsymbol{Z}, \theta_i), \tag{13} $$

where the nominal value of $\theta_i$ is $\theta_i^0$. $\Delta\theta_i = \theta_i - \theta_i^0$ is the zero position offset of the active revolute joint $i$.

The coordinate of $\boldsymbol{B}_i$ in $\{\boldsymbol{W}_{i,2}\}$ is as follows:

$$ {}^{i,2}\boldsymbol{B}_i = [l_{pi}, 0, 0]^T, \tag{14} $$

where the nominal value of the active arm length $l_{pi}$ is $l_p$.

The relationship between ${}^{i,2}\boldsymbol{Z}_{i,3}$ and ${}^{i,2}\boldsymbol{Z}_{i,2}$ is as follows:

$$ {}^{i,2}\boldsymbol{Z}_{i,3} = Rot(\boldsymbol{X}, \Delta\alpha_i)Rot(\boldsymbol{Y}, \Delta\beta_i){}^{i,2}\boldsymbol{Z}_{i,2}, \tag{15} $$

where $\Delta\alpha_i$ and $\Delta\beta_i$ are approximate infinitesimal errors, $\boldsymbol{Z}_{i,2}$ and $\boldsymbol{Z}_{i,3}$ are approximately parallel under error conditions.

The orientation and origin of $\{\boldsymbol{W}_{i,4}\}$ in $\{\boldsymbol{W}_5\}$ are the following:

$$ {}^{5}_{i,4}\boldsymbol{R} = Rot(\boldsymbol{Z}, \varepsilon_i)Rot(\boldsymbol{X}, \tau_i), \tag{16} $$

$$ {}^{5}\boldsymbol{C}_i = [x_{ci}, y_{ci}, z_{ci}]^T, \tag{17} $$

where the nominal values of $\varepsilon_i, \tau_i, x_{ci}, y_{ci}, z_{ci}$ are denoted as $\varepsilon_i^0, \tau_i^0, x_{ci}^0, y_{ci}^0, z_{ci}^0$, respectively, and $\varepsilon_i^0 = \Phi_i, \tau_i^0 = 90°, x_{ci}^0 = r_2\cos\Phi_i, y_{ci}^0 = r_2\sin\Phi_i, z_{ci}^0 = 0$.

The orientation and origin of $\{\boldsymbol{W}_5\}$ in $\{\boldsymbol{W}_0\}$ are the following:

$$ {}^{0}_{5}\boldsymbol{R} = Rot(\boldsymbol{X}, \alpha_5)Rot(\boldsymbol{Y}, \beta_5)Rot(\boldsymbol{Z}, \gamma_5), \tag{18} $$

$$ {}^{0}\boldsymbol{T} = [x_t, y_t, z_t]^T. \tag{19} $$

### 3.2. The Influence of Parallelogram Mechanism Errors

As shown in Figure 3, in the part of the parallelogram mechanism, the nominal points of $\boldsymbol{D}_i$, $\boldsymbol{E}_i$, $\boldsymbol{F}_i$ and $\boldsymbol{G}_i$ are denoted as $\boldsymbol{D}_i^0$, $\boldsymbol{E}_i^0$, $\boldsymbol{F}_i^0$ and $\boldsymbol{G}_i^0$, respectively. The following relations are satisfied: $\|\boldsymbol{D}_i\boldsymbol{E}_i\| = l_{bi}$, $\|\boldsymbol{F}_i\boldsymbol{G}_i\| = l_{ci}$, $\Delta l_{bi} = l_{bi} - l_b^0$, $2\Delta l_{ci} = l_{ci} - l_{bi}$, $\|\boldsymbol{D}_i^0\boldsymbol{F}_i^0\| = \left\|\boldsymbol{E}_i^0\boldsymbol{G}_i^0\right\| = l_n^0$, $\|\boldsymbol{D}_i\boldsymbol{F}_i\| = l_{ni} + \Delta l_i$, $\|\boldsymbol{E}_i\boldsymbol{G}_i\| = l_{ni} - \Delta l_i$, $\Delta l_{ni} = l_{ni} - l_n^0$. In the RPaR configuration, even if there are errors $\Delta l_{bi}$ and $\Delta l_{ci}$, it can be assumed that $\boldsymbol{B}_i$ and $\boldsymbol{C}_i$ are located at the midpoint of $\boldsymbol{D}_i\boldsymbol{E}_i$ and $\boldsymbol{F}_i\boldsymbol{G}_i$, respectively. Establishing a coordinate system $\{\boldsymbol{W}_{i,3}\}$ with $\boldsymbol{B}_i$ as the origin, where its **Z**-axis is $\boldsymbol{Z}_{i,3}$, the $\boldsymbol{X}$ axis is perpendicular to the plane $\boldsymbol{D}_i\boldsymbol{E}_i\boldsymbol{G}_i\boldsymbol{F}_i$ and points outward from the paper, denoted as $\boldsymbol{X}_{i,3}$, and $\boldsymbol{Y}_{i,3} = \boldsymbol{Z}_{i,3} \times \boldsymbol{X}_{i,3}$. The unit vector along $\boldsymbol{B}_i\boldsymbol{C}_i$ is denoted as $\boldsymbol{Y}'_{i,3}$, $\boldsymbol{Z}'_{i,3} = \boldsymbol{X}_{i,3} \times \boldsymbol{Y}'_{i,3}$. The angle between $\boldsymbol{Y}'_{i,3}$ and $\boldsymbol{Y}_{i,3}$ is $\xi_i$.

$$\xi_i = \text{asin}\left\|\boldsymbol{Y}_{i,3} \times \boldsymbol{Y}'_{i,3}\right\| \tag{20}$$

Due to the presence of $\Delta l_{ci}$ and $\Delta l_i$, an angle $\Delta\xi_i$ exists between $\boldsymbol{E}_i\boldsymbol{G}_i$ and $\boldsymbol{Y}'_{i,3}$, and an angle $\Delta\sigma_i$ exists between $\boldsymbol{Z}_{i,4}$ and $\boldsymbol{Z}_{i,3}$. In Figure 3, the new point obtained by rotating $\boldsymbol{G}_i$ around $\boldsymbol{C}_i$ by $\Delta\sigma_i$ is denoted as $\boldsymbol{H}_i$. $\boldsymbol{C}_i\boldsymbol{H}_i$ has the same direction as $\boldsymbol{Z}_{i,3}$. $\boldsymbol{G}'_i$ lies on the line $\boldsymbol{C}_i\boldsymbol{H}_i$, and $\boldsymbol{E}_i\boldsymbol{G}'_i$ is parallel to $\boldsymbol{Y}'_{i,3}$. The intersection point of $\boldsymbol{E}_i\boldsymbol{G}'_i$ and $\boldsymbol{F}_i\boldsymbol{G}_i$ is $\boldsymbol{H}'_i$. In the area enclosed by $\boldsymbol{H}'_i\boldsymbol{G}_i\boldsymbol{H}_i\boldsymbol{G}'_i$, assuming errors $\Delta l_{ci}$, $\Delta\sigma_i$, $\Delta l_i$, $\Delta\xi_i$ are small, and $\sin\Delta\sigma_i \approx \Delta\sigma_i$, $\sin\Delta\xi_i \approx \Delta\xi_i$, then

$$\Delta l_{ci}\boldsymbol{Z}_{i,3} - \Delta\sigma_i l_{ci}\boldsymbol{Y}_{i,3} = -\Delta l_i\boldsymbol{Y}'_{i,3} + \Delta\xi_i(l_{ni} - \Delta l_i)\boldsymbol{Z}'_{i,3}. \tag{21}$$

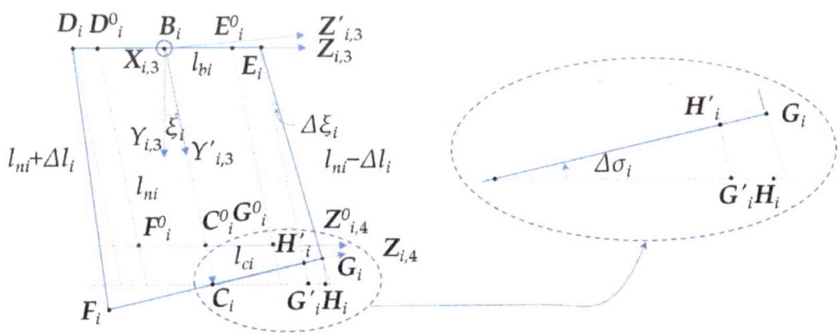


**Figure 3.** The errors of parallelogram mechanism within the kinematic chain.

Equation (21) depicts the extension of $\boldsymbol{C}_i\boldsymbol{G}'_i$ by the distance $\Delta l_{ci}$ to reach point $\boldsymbol{H}_i$, followed by the rotation of $\boldsymbol{H}_i$ around $\boldsymbol{C}_i$ by $\Delta\sigma_i$ to reach point $\boldsymbol{G}_i$ on the left side. Conversely, on the right side, it illustrates the shortening of $l_{ni}$ by $\Delta l_i$ to form segment $\boldsymbol{E}_i\boldsymbol{H}'_i$, followed by the rotation of $\boldsymbol{H}'_i$ around $\boldsymbol{E}_i$ by $\Delta\xi_i$ to reach point $G_i$. Considering the relationship between $\boldsymbol{Y}'_{i,3}$ and $\boldsymbol{Y}_{i,3}$, as well as $\boldsymbol{Z}'_{i,3}$ and $\boldsymbol{Z}_{i,3}$, and $\Delta\sigma_i\Delta l_{ci} \approx 0$, $\Delta\xi_i\Delta l_i \approx 0$, it is derived that,

$$\Delta l_{ci}\boldsymbol{Z}_{i,3} - \Delta\sigma_i l_{bi}\boldsymbol{Y}_{i,3} = -\Delta l_i Rot(\boldsymbol{X}, \xi_i)\boldsymbol{Y}_{i,3} + \Delta\xi_i l_{ni} Rot(\boldsymbol{X}, \xi_i)\boldsymbol{Z}_{i,3}, \tag{22}$$

where $Rot(\boldsymbol{X}, \xi_i)$ represents the rotation matrix around $\boldsymbol{X}$ axis by $\xi_i$. Equations (23) and (24) can be derived along $\boldsymbol{Y}_{i,3}$ and $\boldsymbol{Z}_{i,3}$.

$$\Delta\sigma_i l_{bi} = \Delta l_i \cos\xi_i + \Delta\xi_i l_{ni} \sin\xi_i, \tag{23}$$

$$\Delta l_{ci} = -\Delta l_i \sin\xi_i + \Delta\xi_i l_{ni} \cos\xi_i. \tag{24}$$

Therefore, it can be solved that,

$$\Delta\sigma_i = \frac{\Delta l_i + \Delta l_{ci}\sin\xi_i}{l_{bi}\cos\xi_i}. \tag{25}$$

Equation (25) reveals that the deviations $\Delta l_i$ of links $\boldsymbol{D}_i\boldsymbol{F}_i$ and $\boldsymbol{E}_i\boldsymbol{G}_i$, and the $\Delta l_{ci}$ of links $\boldsymbol{D}_i\boldsymbol{E}_i$ and $\boldsymbol{F}_i\boldsymbol{G}_i$ cause relative pose changes in link $\boldsymbol{F}_i\boldsymbol{G}_i$. The value of $\xi_i$ also affects $\Delta\sigma_i$ when

$\Delta l_i$ and $\Delta l_{ci}$ remain constant. Notably, $l_{bi}$ and $\Delta\sigma_i$ exhibit an inverse relationship, where larger $l_{bi}$ values mitigate the impact of installation and manufacturing errors on link poses.

Although this analysis focuses on point $\boldsymbol{G}_i$, a similar approach can be applied to point $\boldsymbol{F}_i$. Assuming small $\Delta l_{ci}$, $\Delta\sigma_i$, $\Delta l_i$, $\Delta\xi_i$, Equation (25) still holds. Based on these considerations, the Delta robot necessitates the calibration of 54 parameters (Table 1), all of which can be assumed to have small errors.

**Table 1.** Parameters for calibration in Delta robot.

| Index | Parameter | Meaning |
|---|---|---|
| 1 | $x_{ai}$ | the $\boldsymbol{X}$ component of $\boldsymbol{A}_i$ coordinate in $\{\boldsymbol{W}_0\}$ |
| 2 | $y_{ai}$ | the $\boldsymbol{Y}$ component of $\boldsymbol{A}_i$ coordinate in $\{\boldsymbol{W}_0\}$ |
| 3 | $z_{ai}$ | the $\boldsymbol{Z}$ component of $\boldsymbol{A}_i$ coordinate in $\{\boldsymbol{W}_0\}$ |
| 4 | $\varphi_i$ | the $\boldsymbol{Z}$ rotation component transforming from $\{\boldsymbol{W}_0\}$ to $\{\boldsymbol{W}_{i,1}\}$ |
| 5 | $\gamma_i$ | the $\boldsymbol{X}$ rotation component transforming from $\{\boldsymbol{W}_0\}$ to $\{\boldsymbol{W}_{i,1}\}$ |
| 6 | $\theta_i$ | the rotation angle of the active revolute joint $i$ |
| 7 | $l_{pi}$ | the length of the active arm link |
| 8 | $\Delta\alpha_i$ | the $\boldsymbol{X}$ rotation component transforming from ${}^{i,2}\mathbf{Z}_{i,2}$ to ${}^{i,2}\mathbf{Z}_{i,3}$ |
| 9 | $\Delta\beta_i$ | the $\boldsymbol{Y}$ rotation component transforming from ${}^{i,2}\mathbf{Z}_{i,2}$ to ${}^{i,2}\mathbf{Z}_{i,3}$ |
| 10 | $l_{ni}$ | the length of the vector $\boldsymbol{B}_i\boldsymbol{C}_i$ within parallelogram mechanism |
| 11 | $\Delta l_i$ | the lengthening or shortening amount of the links $\boldsymbol{D}_i\boldsymbol{F}_i$ and $\boldsymbol{E}_i\boldsymbol{G}_i$ relative to $l_{ni}$ |
| 12 | $l_{bi}$ | the length of link $\boldsymbol{D}_i\boldsymbol{E}_i$ |
| 13 | $\Delta l_{ci}$ | the half-deviation between the lengths of $\boldsymbol{F}_i\boldsymbol{G}_i$ and $\boldsymbol{D}_i\boldsymbol{E}_i$ |
| 14 | $x_{ci}$ | the $\boldsymbol{X}$ component of $\boldsymbol{C}_i$ coordinate in $\{\boldsymbol{W}_5\}$ |
| 15 | $y_{ci}$ | the $\boldsymbol{Y}$ component of $\boldsymbol{C}_i$ coordinate in $\{\boldsymbol{W}_5\}$ |
| 16 | $z_{ci}$ | the $\boldsymbol{Z}$ component of $\boldsymbol{C}_i$ coordinate in $\{\boldsymbol{W}_5\}$ |
| 17 | $\varepsilon_i$ | the $\boldsymbol{Z}$ rotation component transforming from $\{\boldsymbol{W}_5\}$ to $\{\boldsymbol{W}_{i,4}\}$ |
| 18 | $\tau_i$ | the $\boldsymbol{X}$ rotation component transforming from $\{\boldsymbol{W}_5\}$ to $\{\boldsymbol{W}_{i,4}\}$ |

### 3.3. *The Kinematic Equations under Error Conditions*

Under error conditions, the kinematic chain of the Delta robot satisfies the following equations:

$$\|\boldsymbol{B}_i\boldsymbol{C}_i\| = l_{ni}, \tag{26}$$

$$\|\boldsymbol{Z}_{i,3} \times \boldsymbol{Z}_{i,4}\| = \Delta\sigma_i. \tag{27}$$

Equations (26) and (27) can be further expressed as follows:

$$\left\| \left({}^{0}\boldsymbol{A}_i + {}^{0}_{i,1}\boldsymbol{R}^{i,1}_{i,2}\boldsymbol{R}^{i,2}\boldsymbol{B}_i\right) - \left({}^{0}\boldsymbol{T}_i + {}^{0}_{5}\boldsymbol{R}^{5}\boldsymbol{C}_i\right) \right\| = l_{ni}, \tag{28}$$

$$\left\| \left({}^{0}_{i,1}\boldsymbol{R}^{i,1}_{i,2}\boldsymbol{R} \cdot Rot(\boldsymbol{X}, \Delta\alpha_i)Rot(\boldsymbol{Y}, \Delta\beta_i)^{i,2}\boldsymbol{Z}_{i,2}\right) \times \left({}^{0}_{5}\boldsymbol{R}^{i,4}R_5\right)^{i,4}\boldsymbol{Z}_{i,4} \right\| = \Delta\sigma_i. \tag{29}$$

Expand Equations (28) and (29), and the results can be formally represented as follows:

$$f_i(x_{ai}, y_{ai}, z_{ai}, \varphi_i, \gamma_i, \theta_i, l_p, l_n, x_{ci}, y_{ci}, z_{ci}, x_t, y_t, z_t, \alpha_5, \beta_5, \gamma_5) = 0, \tag{30}$$

$$g_i(\varphi_i, \gamma_i, \Delta\alpha_i, \Delta\beta_i, \varepsilon_i, \tau_i, \Delta\sigma_i, \alpha_5, \beta_5, \gamma_5) = 0. \tag{31}$$

When the Delta robot with the 3-$\underline{\text{R}}$(RPaR) configuration satisfies nominal geometric parameters, according to the modified Chebyshev–Grübler–Kutzbach formula [35], the mechanism's DoF is as follows:

$$M_{3-\underline{R}(RPaR)} = 6(n-g-1) + \sum_{q=1}^{g} f_q + v - \kappa = 6(17-21-1) + 21 + 12 - 0 = 3, \tag{32}$$

where $n$ is the number of bodies, $g$ is the number of kinematic pairs, $f_q$ is the DoF of the $q$th joint, and $\kappa$ represents the internal mobilities or idle DoF. $v = v_{PM} + v_{CJ} = 3 + 9 = 12$;

here, $\nu_{PM}$ is the number of overconstraints assembling the kinematic chains to a parallel manipulator and $\nu_{CJ}$ is the number of overconstraints within the complex joints.

In Figure 3, deviations in parameter errors may lead to non-parallelism between $\boldsymbol{F}_i\boldsymbol{G}_i$ and $\boldsymbol{D}_i\boldsymbol{E}_i$. Connecting the three kinematic chains to the moving and stationary platforms alters the overconstraint situation, resulting in a $v_{PM}$ value of 0. Thus, $v = \nu_{PM} + \nu_{CJ} = 0 + 9 = 9$, and $M_{3\text{-}\underline{R}(RPaR)} = 0$, indicating zero DoF for the Delta robot. Forcefully driving all three active arms induces deformation in the kinematic chains, creating surplus DoF. This paper defines structural deformation in line with the characteristics of the 3-R(RPaR) structure and referencing motion under error in 3-R(2-SS) as the rotation of link $\boldsymbol{F}_i\boldsymbol{G}_i$ (Figure 3) about the $\boldsymbol{Y}_{i,3}$ axis passing $\boldsymbol{C}_i$. To clarify this process, in the planar state depicted in Figure 3, $\boldsymbol{F}_i\boldsymbol{G}_i$ is redefined as $\boldsymbol{F}_i''\boldsymbol{G}_i''$ in Figure 4. Upon rotating $\boldsymbol{F}_i''\boldsymbol{G}_i''$ about the $\boldsymbol{Y}_{i,3}$ axis by $\Delta\eta_i$, the final $\boldsymbol{F}_i\boldsymbol{G}_i$ is obtained.

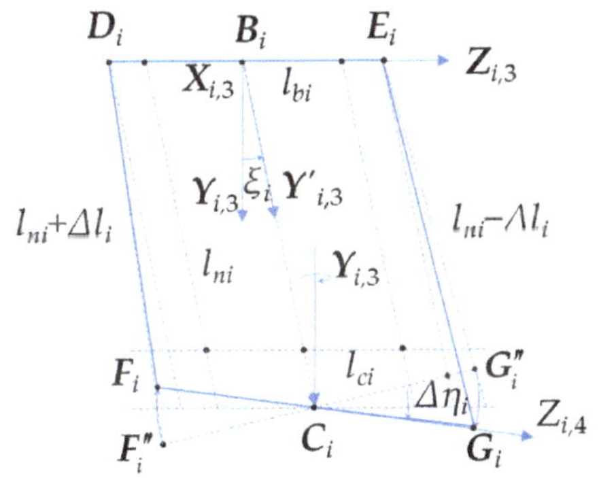


**Figure 4.** The structural deformation of the parallelogram mechanism.

The pose relationship described by Equation (21) is redefined as follows:

$$Rot({}^0\mathbf{Y}_{i,3}, \Delta\eta_i)Rot({}^0\mathbf{X}_{i,3}, \Delta\sigma_i){}^0_{i,1}\mathbf{R}\,{}^{i,1}_{i,2}\mathbf{R} \cdot Rot(\mathbf{X}, \Delta\alpha_i)Rot(\mathbf{Y}, \Delta\beta_i){}^{i,2}\mathbf{Z}_{i,2} = \left({}^0_5\mathbf{R}\,{}^5_{i,4}\mathbf{R}\right){}^{i,4}\mathbf{Z}_{i,4}, \quad (33)$$

where $Rot({}^0\mathbf{Y}_{i,3}, \Delta\eta_i)$ represents the rotation matrix around ${}^0\mathbf{Y}_{i,3}$ by $\Delta\eta_i$ and $Rot({}^0\mathbf{X}_{i,3}, \Delta\sigma_i)$ represents the rotation matrix around ${}^0\mathbf{X}_{i,3}$ by $\Delta\sigma_i$. Both sides of Equation (33) represent unit vectors. Given the small $\boldsymbol{Z}$-component of these unit vectors in $\{\boldsymbol{W}_0\}$ (attributable solely to errors), this paper formulates equations based on their $\boldsymbol{X}$ and $\boldsymbol{Y}$ components. With three chains ($i$ = 1, 2, 3), Equation (33) yields six independent equations, formally expressed as follows:

$$q_{xi}\left(x_{ai}, y_{ai}, z_{ai}, \varphi_i, \gamma_i, \theta_i, l_{pi}, \Delta\alpha_i, \Delta\beta_i, l_{ni}, \Delta\sigma_i, x_{ci}, y_{ci}, z_{ci}, \varepsilon_i, \tau_i, \Delta\eta_i, x_t, y_t, z_t, \alpha_5, \beta_5, \gamma_5\right) = 0, \quad (34)$$

$$q_{yi}\left(x_{ai}, y_{ai}, z_{ai}, \varphi_i, \gamma_i, \theta_i, l_{pi}, \Delta\alpha_i, \Delta\beta_i, l_{ni}, \Delta\sigma_i, x_{ci}, y_{ci}, z_{ci}, \varepsilon_i, \tau_i, \Delta\eta_i, x_t, y_t, z_t, \alpha_5, \beta_5, \gamma_5\right) = 0. \quad (35)$$

Equations (30), (34) and (35) collectively depict the kinematic equations of the Delta robot in the 3-R(RPaR) configuration under error conditions. These equations entail multiple nonlinear terms involving trigonometric functions and higher-order factors, rendering the solution process cumbersome. To streamline these equations, parameters are initially expressed as the sum of nominal values and errors, as presented below:

$$x_{ai} = x_{ai}^0 + \Delta x_{ai},\ y_{ai} = y_{ai}^0 + \Delta y_{ai},\ z_{ai} = z_{ai}^0 + \Delta z_{ai}, \quad (36)$$

$$\varphi_i = \varphi_i^0 + \Delta\varphi_i,\ \gamma_i = \gamma_i^0 + \Delta\gamma_i,\ \theta_i = \theta_i^0 + \Delta\theta_i, \quad (37)$$

$$l_{pi} = l_p + \Delta l_{pi},\ l_{ni} = l_n + \Delta l_{ni}, \quad (38)$$

$$x_{ci} = x_{ci}^0 + \Delta x_{ci},\ y_{ci} = y_{ci}^0 + \Delta y_{ci},\ z_{ci} = z_{ci}^0 + \Delta z_{ci}, \quad (39)$$

$$\varepsilon_i = \varepsilon_i^0 + \Delta\varepsilon_i,\ \tau_i = \tau_i^0 + \Delta\tau_i. \quad (40)$$

The right superscript "0" denotes the nominal value of a parameter, while the prefix $\Delta$ represents the parameter's error. For example, $x_{ai}^0$ and $\Delta x_{ai}$ represent the nominal and error values of $x_{ai}$, respectively. All parameter error values are considered small quantities. Taking $\varphi_i$ as an example, the simplification is as follows:

$$\cos(\Delta\varphi_i) \approx 1,\ \sin(\Delta\varphi_i) \approx \Delta\varphi_i, \tag{41}$$

$$\cos\left(\varphi_i^0 + \Delta\varphi_i\right) \approx \cos\varphi_i^0 - \Delta\varphi_i \sin\varphi_i^0, \tag{42}$$

$$\sin\left(\varphi_i^0 + \Delta\varphi_i\right) \approx \sin\varphi_i^0 + \Delta\varphi_i \cos\varphi_i^0. \tag{43}$$

The other angle parameters $\gamma_i$, $\theta_i$, $\varepsilon_i$, $\tau_i$, follow a similar pattern of $\varphi_i$. Additionally, all second-order and higher small quantities are omitted.

$$\Delta x_{ai} \cdot \Delta x_{ai} \approx 0,\ \Delta x_{ai} \cdot \Delta x_{ai} \cdot \Delta x_{ai} \approx 0, \ldots \tag{44}$$

$$\Delta x_{ai} \cdot \Delta y_{ai} \approx 0,\ \ldots,\ \Delta x_{ai} \cdot \Delta\varphi_i \approx 0, \ldots \tag{45}$$

Equations (30), (34) and (35) can be expressed in matrix form after simplification as follows:

$$\boldsymbol{F}_p \boldsymbol{v} + \boldsymbol{h}_p = 0, \tag{46}$$

$$\boldsymbol{G}_{ox} \boldsymbol{v} + \boldsymbol{h}_{ox} = 0, \tag{47}$$

$$\boldsymbol{G}_{oy} \boldsymbol{v} + \boldsymbol{h}_{oy} = 0, \tag{48}$$

where $\boldsymbol{F}_p = \begin{bmatrix} f_{1,1}, \cdots, f_{1,j}, \cdots, f_{1,18} & 0 & 0 \\ 0 & f_{2,1}, \cdots, f_{2,j}, \cdots, f_{2,18} & 0 \\ 0 & 0 & f_{3,1}, \cdots, f_{3,j}, \cdots, f_{3,18} \end{bmatrix}$, $\boldsymbol{h}_p = \begin{bmatrix} \boldsymbol{h}_{p1} \\ \boldsymbol{h}_{p2} \\ \boldsymbol{h}_{p3} \end{bmatrix}$, $f_{i,j} = f_{i,j}\left(x_{ai}^0, y_{ai}^0, z_{ai}^0, \varphi_i^0, \gamma_i^0, \theta_i^0, l_p, l_n, \varepsilon_i^0, \tau_i^0, x_{ci}^0, y_{ci}^0, z_{ci}^0, x_t, y_t, z_t, \alpha_5, \beta_5, \gamma_5\right)$, $\boldsymbol{h}_{pi} = \boldsymbol{h}_{pi}\left(x_{ai}^0, y_{ai}^0, z_{ai}^0, \varphi_i^0, \gamma_i^0, \theta_i^0, l_p, l_n, \varepsilon_i^0, \tau_i^0, x_{ci}^0, y_{ci}^0, z_{ci}^0, x_t, y_t, z_t, \alpha_5, \beta_5, \gamma_5\right)_{18\times 1}$, $\boldsymbol{v} = \left[\boldsymbol{v}_{c1}, \boldsymbol{v}_{c2}, \boldsymbol{v}_{c3}\right]^T$, $\boldsymbol{v}_{ci} = \left[\Delta x_{ai}, \Delta y_{ai}, \Delta z_{ai}, \Delta\varphi_i, \Delta\gamma_i, \Delta\theta_i, \Delta l_{pi}, \Delta\alpha_i, \Delta\beta_i, \Delta l_{ni}, \Delta l_i, \Delta l_{bi}, \Delta l_{ci}, \varepsilon_i, \tau_i, \Delta x_{ci}, \Delta y_{ci}, \Delta z_{ci}\right]^T$, $\boldsymbol{G}_{ox} = \begin{bmatrix} g_{x1,1}, \cdots, g_{x1,j}, \cdots, g_{x1,18} & 0 & 0 \\ 0 & g_{x2,1}, \cdots, g_{x2,j}, \cdots, g_{x2,18} & 0 \\ 0 & 0 & g_{x3,1}, \cdots, g_{x3,j}, \cdots, g_{x3,18} \end{bmatrix}$, $\boldsymbol{h}_{ox} = \begin{bmatrix} h_{ox1} \\ h_{ox2} \\ h_{ox3} \end{bmatrix}$, $g_{xi,j} = g_{xi,j}\left(x_{ai}^0, y_{ai}^0, z_{ai}^0, \varphi_i^0, \gamma_i^0, \theta_i^0, l_p, l_n, l_b, x_{ci}^0, y_{ci}^0, z_{ci}^0, \varepsilon_i^0, \tau_i^0, \Delta\eta_i, x_t, y_t, z_t, \alpha_5, \beta_5, \gamma_5\right)$, $\boldsymbol{h}_{oxj} = \boldsymbol{h}_{oxj}\left(x_{ai}^0, y_{ai}^0, z_{ai}^0, \varphi_i^0, \gamma_i^0, \theta_i^0, l_p, l_n, l_b, x_{ci}^0, y_{ci}^0, z_{ci}^0, \varepsilon_i^0, \tau_i^0, \Delta\eta_i, x_t, y_t, z_t, \alpha_5, \beta_5, \gamma_5\right)_{18\times 1}$, $\boldsymbol{G}_{oy} = \begin{bmatrix} g_{y1,1}, \cdots, g_{y1,j}, \cdots, g_{y1,18} & 0 & 0 \\ 0 & g_{y2,1}, \cdots, g_{y2,j}, \cdots, g_{y2,18} & 0 \\ 0 & 0 & g_{y3,1}, \cdots, g_{y3,j}, \cdots, g_{y3,18} \end{bmatrix}$, $\boldsymbol{h}_{oy} = \begin{bmatrix} h_{oy1} \\ h_{oy2} \\ h_{oy3} \end{bmatrix}$, $g_{yi,j} = g_{yi,j}\left(x_{ai}^0, y_{ai}^0, z_{ai}^0, \varphi_i^0, \gamma_i^0, \theta_i^0, l_p, l_n, l_b, x_{ci}^0, y_{ci}^0, z_{ci}^0, \varepsilon_i^0, \tau_i^0, \Delta\eta_i, x_t, y_t, z_t, \alpha_5, \beta_5, \gamma_5\right)$, $\boldsymbol{h}_{oyj} = \boldsymbol{h}_{oyj}\left(x_{ai}^0, y_{ai}^0, z_{ai}^0, \varphi_i^0, \gamma_i^0, \theta_i^0, l_p, l_n, l_b, x_{ci}^0, y_{ci}^0, z_{ci}^0, \varepsilon_i^0, \tau_i^0, \Delta\eta_i, x_t, y_t, z_t, \alpha_5, \beta_5, \gamma_5\right)_{18\times 1}$.

$\Delta\eta_i$ satisfies small quantity assumption. Using $\begin{cases} \sum\limits_{j=1}^{18} g_{xi,j} \cdot v_j + h_{oxi} = 0 \\ \sum\limits_{j=1}^{18} g_{yi,j} \cdot v_j + h_{oyi} = 0 \end{cases}$ to eliminate $\Delta\eta_i$, Equations (47) and (48) can be rearranged as:

$$\boldsymbol{G}_o \boldsymbol{v} + \boldsymbol{h}_o = 0, \tag{49}$$

where $G_o = \begin{bmatrix} g_{1,1}, \cdots, g_{1,j}, \cdots, g_{1,18} & 0 & 0 \\ 0 & g_{2,1}, \cdots, g_{2,j}, \cdots, g_{2,18} & 0 \\ 0 & 0 & g_{3,1}, \cdots, g_{3,j}, \cdots, g_{3,18} \end{bmatrix}$, $h_o = \begin{bmatrix} h_{o1} \\ h_{o2} \\ h_{o3} \end{bmatrix}$, $g_{i,j} = g_{i,j}(x_{ai}^0, y_{ai}^0, z_{ai}^0, \varphi_i^0, \gamma_i^0, \theta_i^0, l_p, l_n, l_b, x_{ci}^0, y_{ci}^0, z_{ci}^0, \varepsilon_i^0, \tau_i^0, x_t, y_t, z_t, \alpha_5, \beta_5, \gamma_5)$, $h_{oj} = h_{oj}(x_{ai}^0, y_{ai}^0, z_{ai}^0, \varphi_i^0, \gamma_i^0, \theta_i^0, l_p, l_n, l_b, x_{ci}^0, y_{ci}^0, z_{ci}^0, \varepsilon_i^0, \tau_i^0, x_t, y_t, z_t, \alpha_5, \beta_5, \gamma_5)_{18\times 1}$.

Equations (46) and (49) constitute the linear kinematic equations of the Delta robot with the 3-R(RPaR) configuration under error conditions.

$$\begin{bmatrix} F_p \\ G_o \end{bmatrix} v + \begin{bmatrix} h_p \\ h_o \end{bmatrix} = 0. \tag{50}$$

In theory, utilizing Equations (28) and (33) to derive the partial derivatives of $x_t$, $y_t$, $z_t$, as well as $\alpha_5$, $\beta_5$ and $\gamma_5$ with respect to each error component, yields a similar error model to Equation (50) containing the Jacobian matrix. However, since $x_t$, $y_t$, $z_t$ and $\alpha_5$, $\beta_5$, $\gamma_5$ are implicit in Equations (28) and (33), the solving process becomes relatively complex. This paper adopts the assumption of small errors and follows the principles of small quantity operations for a more intuitive process. Given that the error values of $x_t$, $y_t$, $z_t$ and $\alpha_5$, $\beta_5$, $\gamma_5$ consist of various error components and are relatively large, for enhanced accuracy, the relevant elements in matrices $F_p$ and $G_o$ remain expressed as functions of $x_t$, $y_t$, $z_t$ and $\alpha_5$, $\beta_5$, $\gamma_5$, rather than their corresponding linear error terms.

## 4. The Identification of Parameter Errors

### 4.1. Parameter Error Identifiability

Parameter identifiability comprises two primary aspects: the linear correlation between parameters, potentially impeding accurate individual parameter identification, and parameter sensitivity, indicating the relative impact of parameter errors on overall error. Traditional methods often struggle to effectively address both concerns, resulting in complex analyses. This study employs a simplified approach, the workspace random points + SVD method, to address these challenges. By selecting $n$ random points within the Delta robot's workspace, their actual poses are computed using Equation (50).

In Equation (25), $l_{bi} = l_{bi}^0 + \Delta l_{bi}$; if $\Delta l_{bi}$ is regarded as a first-order small quantity, it will be absorbed in calculations. Consequently, this study overlooks the impact of $\Delta l_{bi}$ and does not account for it. In Equation (50), the component $\Delta l_{bi}$ in the parameter error vector $v$ is consistently treated as zero. To streamline the analysis, $\Delta l_{b1}$, $\Delta l_{b2}$ and $\Delta l_{b3}$ are omitted from $v$. Utilizing Equation (50), an equation set is formulated at each point, yielding the following:

$$Mv + b = 0, \tag{51}$$

where $M$ is the Linear Model Coefficient Matrix (LMCM). Since $\Delta l_{bi}$ is excluded from $v$, $M$ has a size of $6n \times 51$, and the bias $b$ has a size of $6n \times 1$. Performing SVD on matrix $M$, Equation (51) can be rewritten as follows:

$$U\Sigma W^T v + b = 0, \tag{52}$$

where $U$ and $W$ represent orthogonal matrices of size $6n \times 6n$ and $51 \times 51$, respectively. $u_i$ and $w_i$ (for $i = 1, \ldots, 51$) denote the row vectors of $U$ and $W$, respectively. $\Sigma$ is a $6n \times 51$ matrix, with the first 51 rows forming a diagonal matrix consisting of the singular values of $M$ arranged in descending order. The remaining rows of $\Sigma$ are all zeros. To examine the influence of singular values on the parameter solution, Equation (52) can be expressed in terms of $u_i$ and $w_i$:

$$\lambda_1 u_1 w_1^T v + \lambda_2 u_2 w_2^T v + \cdots + \lambda_{51} u_{51} w_{51}^T v + b = 0. \tag{53}$$

The structure of Equation (53) indicates that the smaller $\lambda_i \boldsymbol{u}_i \boldsymbol{w}_i^T \boldsymbol{v}$ is, the lower the impact on the parameter solution. The magnitude of $\lambda_i \boldsymbol{u}_i \boldsymbol{w}_i^T \boldsymbol{v}$ is represented by a vector norm, and

$$\left\| \lambda_i \boldsymbol{u}_i \boldsymbol{w}_i^T \boldsymbol{v} \right\| \leq \lambda_i \|\boldsymbol{u}_i\| \left\| \boldsymbol{w}_i^T \right\| \|\boldsymbol{v}\| = \lambda_i \|\boldsymbol{v}\|. \tag{54}$$

In Equation (54), it is evident that the magnitude of the term $\lambda_i \boldsymbol{u}_i \boldsymbol{w}_i^T \boldsymbol{v}$ depends on the size of $\lambda_i$ when $\boldsymbol{v}$ is given or held constant. When $\lambda_i$ is sufficiently small, $\lambda_i \boldsymbol{u}_i \boldsymbol{w}_i^T \boldsymbol{v}$ can be disregarded in Equation (53). Higher-order small terms are neglected during the derivation of Equation (50) due to the assumption of small errors. Therefore, the accuracy of equations built upon Equation (50) does not surpass the order of $\|\boldsymbol{v}\| \cdot \|\boldsymbol{v}\|$. By comparison with $\lambda_i \|\boldsymbol{v}\|$ in Equation (54), the following criterion can be deduced:

$$\lambda_i < \|\boldsymbol{v}\|. \tag{55}$$

When $\lambda_i$ satisfies Equation (55), the corresponding parameter becomes unidentifiable. The relationship between $\lambda_i$ and the parameters will be illustrated through examples in the simulation section.

The linearity correlation of parameters can be analyzed from the coefficients of the error parameters. For example, in Equation (46), the coefficient of $\Delta x_{ai}$ is as follows:

$$\begin{aligned} f_{i,1} &= 2x_{ai} - 2x_t - 2z_{ci}\sin(\beta_5) - 2x_{ci}\cos(\beta_5)\cos(\gamma_5) \\ &\quad + 2y_{ci}\cos(\beta_5)\sin(\gamma_5) + 2\cos(\varphi_i)\cos(\theta_i)l_p - 2\cos(\gamma_i)l_p\sin(\varphi_i)\sin(\theta_i) \end{aligned} . \tag{56}$$

The coefficient $f_{i,15}$ can be derived as well, but due to its lengthy equation, it is not listed here. Without assuming $\alpha_5$, $\beta_5$, $\gamma_5$ as small quantities, there exists no fixed linear relationship between $f_{i,1}$ and $f_{i,15}$. However, by assuming $\alpha_5$, $\beta_5$, $\gamma_5$ as small quantities and sin() = 0, cos() = 1, then

$$f_{i,1} = -f_{i,15} = 2x_{ai} - 2x_t - 2x_{ci} + 2l_p\cos(\varphi_i)\cos(\theta_i) - 2l_p\cos(\gamma_i)\sin(\varphi_i)\sin(\theta_i). \tag{57}$$

Similarly, the coefficients corresponding to $\Delta y_{ai}$ and $\Delta y_{ci}$, as well as $\Delta z_{ai}$ and $\Delta z_{ci}$, exhibit the same relationship. Given these linear relationships, $\Delta x_{ai}$, $\Delta y_{ai}$ and $\Delta z_{ai}$ can directly replace $\Delta x_{ai} - \Delta x_{ci}$, $\Delta y_{ai} - \Delta y_{ci}$, and $\Delta z_{ai} - \Delta z_{ci}$. This simplifies the dimensionality of $\boldsymbol{v}$. Analyzing or determining linear relationships between parameters directly from lengthy coefficient expressions is often challenging. In the simulation section, these linear relationships will be illustrated graphically with examples.

### *4.2. Measurement Surface Optimization*

This paper assumes that the pose error of the moving platform can be determined through measuring the surface form error of the machined workpiece, termed as the measurement surface. Assuming a fixed form of point acquisition on this surface, its shape directly influences the error parameter identification process. This section aims to identify the optimal measurement surface to enhance the identification process.

First, the measurement surface is parameterized. The coordinate system of the measurement surface is defined as $\{\boldsymbol{W}_6\}$, with $\{\boldsymbol{W}_6\}$ being fixed relative to $\{\boldsymbol{W}_5\}$. The origin of $\{\boldsymbol{W}_6\}$ is offset along the $\boldsymbol{Z}$ axis of $\{\boldsymbol{W}_5\}$ by a distance of $H_q$, with the axes of $\{\boldsymbol{W}_6\}$ parallel and opposite to those of $\{\boldsymbol{W}_5\}$, respectively. $H_q$ is the height of the measurement surface. Within $\{\boldsymbol{W}_6\}$, the measurement surface $\boldsymbol{Q}(s, t)$ is defined. This surface can be extended periodically in both the $s$ and $t$ directions within a finite region, creating an infinitely free-from surface. Alternatively, $\boldsymbol{Q}(s, t)$ can be viewed as a single-period surface on an infinitely periodic surface.

$$\boldsymbol{Q}(s,t) = \boldsymbol{Q}_r(s)\boldsymbol{Q}_h(t), \tag{58}$$

where $\boldsymbol{Q}_r(s)$ and $\boldsymbol{Q}_h(t)$ are both single-period functions, constructed using multiplication to form the $\boldsymbol{Q}(s, t)$ surface. This construction method guarantees that any iso-parameter curve, such as $\boldsymbol{Q}(s_0, t) = \boldsymbol{Q}_r(s_0)\boldsymbol{Q}_h(t)$ or $\boldsymbol{Q}(s, t_0) = \boldsymbol{Q}_r(s)\boldsymbol{Q}_h(t_0)$, remains periodic.

According to the Fourier series expansion theorem, $Q_r(s)$ and $Q_h(t)$ can be further represented as follows:

$$Q_r(s) = \frac{a_{u0}}{2} + \sum_{m=1}^{3} (a_{um}\cos(3ms) + b_{um}\sin(3ms)), \quad (59)$$

$$Q_h(t) = \frac{a_{v0}}{2} + \sum_{n=1}^{3} (a_{vn}\cos(nt) + b_{vn}\sin(nt)), \quad (60)$$

The Fourier series is an infinite sum of trigonometric functions, with high-frequency terms typically having a minimal impact on function values. Here, both $m$ and $n$ are set to 3, a sufficient choice for engineering purposes. Additionally, to capture the symmetry of the three kinematic chains in the $XY$ plane of $\{W_5\}$, coefficients relative to $s$ are represented in a $3m$ form. Consequently, the workpiece's cross-sectional lines in the $XY$ plane of $\{W_5\}$ consist of three identical periodic curves, each spanning 120°.

Subsequently, the optimization problem formulation is established, with Equation (50) serving as the core of parameter identification. At multiple measurement points, an over-constrained equation set can be derived from Equation (50), as detailed in Equation (54) in Section 4.1. The quality of solving Equation (51) is most aptly reflected in the condition number of $M$. Hence, the optimization problem in this paper is defined as follows:

$$\begin{cases} & \min f\left(v_{coff}\right) \\ s.t. & R_q < R_{w\max} \\ & R_q > R_{w\min} \\ & H_q = H_{wm} \end{cases}, \quad (61)$$

where $v_{coff} = [a_{s0}, a_{s1}, a_{s2}, a_{s3}, b_{s1}, b_{s2}, b_{s3}, a_{t0}, a_{t1}, a_{t2}, a_{t3}, b_{t1}, b_{t2}, b_{t3}]$, $f\left(v_{coff}\right) = cond(\boldsymbol{M})$, and cond($\boldsymbol{M}$) denotes the condition number of $\boldsymbol{M}$ in Equation (51). $R_q$ denotes the maximum radius of the measurement surface in the $\boldsymbol{X}_5\boldsymbol{Y}_5$ plane. $R_q$ must be smaller than the maximum radius $R_{wmax}$ of the cylindrical workspace of the Delta robot. Simultaneously, $R_q$ should not be excessively small, as this could compromise the strength of the workpiece, leading to fractures, or result in deep grooves that are challenging to measure. The lower limit of $R_q$ is set to $R_{wmin}$. The height $H_q$ of the measurement surface is equal to the maximum height $H_{wmax}$ of the robot cylindrical workspace by replacing $t$ in Equation (60) with $2\pi t / H_{w\max} - \pi$.

The process of measurement surface optimization and parameter identification is shown in Figure 5. Specific implementation details are described in Section 5. The "triangle" in the upper left corner indicates the start, and the "circle" in the lower right corner indicates the end.

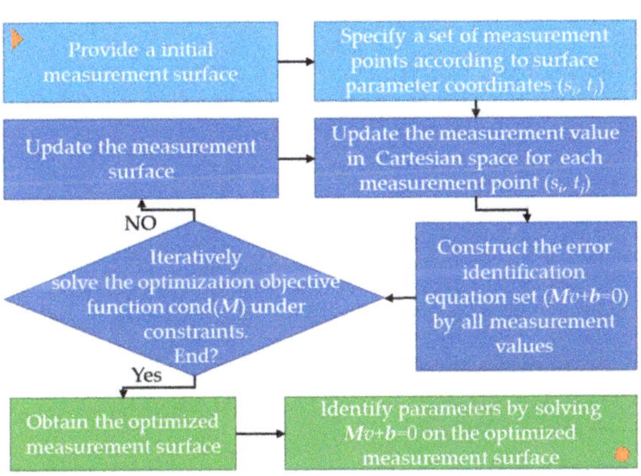


**Figure 5.** The process of measurement surface optimization and parameter identification.

## 5. Simulation

This section validates the proposed kinematic calibration method for the Delta robot with 3-R(RPaR) configuration through computational cases. Initially, nominal geometric parameters are specified: $r_1$ = 0.127 m, $r_2$ = 0.046 m, $l_p$ = 0.22751 m, $l_n$ = 0.55361 m, and $l_{bi}^0$ = 0.1 m. Eliminating $\Delta l_{bi}$, the other 51-dimensional (or positional) error terms are set to $1 \times 10^{-4}$ m, and angular (or orientation) error terms are set to 0.01° ($1.745 \times 10^{-4}$ rad). For instance, $\Delta x_{ai} = 1 \times 10^{-4}$ m and $\Delta\theta_i = 0.01°$.

1. Verification of parameter identifiability. A total of 50 points are randomly selected in the Delta robot's workspace, as detailed in Section 4.1 with $n = 50$. Figure 6 illustrates the singular values of the LMCM $M$ from Equation (51), plotted in descending order on a log10 scale. The singular values of $M$ form three distinct tiers, labeled I, II, and III, with magnitudes of approximately $10^0$, $10^{-7}$, and $10^{-15}$, respectively. Given that position and angle errors are on the order of $10^{-4}$, the magnitude of $\|v\|$ is also $10^{-4}$. According to inequality (55), parameters can be ignored when $\lambda_i < 10^{-4}$. Therefore, parameters corresponding to tiers other than tier I will be ignored. The lowest point in tier I corresponds to the singular value $\lambda_{24}$.

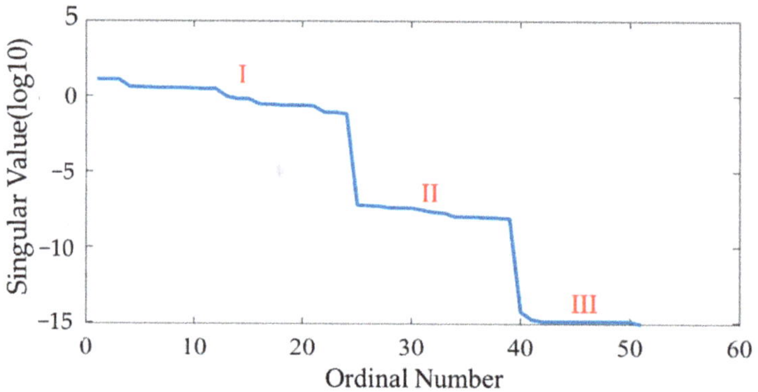


**Figure 6.** Singular values of the LMCM $\boldsymbol{M}$ (Tier I: $10^0$, Tier II: $10^{-7}$, Tier III: $10^{-15}$).

Rewriting Equation (52) yields the following:

$$\sum_{i=1}^{24} \lambda_i \boldsymbol{w}_i^T \boldsymbol{v} = -\boldsymbol{U}^T \boldsymbol{b}, \tag{62}$$

where $\boldsymbol{v}$ denotes the 51-component error vector excluding $\Delta l_{b1}$, $\Delta l_{b2}$, and $\Delta l_{b3}$. Figure 7 displays a scatter plot of the components of the column vectors $\boldsymbol{w}_i$, with each color corresponding to the same $i$ value. The $j$th component of $\boldsymbol{w}_i$ also serves as the coefficient for the $j$th error component of $\boldsymbol{v}$. The horizontal axis in Figure 7 represents each of the 17 error components in each chain, ordered as chains 1, 2, and 3. Notably, the component values in $\boldsymbol{w}_i$ have been preprocessed based on the error scale; specifically, if the absolute value of a component in $\boldsymbol{w}_i$ is less than $10^{-4}$, it is set to zero.

From Figure 7, it is evident that the coefficients corresponding to components 8, 9, 11, 12, 13, and 14 in the error vectors within each chain consistently evaluate to zero. Consequently, these parameters are deemed unidentifiable. Furthermore, the coefficients of components 1, 2, and 3 within each chain exhibit sign symmetry in relation to the coefficients of components 15, 16, and 17, respectively. This indicates that either the error components 1, 2, and 3, or the error components 15, 16, and 17, cannot be individually discerned. What can be discerned are $\boldsymbol{v}_{ci}(1)$–$\boldsymbol{v}_{ci}(15)$, $\boldsymbol{v}_{ci}(2)$–$\boldsymbol{v}_{ci}(16)$, and $\boldsymbol{v}_{ci}(3)$–$\boldsymbol{v}_{ci}(17)$, where $\boldsymbol{v}_{ci}(j)$ represents the $j$th component of the $\boldsymbol{v}_{ci}$ vector. In subsequent optimization, $\boldsymbol{v}_{ci}(1)$,

$v_{ci}(2)$ and $v_{ci}(3)$ will be utilized to denote $v_{ci}(1)$–$v_{ci}(15)$, $v_{ci}(2)$–$v_{ci}(16)$, and $v_{ci}(3)$–$v_{ci}(17)$, respectively. The remaining identifiable parameters for each chain are as follows:

$$v_{ci} = \left[\Delta x_{ai}, \Delta y_{ai}, \Delta z_{ai}, \Delta \varphi_i, \Delta \gamma_i, \Delta \theta_i, \Delta l_{pi}, \Delta l_{ni}\right]^T. \tag{63}$$

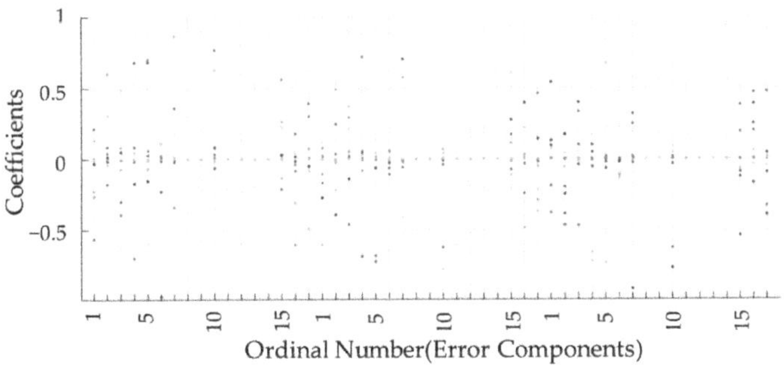


**Figure 7.** Coefficient distribution of error components.

2. Determining the optimal measurement surface. The cylindrical workspace of the Delta robot is characterized by a maximum radius $R_{wmax}$ = 0.265 m, a given minimum radius $R_{wmin}$ = 0.100 m, and a height $H_{wm}$ = 0.2 m, with its zero point in $\{W_0\}$ along the Z axis positioned at $Z_{home}$ = 0.3795 m. When evaluating $f\left(v_{coff}\right)$ in Problem (61), the surface coefficients in Equations (59) and (60) are considered given, and a point acquisition method on the surface needs to be specified. Here, $s \in [0, 2\pi]$, $t \in [0, H_{wm}]$, and $s$ and $t$ are uniformly divided into 36 and 12 parts, respectively, generating a grid of points. Since the $s$ direction forms a closed curve, it is iterated from $i$ = 0 to 35 and $j$ = 0 to 12, yielding a total of 36 × 13 non-repeating grid points in space. While more points theoretically enhance the accuracy of the grid, an excessive number would escalate measurement and computation times. In this study, the choice of points is balanced; even for trigonometric components with $m$, and $n$ = 3 in Equations (59) and (60), a minimum of four sampling points per cycle is ensured.

This paper employs a hybrid solving algorithm that combines the Hook–Jeeves Direct Search Algorithm (DSA) with the Inner Point Method (IPM) for the optimization problem outlined in Problem (61). Initially, the constrained optimization problem is converted into an unconstrained one through the application of the interior point method. Here, the constraints refer to "<" and ">" constraints, while equality constraints can be ensured through the definition of the measurement surface. The augmented objective function needed for the interior point method is formulated as follows:

$$f_a\left(v_{coff}, \tau\right) = f\left(v_{coff}\right) + \tau f_b\left(v_{coff}\right), \tag{64}$$

where $\tau > 0$ is the penalty factor.

The essence of the interior point method is to transform a constrained optimization problem into a sequence of unconstrained optimization problems $\min f_a\left(v_{coff}, \tau_k\right)$, where $\tau_{k+1} = \rho\tau_k$, as shown in "IPM" part of Figure 8. In the simulation process, $\rho$ = 0.5, $\tau_0$ = 1. The barrier function $f_b\left(v_{coff}\right)$ is defined as follows:

$$f_b\left(v_{coff}\right) = \frac{1}{R_{w\max} - R_q} + \frac{1}{R_q - R_{w\min}}, \tag{65}$$

where $R_q$ is achieved by separately solving the maximum value of $Q_r(s)$ and $Q_h(t)$, i.e., $R_q = \max(Q_r(s)) \cdot \max(Q_h(t))$. As $k$ increases, $\tau_k f_b\left(v_{coff}\right)$ will gradually decrease. The termination criterion for the iterative process is $\tau_k f_b\left(v_{coff}\right) < \varepsilon_{IM}$. In simulation process, $\varepsilon_{IM} = 0.01$.

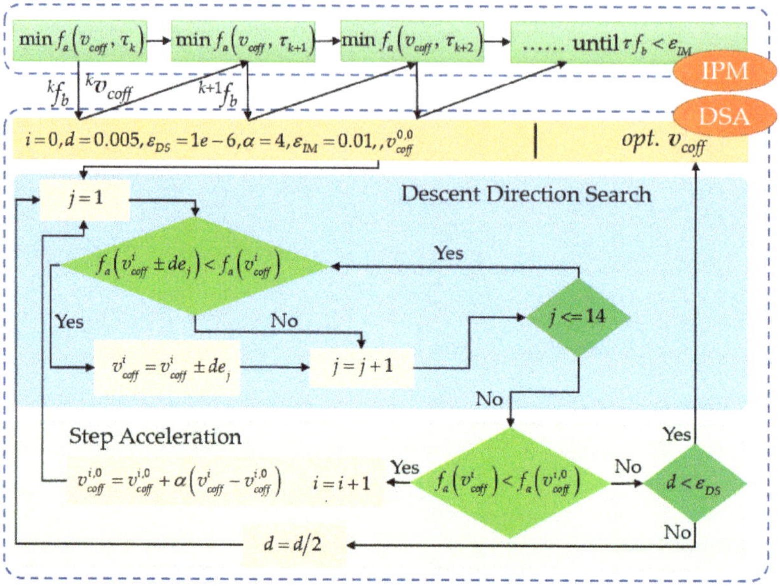


**Figure 8.** IPM + DSA hybrid optimization algorithm.

Each $f_a\left(v_{coff}, \tau_k\right)$ undergoes optimization using the Hook–Jeeves DSA to determine the minimum value. While DSAs may not converge as rapidly as derivative-based methods like gradient descent or Newton's method, the latter are susceptible to local optima. The DSA, being derivative-free, offers global optimality and requires fewer parameters than heuristic algorithms, making it well-suited for engineering applications. The Hook–Jeeves algorithm comprises two key components: Descent Direction Search (DDS) and Step Acceleration (SA), depicted in the "DSA" section of Figure 8. In DDS, the search progresses along the positive and negative directions of each unit component $e_j$ of the $v_{coff}$ vector sequentially, with a step size of $d$. The initial value of $v_{coff}$ is denoted as $v^{0,0}_{coff}$. Upon identifying a direction of function reduction, the SA phase is initiated. During SA, the step length is accelerated using the acceleration factor $\alpha\left(v^{i}_{coff} - v^{i,0}_{coff}\right)$ to update $v_{coff}$ and commence a new cycle of DDS. In the simulation process, $\alpha = 4$. If a direction of function reduction is not detected, $d = d/2$, and DDS continues. The termination criterion for the Hook–Jeeves DSA is d < $\varepsilon_{DS}$. In simulations, the initial value of $d$ is set to 0.005, with $\varepsilon_{DS} = 10^{-6}$.

Figure 9 illustrates the iterative convergence process of the augmented objective function. The horizontal axis represents the number of iterations of the Hook–Jeeves DSA, which is the ordinal of the sequence min$f_a(v_{coff}, \tau_k)$. The convergence process is very rapid, with the augmented objective function entering a bottom plateau phase after approximately five iterations. The value of the augmented objective function decreases from 1239.6853 to 236.8953, a decrease of more than five times. Figure 10a and 10b, respectively, depict the iterative convergence process of the surface coefficients $a_{s0}$, $a_{s1}$, $a_{s3}$, $b_{s1}$, $b_{s2}$, $b_{s3}$ and $a_{t0}$, $a_{t1}$, $a_{t2}$, $a_{t3}$, $b_{t1}$, $b_{t2}$, $b_{t3}$ with the same horizontal axis as Figure 9. From Figure 10, it can be observed that the amplitudes of $a_{s0}$ and $b_{s0}$ are significantly larger compared to other trigonometric coefficients, and their initial values change noticeably during the initial iterations.

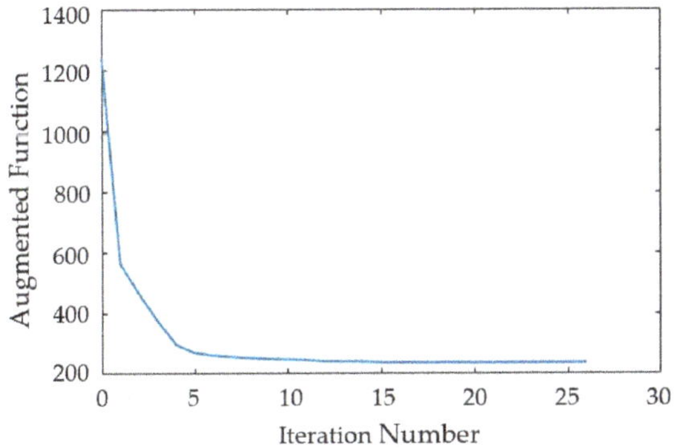


**Figure 9.** Iterative convergence of augmented function.

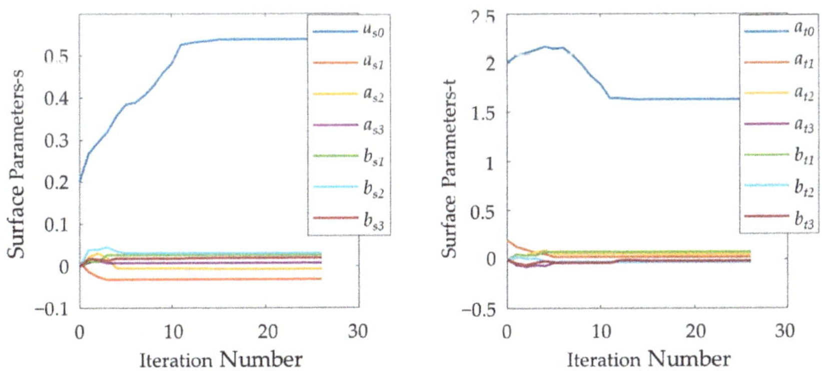


(**a**) (**b**)

**Figure 10.** Iterative convergence of surface parameters. (**a**): $Q_r(s)$. (**b**): $Q_h(t)$.

Figure 11 depicts the optimized surface profile. In the 3D representation shown in Figure 11a, the surface demonstrates pronounced periodicity along both the $s$ and $t$ directions. Moreover, the surface's position and height in the $\boldsymbol{Z}$ direction align with the specified values. As illustrated in Figure 11b, the optimized surface profile adheres to the "<" and ">" constraints outlined in Problem (60), with both constraint boundaries being attained.

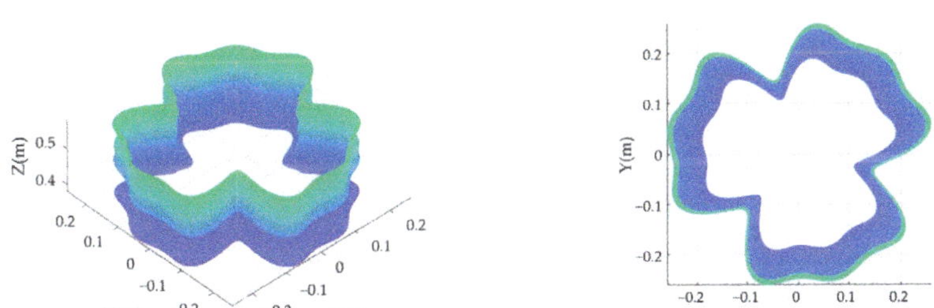


(**a**) (**b**)

**Figure 11.** The optimized surface profile. (**a**): 3D view. (**b**): $XY$-plane view.

Table 2 presents the conditions used in the optimization process, the values of the surface coefficients before and after optimization, and the values of the augmented objective function before and after optimization.

**Table 2.** The optimized result for measurement surface.

| Title 1 | Initial Values | Optimized Values | Conditions |
|---|---|---|---|
| $a_{s0}/a_{t0}$ | 0.2/2 | 0.53845/1.6278 | |
| $a_{s1}/a_{t1}$ | 0.01/0.2 | −0.031114/0.025262 | |
| $a_{s2}/a_{t2}$ | 0/0 | −0.0066394/0.054048 | |
| $a_{s3}/a_{t3}$ | 0/0 | 0.0078314/−0.025262 | $D = 0.005$, $\varepsilon_{DS} = 10^{-6}$, $\rho = 0.5$, $\tau_0 = 1$, $\varepsilon_{IM} = 0.01$ |
| $b_{s1}/b_{t1}$ | 0/0 | 0.026959/0.078273 | |
| $b_{s2}/b_{t2}$ | 0/0 | 0.031114/−0.028414 | |
| $b_{s3}/b_{t3}$ | 0/0 | 0.020146/−0.015076 | |
| $f_a$ | 1239.6853 | 236.8953 | |

3. Identify the parameter errors. Firstly, simulation error data are generated for the measurement surface. Assuming a uniform error distribution, $\Delta l_{bi}$, $\Delta x_{ci}$, $\Delta y_{ci}$ and $\Delta z_{ci}$ are set to zero, and the remaining 42 positional and angular errors are set to $1 \times 10^{-4}$ m and 0.01°, respectively. These errors are then added to the nominal parameter values accordingly. Utilizing the actual parameter values, the pose vectors for the points $Q(s_i, t_j)$ on the measurement surface are computed directly by Equations (30), (34), and (35), without introducing linearization errors. This method ensures a more precise assessment of the comprehensive error magnitude. Additionally, besides acquiring the pose components of the moving platform control point $T$, the non-planar deformation angles for the three parallelogram mechanisms are also calculated. Notably, in the linearized Equation (50), $\Delta\eta_1^{i,j}$, $\Delta\eta_2^{i,j}$ and $\Delta\eta_3^{i,j}$ have been eliminated as intermediate variables.

Subsequently, parameter identification is performed based on the simulated error data. By employing the linearized Equation (51) and the identifiable error vector in Equation (63), a set of linear over-constrained equations is established. The identifiable error components in Equation (63) are then resolved using the least squares method, and the outcomes are presented in Table 3. In Table 3, "GV" denotes the error values provided during the construction of the simulation data, "CV" represents the error components computed based on Equations (51) and (63), and "Error" indicates the relative deviation between the two. Notably, Table 3 illustrates that the maximum absolute relative deviation is 0.9%, thereby affirming the efficacy of the proposed method for parameter error identification. This also corroborates the parameter identifiability judgment. Furthermore, it suggests that the error components not listed in Table 3 have a negligible impact on the overall error.

**Table 3.** The calibration results of Delta robot's geometric parameter errors.

| | GV (mm or °) | CV $i = 1$ (mm or °) | Error $i = 1$ (%) | CV $i = 2$ (mm or °) | Error $i = 2$ (%) | CV $i = 3$ (mm or °) | Error $i = 3$ (%) |
|---|---|---|---|---|---|---|---|
| $\Delta x_{ai}$ | 0.1 | 0.099873 | −0.13 | 0. 099746 | −0.25 | 0.100294 | 0.29 |
| $\Delta y_{ai}$ | 0.1 | 0.100206 | 0.20 | 0.099781 | −0.22 | 0.099983 | −0.02 |
| $\Delta z_{ai}$ | 0.1 | 0.100100 | 0.10 | 0.100066 | 0.07 | 0.100170 | 0.17 |
| $\Delta\varphi_i$ | 0.01 | 0.009941 | −0.59 | 0.009910 | −0.90 | 0.009924 | −0.76 |
| $\Delta\gamma_i$ | 0.01 | 0.009999 | −0.01 | 0.010009 | 0.09 | 0.010005 | 0.05 |
| $\Delta\theta_i$ | 0.01 | 0.010012 | 0.12 | 0.009994 | −0.06 | 0.010022 | 0.22 |
| $\Delta l_{pi}$ | 0.1 | 0.100067 | 0.07 | 0.100059 | 0.06 | 0.100026 | 0.03 |
| $\Delta l_{ni}$ | 0.1 | 0.099801 | −0.20 | 0.099937 | −0.06 | 0.099735 | −0.27 |

In Ref. [12], Table 3 presented the simulation results of kinematic parameter identification for a Delta robot. It showed that the accuracy of the identification results varied with different spatial distributions of measurement points, with Set 1 achieving the best accuracy. The maximum deviation between the identified and given errors in Ref. [12] is

1.6%, which is larger than 0.9%. In our research, the measurement surface or the distribution of measurement points is optimized, allowing the algorithm to determine a unique set of measurement points.

To verify the parameter identification performance of the algorithm under error conditions, random errors of $\pm 5\%\ \Delta T_{i,j}$ are added to the pose error vector $\Delta T_{i,j}$, and the error components are recalculated. Table 4 presents a typical result of parameter identification under random measurement errors. The maximum identification error in the identification results is $-10.36\%$. Although this is considerably larger than the results in Table 3, it is still sufficient for engineering applications.

**Table 4.** Parameter error calibration results with random measurement noises.

| | GV (mm or °) | CV $i$ = 1 (mm or °) | Error $i$ = 1 (%) | CV $i$ = 2 (mm or °) | Error $i$ = 2 (%) | CV $i$ = 3 (mm or °) | Error $i$ = 3 (%) |
|---|---|---|---|---|---|---|---|
| $\Delta x_{ai}$ | 0.1 | 0.102633 | 2.63 | 0.095674 | −4.33 | 0.096670 | −3.33 |
| $\Delta y_{ai}$ | 0.1 | 0.099947 | −0.05 | 0.101908 | 1.91 | 0.097117 | −2.88 |
| $\Delta z_{ai}$ | 0.1 | 0.093943 | −6.06 | 0.093356 | −6.64 | 0.094659 | −5.34 |
| $\Delta \varphi_i$ | 0.01 | 0.010051 | 0.51 | 0.009332 | −6.68 | 0.010415 | 4.15 |
| $\Delta \gamma_i$ | 0.01 | 0.010105 | 1.05 | 0.009738 | −2.62 | 0.010226 | 2.26 |
| $\Delta \theta_i$ | 0.01 | 0.008964 | −10.36 | 0.009779 | −2.21 | 0.009993 | 0.07 |
| $\Delta l_{pi}$ | 0.1 | 0.100067 | 1.27 | 0.100059 | 0.059 | 0.100026 | 0.026 |
| $\Delta l_{ni}$ | 0.1 | 0.102633 | 2.63 | 0.095674 | −4.33 | 0.096670 | −3.33 |

## 6. Experiments

The dynamic calibration method is validated by experiments. The nominal geometric parameters are specified: $r_1$ = 0.065 m, $r_2$ = 0.035 m, $l_p$ = 0.1042 m, $l_n$ = 0.1650 m. The motor used has a rated current of 3.47 A, a torque constant of 25.9 mNm/A, a no-load speed of 8810 rpm at 24 V voltage, and a reduction ratio of 12. In contrast to the top-mounted configuration illustrated in Figure 1, in the experiment, the Delta robot is laterally installed, and a preloaded spring in the upper kinematic chain compensates for the torque of the integrated joint. The spring torque is defined as follows:

$$\tau_s = k_s \theta_1 + \tau_{s0}, \tag{66}$$

where $k_s$ and $\tau_{s0}$ are the proportional coefficient and initial torque of the spring, respectively. During the grinding task, the robot usually follows a smooth and slow curve with very small acceleration and deceleration. Therefore, the parameters in the simplified dynamic model in this paper mainly include friction and mass, and the inertia parameters closely related to acceleration are not reflected, i.e., $I\ddot{\Theta}$, $J^T F_A$, and $\Gamma_{CM}$ are negligible. Consequently, these three factors will be disregarded in the identification process. Additionally, accounting for the frictional forces of integrated joints, the motor torque $\Gamma$ will be adjusted to incorporate the effect of the frictional torque $\Gamma_c$. The Coulomb model is employed as the friction model,

$$\tau_{c1}\left(\dot{\theta}_i\right) = \tau_{ci0} sign(\dot{\theta}_i), \tag{67}$$

where $\Gamma_c = [\tau_{c1}, \tau_{c2}, \tau_{c3}]^{\mathrm{T}}$. Therefore, the dynamic equation can be re-defined as follows:

$$\Gamma = -J^T F_G - \Gamma_c - [\tau_s, 0, 0]^T \tag{68}$$

The excitation trajectory adopts a uniform synchronous curve in joint space, i.e., $\theta_1(t) = \theta_2(t) = \theta_3(t) = vt + \theta_{min}$, where $v$ is the moving speed, with a motion range of $\theta_{min} = -32°$ and $\theta_{max} = 72°$. A suitable velocity, such as 5°/s, is selected. Under the servo driver's CSP mode, the system moves from the starting point to the ending point at a constant speed. The joint angle is obtained through the encoder at the rear of the motor, and the torque is acquired by reading the motor's current value and combining it with the torque constant and reduction ratio. Both the encoder values and current values are read in real-time at a 1 ms cycle from the corresponding variables inside the driver using the TwinCAT software (v3.1.4024.11). The same procedure is repeated from the ending point back to the starting point. The joint angle is obtained through the encoder at the rear of the motor, and the torque is acquired by reading the motor's current value and combining it with the torque constant and reduction ratio. Both the encoder values and current values are read in real-time at a 1 ms cycle from the corresponding variables inside the driver using the TwinCAT software.

This simple curve form can fully meet the identification needs of the dynamic model in this paper. Additionally, this curve can be implemented at the early stage of robot controller development, without requiring interpolation in Cartesian space or worrying whether the $T$ point after synthesizing the three joint motions always remains within the robot workspace.

Figure 12 display the raw torque-angle data and the data after applying moving average filtering for positive and negative movements. Based on the filtered data, nonlinear least squares fitting was conducted to obtain the calibration results of the dynamic parameters, as shown in Table 5. Figure 12 also depicts the model prediction curves based on the identified dynamic parameters. The torque RMSEs between the filtered data and model-predicted data for the three joints are 14.5529 mNm, 26.2232 mNm, and 31.7745 mNm, which are approximately 1.42%, 2.55%, and 3.09% of the joints' rated torques, respectively. Apart from slightly larger deviations at the beginning and end, the prediction curves fit the actual curves very well in most regions.

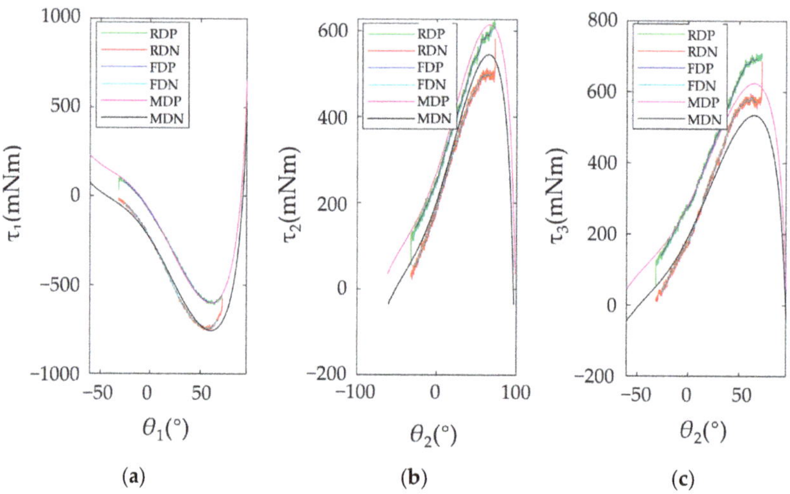


**Figure 12.** The curves of joint torque vs. joint position, RDP: raw data for positive movement, RDN: raw data for negative movement, FDP: filtered data for positive movement, FDN: filtered data for negative movement, MDP: model prediction data for positive movement, MDN: model prediction data for negative movement. (**a**): For joint 1. (**b**): For joint 2. (**c**): For joint 3.

**Table 5.** Dynamic parameter calibration results.

| Index | Parameter | Value | Unit | Meaning |
|---|---|---|---|---|
| 1 | $m_e$ | 0.9735 | kg | the mass of moving platform |
| 2 | $\tau_{c10}$ | 78.9 | mNm | the friction of joint 1 |
| 3 | $\tau_{c20}$ | 35.3 | mNm | the friction of joint 2 |
| 4 | $\tau_{c30}$ | 45.1 | mNm | the friction of joint 3 |
| 5 | $k_s$ | −152.9 | mNm/rad | the proportional coefficient of the spring |
| 6 | $\tau_{s0}$ | −313.0 | mNm | the intial torque of the spring |

In Ref. [28], Table 6 provided the simulation results of RMSE between the given torques and the predicted torques of the identified model for a Delta robot. Ref. [28] did not provide the rated torques of robot joints. To compare with the results in our research, it is assumed that the maximum output torque in Ref. [28] is 0.7 times the rated torque, making it consistent with the situation our research. Additionally, the units in Ref. [28] are converted to mNm. First, looking at the offline parameter identification results in Table 6 of Ref. [28], the RMSE was 632.7 mNm, which is 4.4% of the rated torque. The identification process in our paper is also offline, and the maximum RMSE between the model-predicted torque and the actual torque is 3.09% of the rated torque, which is better than the offline identification results in Ref. [28]. Additionally, Ref. [28] provides online identification results, with the corresponding torque RMSE being 31.5 mNm, or 0.22% of the rated torque. It is worth noting that the results in Table 6 of Ref. [28] are simulation results, while our research provides experimental results. The results in our research include data noise and other unknown factors.

## 7. Conclusions

This paper focused on the calibration of kinematic errors and dynamic parameters for Delta robots driven by integrated joints with 3-R(RPaR) configuration in machining tasks, including position errors and frictions of integrated joints.

The influence of parallelogram mechanism dimension errors was analyzed based on the vector loop method for the 3-R(RPaR) configuration. The modeling of the in-plane deviation angle of the connecting platform link was conducted. This paper also defined the spatial deformation angle of the parallelogram mechanism constrained by the robot's DoF, ultimately forming a 54-parameter kinematic error model and its corresponding linearized model.

By employing the SVD of LMCM, this paper established criteria for the identifiability of error component combinations based on the inequality between the singular value and the error vector norm. A measurement surface with Fourier series form containing 14 coefficients was conducted. The condition number of LMCM derived from the measurement surface was defined as the optimization objective. A hybrid algorithm that combines the Hook–Jeeves DSA with IPM was successfully applied to solve the optimal measurement surface under constraints.

The friction coefficients of integrated joints and other dynamic parameters were obtained by fitting the joint torque-angle pairs measured along specific trajectories using nonlinear least squares regression.

The simulation and experiment results validated the proposed kinematic and dynamic calibration methods, providing a foundation for precise control of integrated joints and high-precision motion of robots.

Based on the actual conditions of the grinding process, this study assumed that the motion process is always smooth and uniform, and based on this, the simplified dynamic model was established. If the motion behavior of the robot during the grinding process differs significantly from the assumption, the simplified model proposed in this paper will no longer be applicable, and the calibration accuracy will also be affected.

Our next step is to implement force–position hybrid control during the grinding process, examine the quality of the actual machined surface, and further optimize the dynamic performance through online optimization.

**Author Contributions:** Conceptualization, Z.J. and T.S.; methodology, Z.J.; software, Y.W.; writing—original draft, Y.W. and D.L.; writing—review and editing, Z.J. All authors will be informed about each step of manuscript processing including submission, revision, revision reminder, etc. via emails from our system or assigned Assistant Editor. All authors have read and agreed to the published version of the manuscript.

**Funding:** This research was funded by National Natural Science Foundation of China, grant number 52105510, Shanghai Science and Technology Innovation Action Plan (High-tech Field Project), grant number 22511102102, Shanghai Science and Technology Planning Project, grant number 20DZ2251400.

**Data Availability Statement:** The datasets used and/or analyzed during the current study are available from the corresponding author on reasonable request.

**Conflicts of Interest:** The authors declare no conflicts of interest.

## References

1. Li, Y.; Huang, T.; Chetwynd, D.G. An approach for smooth trajectory planning of high-speed pick-and-place parallel robots using quintic B-splines. *Mech. Mach. Theory* **2018**, *126*, 479–490. [CrossRef]
2. Moustris, G.P.; Tzafestas, C.S. Modelling and analysis of a parallel double delta mechanism for robotic surgery. In Proceedings of the 2022 30th Mediterranean Conference on Control and Automation (MED), Vouliagmeni, Greece, 28 June–1 July 2022; pp. 861–866.
3. Chen, J.; San, H. Workspace analysis for a five degrees of freedom hybrid engraving plotter. *J. Eng.* **2019**, *13*, 278–283. [CrossRef]
4. Kelaiaia, R. Improving the pose accuracy of the Delta robot in machining operations. *Int. J. Adv. Manuf. Technol.* **2017**, *91*, 2205–2215. [CrossRef]
5. Ma, J.; Shen, Y.; Zhang, S.; Yan, H.; Jia, Z. A New error compensation method for Delta robots combining geometric error modeling with spatial interpolating. *J. Mech. Robot.* **2024**, *16*, 051012. [CrossRef]
6. Bentaleb, T.; Iqbal, J. On the improvement of calibration accuracy of parallel robots–modeling and optimization. *Theor. Appl. Mech. Lett.* **2020**, *58*, 261–272. [CrossRef]
7. Denavit, J.; Hartenberg, R.S. A kinematic notation for lower pair mechanisms based on matrices. *J. Appl. Mech.* **1995**, *77*, 215–221. [CrossRef]
8. Okamura, K.; Park, F.C. Kinematic calibration using the product of exponentials formula. *Robotica* **1996**, *14*, 415–421. [CrossRef]
9. Ottaviano, E.; Gosselin, C.M.; Ceccarelli, M. Singularity analysis of CaPaMan: A three-degree of freedom spatial parallel manipulator. In Proceedings of the 2001 ICRA. IEEE International Conference on Robotics and Automation (Cat. No. 01CH37164), Seoul, Republic of Korea, 21–26 May 2001; pp. 1295–1300.
10. Vischer, P.; Clavel, R. Kinematic calibration of the parallel Delta robot. *Robotica* **1998**, *16*, 207–218. [CrossRef]
11. Li, Y.; Shang, D.; Liu, Y. Kinematic modeling and error analysis of Delta robot considering parallelism error. *INT J. Adv. Robot. Syst.* **2019**, *16*, 1729881419878927. [CrossRef]
12. Shen, H.; Meng, Q.; Li, J.; Deng, J.; Wu, G. Kinematic sensitivity, parameter identification and calibration of a non-fully symmetric parallel Delta robot. *Mech. Mach. Theory* **2021**, *161*, 104311. [CrossRef]
13. Tian, W.; Huo, M.; Zhang, X.; Song, Y.; Wang, L. A general approach for robot pose error compensation based on an equivalent joint motion error model. *Measurement* **2022**, *203*, 111952. [CrossRef]
14. Shi, J.; Yu, C.; Li, Z. Kinematic model identification of planar delta manipulator using Random Levenberg-Marquardt algorithm. In Proceedings of the 2011 9th World Congress on Intelligent Control and Automation, Taipei, Taiwan, 21–25 June 2011; pp. 1097–1102.
15. Caro, S.; Binaud, N.; Wenger, P. Sensitivity analysis of 3-RPR planar parallel manipulators. *J. Mech. Des.* **2009**, *131*, 121005. [CrossRef]
16. Wang, L.; Li, M.; Yu, G. A Novel Sensitivity Analysis Method for Geometric Errors of a Parallel Spindle Head. In Proceedings of the International Conference on Intelligent Robotics and Applications, Hangzhou, China, 5–7 July 2023; pp. 202–211.
17. Kim, H.S.; Choi, Y.J. The kinematic error bound analysis of the Stewart platform. *J. Robot. Syst.* **2000**, *17*, 63–73. [CrossRef]
18. Tannous, M.; Caro, S.; Goldsztejn, A. Sensitivity analysis of parallel manipulators using an interval linearization method. *Mech. Mach. Theory* **2014**, *71*, 93–114. [CrossRef]
19. Swevers, J.; Ganseman, C.; Tukel, D.B.; De Schutter, J.; Van Brussel, H. Optimal robot excitation and identification. *IEEE Trans. Robot. Autom.* **1997**, *13*, 730–740. [CrossRef]
20. Park, K.J. Fourier-based optimal excitation trajectories for the dynamic identification of robots. *Robotica* **2006**, *24*, 625–633. [CrossRef]

21. Bingül, Z.; Karahan, O. Dynamic identification of Staubli RX-60 robot using PSO and LS methods. *Expert Syst. Appl.* **2011**, *38*, 4136–4149. [CrossRef]
22. Grotjahn, M.; Heimann, B.; Abdellatif, H. Identification of friction and rigid-body dynamics of parallel kinematic structures for model-based control. *Multibody Syst. Dyn.* **2004**, *11*, 273–294. [CrossRef]
23. Han, Y.; Wu, J.; Liu, C.; Xiong, Z. An iterative approach for accurate dynamic model identification of industrial robots. *IEEE Trans. Robot.* **2020**, *36*, 1577–1594. [CrossRef]
24. Dong, J.; Xu, J.; Zhou, Q.; Zhu, J.; Yu, L. Dynamic identification of industrial robot based on nonlinear friction model and LS-SOS algorithm. *IEEE Trans. Instrum. Meas.* **2021**, *70*, 1–12. [CrossRef]
25. Song, S.; Dai, X.; Huang, Z.; Gong, D. Load parameter identification for parallel robot manipulator based on extended Kalman filter. *Complexity* **2020**, *2020*, 8816374. [CrossRef]
26. Ohno, M.; Takeda, Y. Design of target trajectories for the detection of joint clearances in parallel robot based on the actuation torque measurement. *Mech. Mach. Theory* **2021**, *155*, 104081. [CrossRef]
27. Diaz-Rodriguez, M.; Valera, A.; Mata, V.; Valles, M. Model-based control of a 3-DOF parallel robot based on identified relevant parameters. *IEEE-ASME Trans. Mechatron.* **2012**, *18*, 1737–1744. [CrossRef]
28. Abed Azad, F.; Ansari Rad, S.; Hairi Yazdi, M.R.; Tale Masouleh, M.; Kalhor, A. Dynamics analysis, offline–online tuning and identification of base inertia parameters for the 3-DOF Delta parallel robot under insufficient excitations. *Meccanica* **2022**, *57*, 473–506. [CrossRef]
29. Liu, Y.; Xi, N. Low-cost and automated calibration method for joint offset of industrial robot using single-point constraint. *Ind. Robot.* **2011**, *38*, 577–584. [CrossRef]
30. Li, M.; Wang, L.; Yu, G.; Li, W. A new calibration method for hybrid machine tools using virtual tool center point position constraint. *Measurement* **2021**, *181*, 109582. [CrossRef]
31. Bai, P.; Mei, J.; Huang, T.; Chetwynd, D.G. Kinematic calibration of Delta robot using distance measurements. *Proc. Inst. Mech. Eng. Part C J. Mech. Eng. Sci.* **2016**, *230*, 414–424. [CrossRef]
32. Li, M.; Wang, L.; Yu, G.; Li, W.; Kong, X. A multiple test arbors-based calibration method for a hybrid machine tool. *Robot. Cim-Int. Manuf.* **2023**, *80*, 102480. [CrossRef]
33. Xiao, B.; Alamdar, A.; Song, K.; Ebrahimi, A.; Gehlbach, P.; Taylor, R.H.; Iordachita, I. Delta robot kinematic calibration for precise robot-assisted retinal surgery. In Proceedings of the 2022 International Symposium on Medical Robotics (ISMR), Atlanta, GA, USA, 13–15 April 2022; pp. 1–7.
34. Daney, D.; Papegay, Y.; Madeline, B. Choosing measurement poses for robot calibration with the local convergence method and Tabu search. *Int. J. Robot. Res.* **2005**, *24*, 501–518. [CrossRef]
35. Brinker, J.; Corves, B.; Takeda, Y. Kinematic performance evaluation of high-speed Delta parallel robots based on motion/force transmission indices. *Mech. Mach. Theory* **2018**, *125*, 111–125. [CrossRef]

*actuators*

MDPI

*Article*

# Research on Micro-/Nano-Positioning System Driven by a Stepper Motor

**Minjie Liu [1], Yangyang Yu [1,*], Liangyu Cui [2], Ning Ji [1,3,*] and Xiaofan Deng [2]**

[1] School of Mechanical Engineering, Tianjin Renai College, Tianjin 301636, China; 093015@tjrac.edu.cn
[2] School of Mechanical Engineering, Tianjin University of Technology and Education, Tianjin 300222, China; cuily@tju.edu.cn (L.C.); dengxiaofan@tju.edu.cn (X.D.)
[3] State Key Laboratory of Engine, Tianjin University, Tianjin 300072, China
* Correspondence: yuyangyang@tju.edu.cn (Y.Y.); jining@tju.edu.cn (N.J.)

**Abstract:** To achieve cost-effective micro-/nano-displacement adjustment, this paper integrates the advantages of flexible hinge micro-/nano-displacement transmission. A linear stepper motor is utilized as the driving component to design and develop a high-precision, low-cost micro-/nano-positioning system. The structure, design, and working principles of the micro-/nano-positioning platform are introduced. The scaling factor model between micro-positioning platforms and nano-positioning platforms is analyzed. Static and dynamic models of flexible mechanisms have been established. The dimensions of the mechanical structure and the selection of motors are determined. The mechanical characteristics of the micro-/nano-positioning platform are validated through finite element analysis. To address the characteristic of increasing loads during the transmission process, an intelligent control system based on current feedback is designed and developed. The integration of drive and control provides a high level of system integration. Finally, experimental calibration was conducted to test the motion characteristics of the linear stepper motor-driven micro-/nano-positioning platform. It achieved a minimum displacement control resolution of 100 nm and demonstrated a certain level of stability.

**Keywords:** micro-/nano-positioning platform; flexible hinge; stepper motor; low-cost

## 1. Introduction

Micro-/nano-positioning technology represents a sophisticated discipline in displacement measurement and control, targeting precise manipulation and localization at the micron and nanoscale levels [1]. Synthesizing expertise from microelectronics, precision engineering, materials science, control theories, and systems, it stands as a cornerstone in contemporary high-tech applications [2]. Its prevalence spans across avant-garde sectors such as semiconductor fabrication, biomedical diagnostics, precision metrology, material characterization, and advanced manufacturing processes [3–6]. Fundamental to its efficacy are advancements in precision actuation, high-resolution sensing, control algorithms, environmental isolation, and vibration mitigation technologies [7], rendering the design and development of innovative micro-/nano-positioning systems crucial for technological progress and catering to diverse sectoral needs, with cost-effectiveness being a paramount consideration for users of micro-/nano-positioning platforms.

A prevalent strategy in micro-positioning involves piezoelectric ceramic actuators combined with compliant mechanisms for guidance. Flexible hinges leverage elastic deformation for guidance, offering a non-contact, frictionless, lubrication-free, and assembly free transmission and steering mechanism. Piezoceramic-driven systems can thereby attain nanoscale positioning resolution [8]. Nonetheless, their restricted stroke confines their utility in millimeter-scale operations, and the supplementary high-cost sensor systems necessary for enhancing motion precision pose adoption barriers.

Voice Coil Motors (VCMs), capitalizing on their unique performance attributes such as rapid response times, high-precision positioning capabilities, and the capacity to deliver substantial thrust within confined spaces, occupy a pivotal role in micro-/nano-positioning systems. VCM positioning systems can employ guidance mechanisms including sliding guides, hydrostatic guides, or flexible hinge architectures, with their positioning accuracy being utterly contingent upon feedback and control systems, exemplified by optical encoders, laser displacement sensors, and strain gauges [9–11]. Kihyun Kim introduced a novel, precision-grade 3 Degrees of Freedom (X-Y-θ) configuration utilizing an array of four coplanar VCMs [12]. Xu has designed a series of large-stroke micro-positioning platforms integrating VCMs with flexible hinges [13–15]. Building upon VCM technology, Hashizume developed a hybrid linear motor and validated a high-speed nano-positioning platform under hydrostatic rail guidance [16]. Liang proposed a novel moving-coil VCM designed for actuating nano-positioning devices, which, supported by compliant mechanisms, achieves both large strokes and nanoscale positioning precision [17]. Nevertheless, VCM systems, with their inherently low damping, are prone to mechanical vibrations, thereby limiting the operational speed of precision positioning systems [18]. Research has illuminated that through ingeniously designed flexible hinge structures, millimeter-scale extensive travels can be attained. Moreover, the synergy of VCM actuation and sensor feedback mechanisms permits millimeter-scale strokes with nanoscale resolution positioning. However, the necessity for VCMs to operate in tandem with high-accuracy sensors significantly escalates the overall cost of positioning systems.

Linear stepping motors are an excellent solution for positioning applications that require rapid acceleration and high-speed moves with low mass payloads. Mechanical simplicity and precise open-loop operation are additional features of the micro stepping linear motor systems. However, there is a scarcity of literature investigating micro-/nano-positioning systems employing linear stepper motors.

In summary, of the aforementioned analysis, this study endeavors to realize low-cost micro-/nano-scale displacement orientation. Leveraging the advantages of flexible hinge transmission, we employ linear stepper motors as driving components to engineer a high-precision, cost-effective micro-/nano-positioning system. Additionally, we devise a stepper motor drive and control system tailored for nano-positioning. Moreover, we introduce a PID-based current loop closed-loop control mechanism. This enables fine-tuning of the stepper motor's operational current via upper-computer software, thereby modulating motor thrust and mitigating input errors on the micrometer stage.

## 2. Micro-/Nano-Positioning Platform: Structural Design and Modeling

### 2.1. Micro-/Nano-Positioning Platform Design

The overall composition of the micro-/nano-positioning platform is illustrated in Figure 1, consisting of the main structure based on a flexible hinges, a linear stepper motor, couplings, and a motor mounting plate. The micro stage utilizes long-stroke flexible beams for translational guidance, achieving micrometer-level positioning accuracy through direct coupling with a linear stepper motor. Conversely, the nano stage employs short flexible beams for guidance. By exploiting the stiffness difference between the large-stroke and short-stroke flexible beams, the micro-motion stage's micrometer resolution is effectively downscaled to nanometer resolution, consequently attaining nanometer-level positional precision. The overall dimensions of the motion platform are 110 × 90 × 42 mm, with the linear stepper motor installed inside the structure. There are two motion platforms, including the micro-positioning stage and the nano-positioning stage. The linear stepper motor is connected to the micro-positioning platform via a coupling, as shown in Figure 1c. The coupling utilizes flexible hinge transmission technology, providing certain flexibility in the axial direction to mitigate the impact of installation errors.

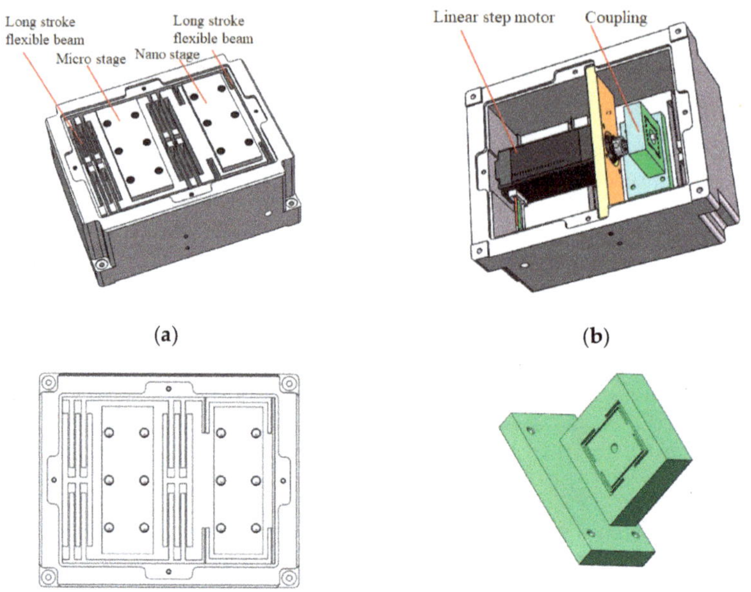


(a) (b) (c) (d)

**Figure 1.** Micro-/nano-positioning system based on a stepper motor. (**a**) Overall view of the micro-/nano-positioning platform; (**b**) interior view of the micro-/nano-positioning platform; (**c**) top view of the micro-/nano-positioning platform based on a flexible hinge; (**d**) coupling based on a flexible hinge.

*2.2. Design of the Flexible Hinge-Based Transmission Mechanism*

A flexible mechanism utilizes material deformation to transmit or convert motion, force, or energy. Flexible mechanisms implement motion through flexible hinge joints, commonly referred to as flexible hinge mechanisms. Compared to rigid mechanisms, flexible hinge mechanisms offer several advantages: (1) they can be integrated into the overall design and manufacturing process, enabling miniaturization, assembly free construction, cost reduction, and increased reliability; (2) they exhibit no backlash or friction, allowing for high-precision motion; (3) they are free from friction wear and do not require lubrication.

Based on the principles of materials mechanics, it is known that utilizing symmetrically distributed flexible beams as guiding structures can yield significant displacement output. Typically, by adjusting the length of the flexible beams, various stiffness levels can be achieved, enabling both micron- and nanometer-level displacement ranges. Based on the difference in guiding stiffness, the transition from micron-level displacement to nanometer-level displacement can be achieved. The transmission principle of the micro-motion platform, as shown in Figure 2, consists of a micrometer platform, a nanometer platform, and two different types of straight-beam flexible hinges. The micrometer platform is guided by symmetrically distributed long flexible beams, resulting in lower stiffness, while the guiding flexible beams of the nanometer platform are shorter, resulting in higher stiffness.

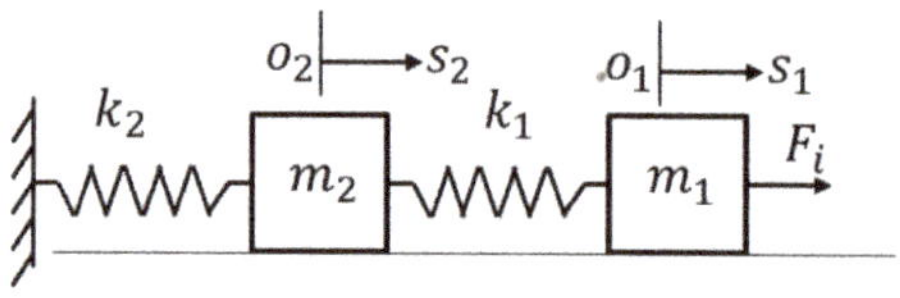


**Figure 2.** Design principle of the micro-/nano-positioning platform.

As shown in Figure 2, under the action of driving force $F_i$, the displacement of the micro stage is $s_1$, which represents the input displacement of the system. According to Hooke's Law, $s_1$ can be expressed as:

$$s_1 = \frac{F_i(k_1 + k_2)}{k_1 k_2} \tag{1}$$

$k_1$ represents the stiffness of the micrometer platform along the driving direction, while $k_2$ represents the stiffness of the nanometer platform along the driving direction.

The output displacement $s_2$ of the nanometer platform can be expressed as:

$$s_2 = \frac{F_i}{k_2} \tag{2}$$

The proportional relationship between the driving input and the platform output can be expressed as:

$$\lambda = \frac{s_1}{s_2} = \frac{k_1 + k_2}{k_1} \tag{3}$$

In the equation, $\lambda$ represents the displacement scaling factor.

When $\lambda$ is larger, the output displacement of the platform is smaller under the same driving input, resulting in higher platform resolution. This means that through the flexible beam guiding mechanism, lower-precision actuators can achieve high-resolution and high-precision output displacements. As a result, the cost of precision positioning platforms is significantly reduced.

### 2.3. Stiffness Modeling Analysis of the Flexible Mechanism of the Micro-Positioning Platform

The basic elastic elements in the structure are mainly long, straight flexible beams, and the flexibility modeling model of a single cantilever beam is shown in Figure 3. First, a generalized force $W_E = \left[M_y F_x F_z\right]^T$ is applied to the reference point at the end of the flexible beam on the moving platform. Then, the deformation of the end of the flexible beam on the moving platform under the action of each component force $M_y$, $F_x$ and $F_z$ is calculated, and the deformation is $T_{1E}$, $T_{2E}$ and $T_{3E}$ respectively.

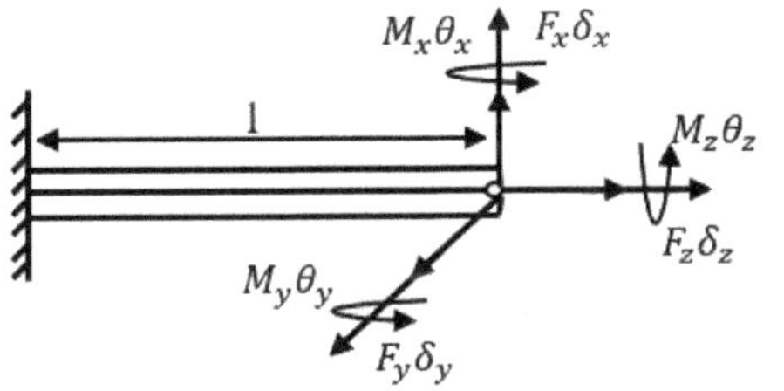


**Figure 3.** Model for the flexibility of a single cantilever beam.

According to the knowledge of material mechanics, it is easy to obtain the deformation at the reference point of the flexible beam end when $M_y$, $F_x$ and $F_z$ are acted on the end of the flexible beam on a moving platform.

When acting alone, $M_y$, the deformation $T_{1E}$ is given by:

$$T_{1E} = \begin{bmatrix} \frac{M_y l}{EI_y} & \frac{M_y l^2}{2EI_y} & 0 \end{bmatrix}^T \tag{4}$$

When acting alone, $F_x$, the deformation $T_{2E}$ is given by:

$$T_{2E} = \begin{bmatrix} \frac{F_x l^2}{2EI_y} & \frac{F_x l^3}{3EI_y} & 0 \end{bmatrix}^T \tag{5}$$

When acting alone, $F_z$, the deformation $T_{3E}$ is given by:

$$T_{3E} = \begin{bmatrix} 0 & 0 & \frac{F_z l}{EA} \end{bmatrix}^T \tag{6}$$

In the equation, $I_x$ and $I_y$ are the moments of inertia about the $x$-axis and $y$-axis, respectively; $J = I_x + I_y$ is the polar moment of inertia; $E$ is the Young's elastic modulus; and $A$ is the cross-sectional area of the homogeneous flexible beam.

The total deformation at the reference point $E$ of the flexible beam on the moving platform is given by:

$$T_E = T_{1E} + T_{2E} + T_{3E} \tag{7}$$

$$T_E = \begin{bmatrix} \frac{F_x l^2}{2EI_y} + \frac{M_y l}{EI_y} \\ \frac{F_x l^3}{3EI_y} + \frac{M_y l^2}{2EI_y} \\ \frac{F_z l}{EA} \end{bmatrix} = \begin{bmatrix} \frac{l}{EI_y} & \frac{l^2}{2EI_y} & 0 \\ \frac{l^2}{2EI_y} & \frac{l^3}{3EI_y} & 0 \\ 0 & 0 & \frac{l}{EA} \end{bmatrix} \cdot \begin{bmatrix} M_y \\ F_x \\ F_z \end{bmatrix} = C_E W_E \tag{8}$$

For a flexible beam guide platform consisting of two pairs of symmetrical long straight flexible beams, as shown in Figure 4, for a symmetrical arrangement, the platform is translational under the action of thrust.

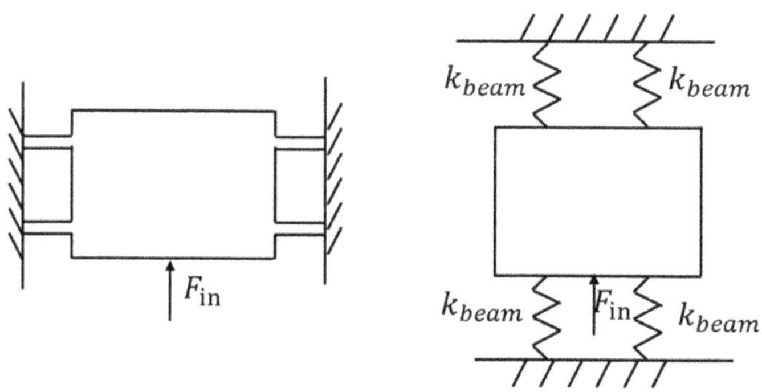


**Figure 4.** Parallel four-bar guide flexible beam mechanism and equivalent diagram.

Due to the symmetric flexible hinge joint, the moving platform undergoes translational motion under thrust, resulting in zero rotation at the end of the flexible beam. Therefore, the rotation angle of the flexible beam under the action of $M_y$ and $F_x$ is zero.

$$\frac{F_x l^2}{2EI_y} + \frac{M_y l}{EI_y} = 0 \tag{9}$$

Thus, we have:

$$M_y = -\frac{F_x l}{2} \tag{10}$$

Therefore, the displacement at the end of the flexible beam is:

$$d = \frac{F_x l^3}{12EI} \tag{11}$$

Hence, the stiffness $K_{beam}$ of a single-guided flexible beam is:

$$K_{beam} = \frac{Eht^3}{l^3} \tag{12}$$

The equivalent stiffness of the platform is illustrated in Figure 3. The input stiffness of the dynamic platform, with four guided beams, is denoted as $F_x$. Through force analysis, it is known that:

$$K_x \cdot d = F_{in} \tag{13}$$

$$F_{in} = 4F_x \tag{14}$$

Thus, the stiffness $K_x$ of the platform in the micro-feeding direction is:

$$K_x = \frac{4Eht^3}{l^3} \tag{15}$$

The elastic deformation of the flexible beam can serve as the micro-feeding stroke, but typically, the stroke is limited. To increase the stroke of the flexible beam, the length of the flexible beam can be increased. To maintain compact dimensions, a series-connected flexible beam guiding mechanism is designed, as shown in Figure 5. Due to the symmetry of the overall system, the moving platform achieves feed motion in the thrust direction. Its serial-parallel compound equivalent model is depicted in Figure 5.

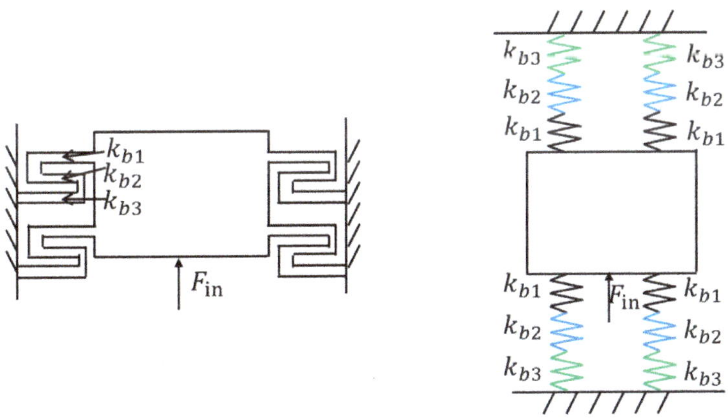


**Figure 5.** Long-stroke parallel-guided flexible beam mechanism and its equivalent diagram.

Based on the above analysis, it can be deduced that the stiffness $K_{nbeam}$ of multiple serially connected flexible beams in their micro-displacement direction is:

$$K_{nbeam} = \left( \sum_{n} \left( \frac{Eht^3}{{l_{bn}}^3} \right)^{-1} \right)^{-1} \tag{16}$$

Therefore, the stiffness $K_{nx}$ of a platform guided by four symmetrically arranged serial flexible beams in its micro-displacement direction is:

$$K_{nx} = 4 \left( \sum_{n} \left( \frac{Eht^3}{{l_{bn}}^3} \right)^{-1} \right)^{-1} \tag{17}$$

### 2.4. *Dynamic Modeling of Flexible Mechanisms in the Micro-Positioning Platform*

To analyze the system's dynamic performance, as shown in Figure 2, generalized coordinates $o_1$ and $o_2$ are chosen. The system's kinetic energy is expressed as:

$$T = \frac{1}{2} m_1 {x_1}^2 + \frac{1}{2} m_2 {x_2}^2 \tag{18}$$

The potential energy of the system is:

$$U = \frac{1}{2}K_1(x_2 - x_1)^2 + \frac{1}{2}K_2{x_2}^2 \tag{19}$$

By applying the Lagrange equations, the system's equations of motion can be directly obtained:

$$M\begin{bmatrix} \ddot{x}_1 \\ \ddot{x}_2 \end{bmatrix} + K\begin{bmatrix} x_1 \\ x_2 \end{bmatrix} = \begin{bmatrix} F \\ 0 \end{bmatrix} \tag{20}$$

Thus, the natural frequency expression of the platform is:

$$f_n = \frac{1}{2\pi}\sqrt{\frac{K}{M}} \tag{21}$$

## 3. Finite Element Simulation Verification

### *3.1. Scaling Analysis of the Micro-/Nano-Positioning Platform*

Finite element simulation is a mature and effective method for analyzing the static and dynamic characteristics of mechanical systems. The structural material of the mechanical system is chosen as aluminum alloy 7075, with a density of 2810 Kg/m$^3$, Poisson's ratio of 0.3, and elastic modulus of 71 GPa. The simulation is conducted using the finite element simulation software ANSYS.

When a driving displacement of 5 mm is applied in the translational direction of the micro-positioning platform, the simulated displacement variation results, as shown in Figure 6a, indicate an output displacement of 0.18 mm for the nano-positioning platform. The displacement scaling ratio is 27.8, meaning that when the range of motion for the micro-positioning platform is (−5 mm, 5 mm), the output range for the nano-positioning platform is (−180 μm, 180 μm).

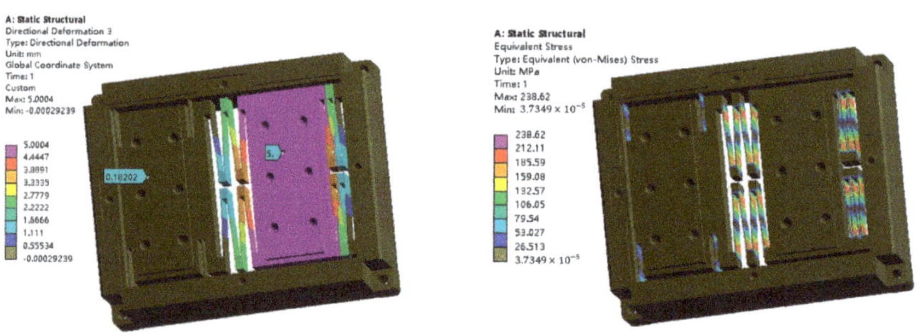


(a) (b)

**Figure 6.** Simulation analysis of the micro-/nano-positioning platform. (**a**) Micro-/nano-positioning platform displacement analysis; (**b**) micro-/nano-positioning platform stress analysis.

Similarly, when a driving displacement of 5 mm is applied in the translational direction of the micro-positioning platform, the simulated stress results, as shown in Figure 6b, indicate a maximum stress of 238.6 MPa. This value is significantly lower than the allowable stress of the material, satisfying the material's usage requirements. The maximum stress occurs at the root of the flexible guiding beam of the micro-positioning platform.

### *3.2. Stiffness Analysis*

When a driving force of 50 N is applied to the micro-positioning platform, it can be observed from Figure 7 that the output displacement of the micro-positioning platform is 5.75 mm and the output displacement of the nano-positioning platform is 0.208 mm. Thus, the input stiffness and output stiffness of the platform are determined to be 8.7 N/mm and 240.4 N/mm, respectively. Due to the limitations imposed by the motor thrust and stroke, there are maximum requirements for the platform's stiffness. Therefore, the design stiffness of the flexible hinge joints is relatively small.

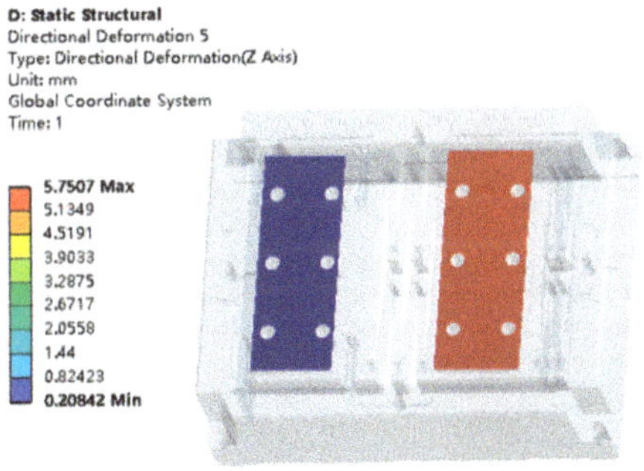


**Figure 7.** Displacement analysis of the micro-/nano-positioning platform under a 50 N driving force.

### *3.3. Modal Analysis*

Modal analysis of the micro-/nano-positioning platform was conducted using the finite element software ANSYS. The first two modal frequencies are 57.4 Hz and 114.3 Hz, respectively. Their mode shapes are depicted in Figure 8, corresponding to the translational and rotational modes of the micro-positioning platform. Due to the requirement for a compact structure, combined with the dimensions and thrust of the selected motor, the designed hinge structure has relatively low stiffness, thereby limiting the operating frequency of the platform to around 20 Hz. By reducing the mass of the working platform, the natural frequency of the positioning platform can be increased, thereby achieving higher natural frequencies.

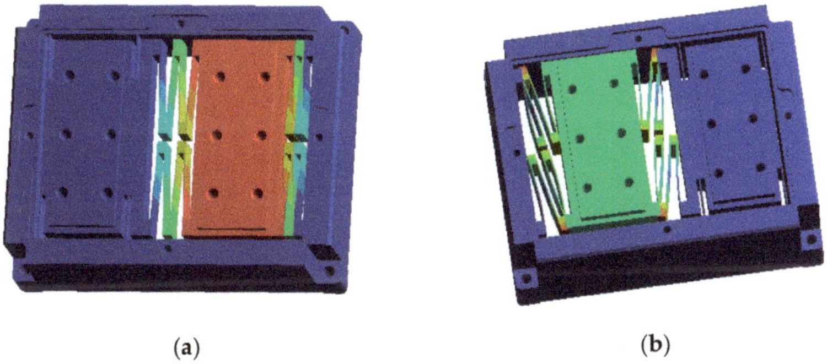

(**a**) (**b**)

**Figure 8.** Modal analysis of the micro-positioning platform. (**a**) Translational modal analysis; (**b**) rotational modal analysis.

## 4. Integrated Design of Stepper Motor Drive and Control

In the proposed micro-/nano-positioning platform driven by a stepper motor, the driving force of the motor increases with the stroke, necessitating dynamic adjustment of the motor's driving force during operation. To address this, an integrated control system based on current feedback is designed and developed, as illustrated in Figure 9. The control board consists of communication, control, drive, power supply, encoder, and current feedback modules. The host computer sends control commands to the control module via the communication module, prompting it to send pulse commands to the drive module, which then sends pulses to the stepper motor. The motor provides speed, angle signals, and current feedback to the host computer through the encoder and current feedback modules, forming a closed-loop control system.

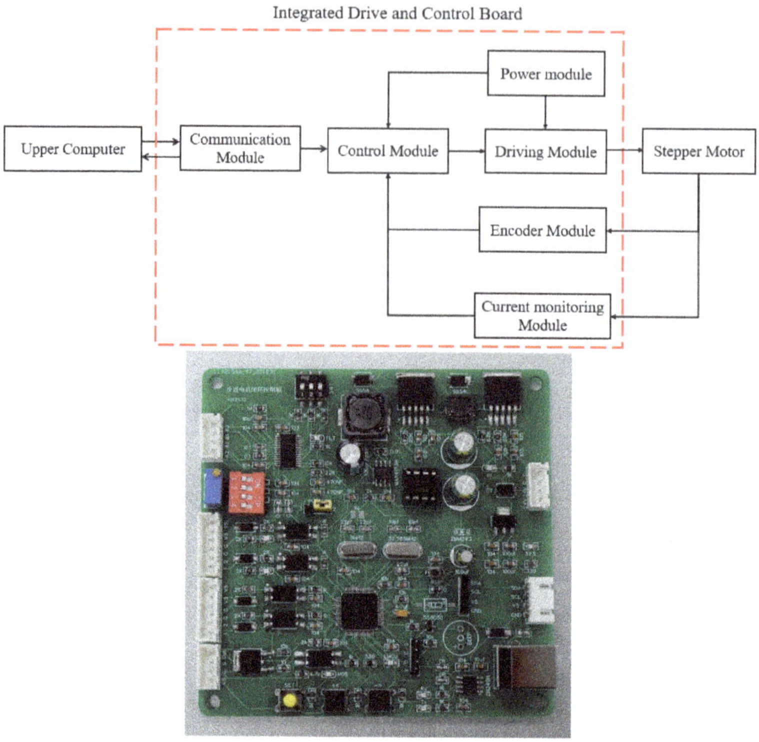


**Figure 9.** Schematic diagram and physical diagram.

## 5. Experimental Testing

The micro-/nano-positioning platform designed in this paper is made of aluminum alloy and processed using advanced wire electrical discharge machining technology. Figure 10 shows the overall experimental setup, consisting primarily of the micro-/nano-positioning platform, stepper motor, integrated drive and control board, 24 V switch power supply, capacitive displacement sensor, host computer, dSPACE, and calibration block.

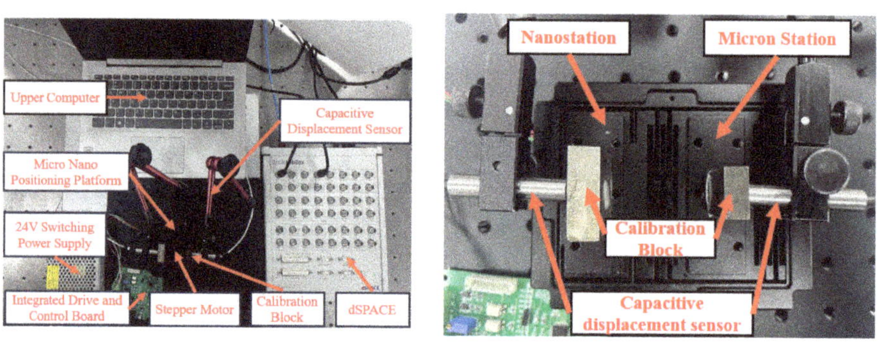


**Figure 10.** Overall experimental setup.

### *5.1. Micro-Positioning Platform Input Displacement Calibration Analysis*

Due to the selected motor's step angle of 1.8° and the lead screw pitch of 0.609 mm, the displacement per pulse sent to the stepper motor should be 3.04 μm, resulting in a resolution of 3.04 μm. After setting up the platform, 22 pulses were sent via the host computer software for both unloaded testing and testing while driving the micro-positioning platform. The data were processed using MATLAB R2017 to obtain the curve depicting the unloaded displacement of the stepper motor versus the input displacement of the micro-positioning platform, as shown in Figure 11.

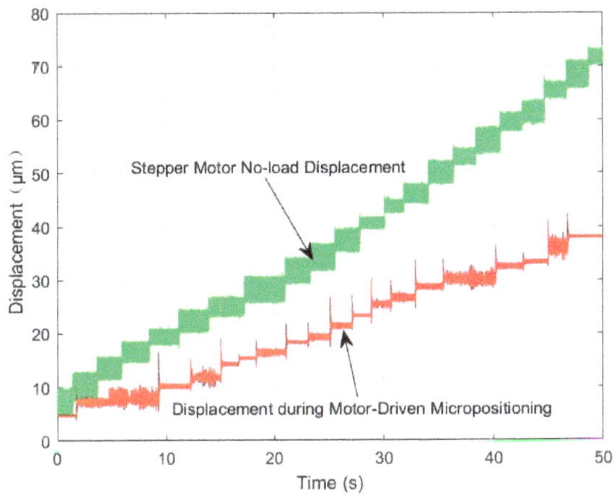


**Figure 11.** Stepper motor unloaded—micro stage input displacement curve.

In Figure 11, it can be observed that as the number of pulses increases, the actual input displacement curve of the micro-positioning platform driven by the stepper motor gradually deviates from the unloaded curve, and the deviation becomes larger over time. This discrepancy arises from the resistance of the micro-/nano-positioning platform. When the motor drives the micro-positioning platform to generate nanometer-level output along with the nano-positioning platform, the resistance increases as the motor stroke increases, resulting in a deviation between the input displacement of the micro-positioning platform and the theoretical value.

### *5.2. Nano-Positioning Platform Output Displacement and Scaling Ratio Calibration Analysis*

The capacitive displacement sensor probe is placed against the surface of the calibration block on the nano-positioning platform of the micro-/nano-positioning platform. After the stepper motor drives the micro-positioning platform to generate micrometer-level displacement, the actual output displacement of the nano-positioning platform is obtained via dSPACE. The displacement curve is depicted in Figure 12.

It can be observed that both the displacement curves of the micro-positioning platform driven by the stepper motor and the displacement curves of the nano-positioning platform exhibit a linear trend, indicating that as the stroke of the stepper motor increases, the output displacement of the nano-positioning platform also increases. From the graph, it is evident that both the micro-positioning platform and the nano-positioning platform's displacement curves show a step-like increase.

Figure 13 illustrates the relationship between the displacement generated by driving the micro-positioning platform with the stepper motor and the output displacement of the nano-positioning platform. Based on the stiffness of the micro-positioning platform (8.4 N/mm) and the nano-positioning platform (236.8 N/mm), the estimated scaling ratio between the micro-positioning platform and the nano-positioning platform is 29.19. Through simulation, the displacement scaling ratio of the micro-/nano-positioning platform was found to be 27.8. However, the calibration experiment reveals that the actual scaling ratio of the platform displacement is 20.76.

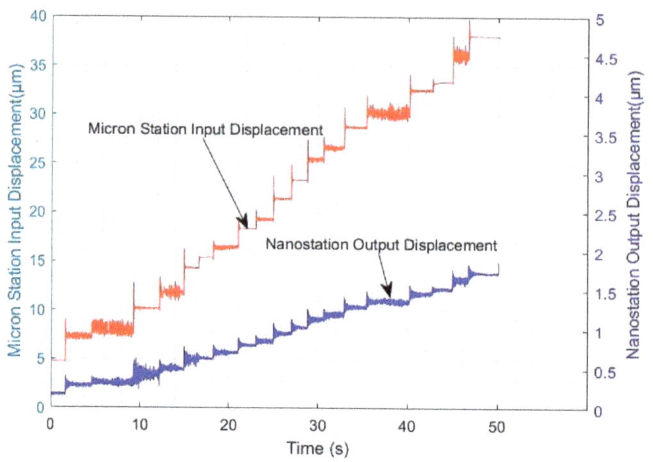


**Figure 12.** Displacement curve.

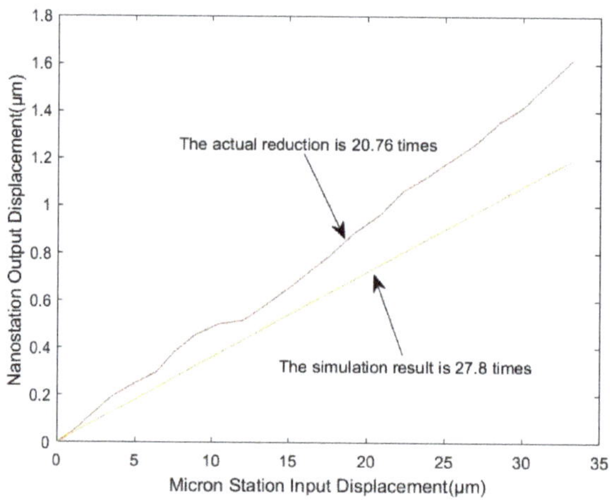


**Figure 13.** Curve depicting the relationship between the input displacement of the micrometer stage and the output displacement of the nanometer stage.

### 5.3. Stability and Round-Trip Testing of the Micro-/Nano-Positioning Platform

(1) Stability Testing

The stability of the micro-/nano-positioning platform is a crucial indicator of the micro-/nano-positioning system and requires testing experiments. Using the host computer software, a pulse is sent to the stepper motor every 2 s, totaling 32 pulses. The platform is paused for approximately one minute at the 2nd, 12th, 22nd, and 32nd pulses, and the stability of the input displacement and output displacement is observed. Figure 14 illustrates the stability testing of the micro-positioning platform and the nano-positioning platform.

From Figure 14, it can be observed that both the input displacement of the micro-positioning platform and the output displacement of the nano-positioning platform stabilize after pausing for approximately one minute at the 2nd, 12th, 22nd, and 32nd pulses. It is evident from the graph that after pausing for about one minute at the 12th and 22nd pulses, both the micro-positioning platform and the nano-positioning platform exhibit relatively stable and step-like increasing curves. This indicates that the stability of the micro-/nano-positioning system is good.

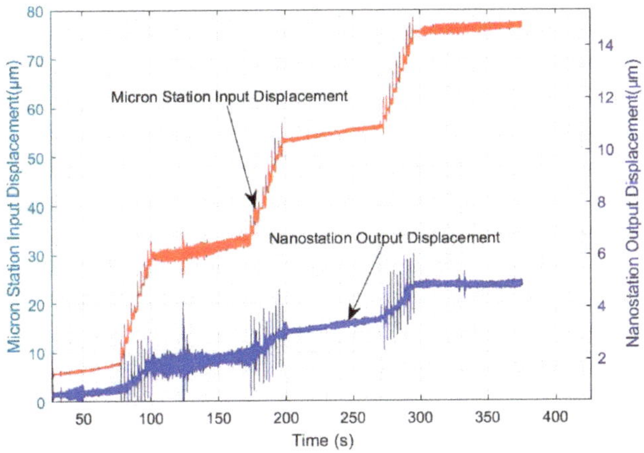


**Figure 14.** Stability testing of the micro and nano stage displacements.

(2) Round-Trip Testing

Using the host computer software, control the stepper motor to drive the micro /nano-positioning platform forward in the positive direction and then reverse to return. During the round-trip motion, send 22 pulses to the stepper motor for each direction, totaling 44 pulses. Measure the displacement using the capacitive displacement sensor, send the data to dSPACE, and obtain the round-trip curve as shown in Figure 15. From Figure 15, it can be observed that the displacement curve generated by driving the micro-positioning platform with the stepper motor is essentially symmetrical with the round-trip operation curve of the nano-positioning platform. The resolution of the nano-positioning platform is 100 nm, indicating good operation of the micro-/nano-positioning system.

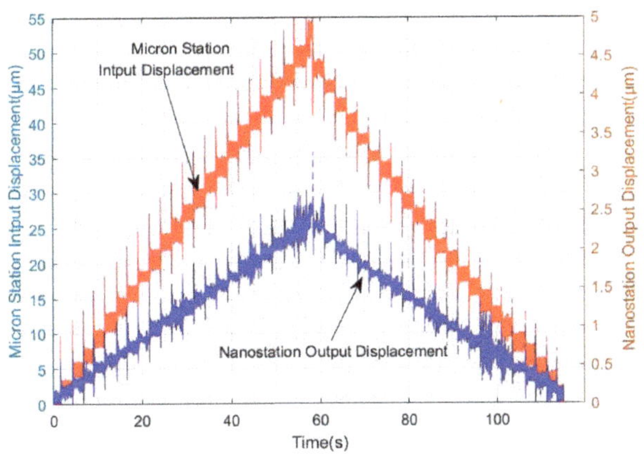


**Figure 15.** Micro-/Nano-positioning Platform Round-trip Test.

The aforementioned experiments illustrate that the micrometer-nanometer positioning platform, driven by a linear stepping motor and guided by a flexible hinge mechanism, successfully achieves the anticipated nanoscale positioning functionality. The deformation range of the serially connected flexible hinges and the thrust generated by the stepping motor influence the effective stroke of the platform. The guidance stiffness ratio between the micrometer and nanometer stages, along with the micro-stepping displacement of the stepping motor, constitute the primary factors affecting the platform's resolution. An increase in travel range leads to a larger system structure and a decrease in system stiffness, which in turn impacts stability. Consequently, stepping motors with greater thrust capacities and more sophisticated, stable control strategies become imperative.

## 6. Conclusions

In this study, a micronano-positioning system with micron-level travel and nanometer-level resolution was designed based on the principles of open-loop driving of linear stepper motors and flexible hinge guiding. A stiffness model of the flexible mechanism was established, and the scaling ratio model between the micro-positioning platform and the nano-positioning platform was analyzed. Based on the operational requirements, the dimensions of the mechanical structure and the selection of motors were determined. The mechanical properties of the micronano-positioning platform were verified using finite element analysis software. Additionally, an integrated design of stepper motor drive and control was implemented, with a closed-loop current control system designed based on PID control theory. An experimental platform was constructed to test and analyze the system. The results demonstrate that the micronano-positioning platform designed in this study achieves a single-pulse resolution of 100 nm, with a compact overall structure and simple control method, enabling precise adjustment of micro-/nano-scale displacements.

**Author Contributions:** Conceptualization, M.L. and L.C.; methodology, Y.Y.; validation, N.J. and X.D.; writing—M.L., Y.Y., L.C., and N.J. All authors have read and agreed to the published version of the manuscript.

**Funding:** This research was funded by the Tianjin Natural Science Foundation, NO. 23JCQNJC00990 and Tianjin Education Committee Scientific Research Project, grant number 2023KJ189.

**Data Availability Statement:** Data are contained within the article.

**Conflicts of Interest:** The authors declare no conflicts of interest.

## References

1. Wu, Z.; Xu, Q. Survey on Recent Designs of Compliant Micro-/Nano-Positioning Stages. *Actuators* **2018**, *7*, 5. [CrossRef]
2. Ukraintsev, V.; Banke, B. Review of reference metrology for nanotechnology: Significance, challenges, and solutions. *J. Micro/Nanolithography MEMS MOEMS* **2012**, *11*, 011010. [CrossRef]
3. Istriţeanu, S.; Gheorghe, I.G. *Research Work and Study on Ultraprecise High-Tech Robotic Micro-Nano- Systems for Measurement, Positioning and Alignment Used in the Fields of Mechatronics and Integronics*; The Scientific Bulletin of VALAHIA University: Waltham, MA, USA, 28 May 2024.
4. Lin, H.B.; Yang, G.Z. Design and Analysis for a Nano-Positioning Micro-Motion Platform Driven by Piezoelectric Ceramic Linear Motors. *Adv. Mater. Res.* **2010**, *102*, 402–406. [CrossRef]
5. Jha, N.; Singh Naruka, G.; Dutt Sharma, H. Design of Embedded Control System Using Super-Scalar ARM Cortex-A8 for Nano-Positioning Stages in Micro-Manufacturing. *Signal Image Process. Int. J.* **2013**, *4*, 71–82. [CrossRef]
6. Hyde, H.C.; Sandtner, W.; Vargas, E.; Dagcan, A.T.; Robertson, J.L.; Roux, B.; Correa, A.M.; Bezanilla, F. Nano-Positioning System for Structural Analysis of Functional Homomeric Proteins in Multiple Conformations. *Structure* **2012**, *20*, 1629–1640. [CrossRef] [PubMed]
7. Ru, C.; Liu, X.; Sun, Y. Nanopositioning Technologies. In *Piezoelectric Motor Technology: A Review*; Springer: Cham, Switzerland, 2016; Chapter 2; pp. 33–59. [CrossRef]
8. Javed, A.; Rout, B.K.; Mittal, R.K. A Review on Design and Synthesis of Compliant Mechanism for Micro Actuation. In Proceedings of the 2nd ISSS National Conference on MEMS, Microsensors, Smart Materials, Structures and Systems, Pilani, India, 16–17 November 2007.
9. Jeong, J.H.; Kim, M.H.; Woo, S.W.; Gweon, D.G.; Hong, D.P. Optimal Design of Voice Coil Motor for Micro Stage. *Appl. Mech. Mater.* **2014**, *607*, 507–510. [CrossRef]
10. Lin, S.; Jia, Y.; Lei, I.P.; Xu, Q. Design and Optimization of a Long-Stroke Compliant Micropositioning Stage Driven by Voice Coil Motor. In Proceedings of the 2012 12th International Conference on Control Automation Robotics & Vision (ICARCV), Guangzhou, China, 5–7 December 2012; IEEE: Piscataway, NJ, USA, 2012.
11. Shan, G.; Li, Y.; Zhang, L.; Wang, Z.; Zhang, Y.; Qian, J. Contributed Review: Application of voice coil motors in high-precision positioning stages with large travel ranges. *Rev. Sci. Instrum.* **2015**, *86*, 101501. [CrossRef] [PubMed]
12. Kim, K.H.; Gweon, D.G.; Jung, H.S.; Lee, S.H.; Hong, M.S.; Lee, M.G. A Stage Based on Voice Coil Motor with High Speed and Long Range for Laser Micro/Nano Fabrication. *Key Eng. Mater.* **2007**, *345–346 Pt 1*, 757–760. [CrossRef]
13. Xu, Q. Design and implementation of a novel rotary micropositioning system driven by linear voice coil motor. *Rev. Sci. Instrum.* **2013**, *84*, 063701. [CrossRef] [PubMed]

14. Liu, Y.; Xu, Q. Design of a flexure-based auto-focusing device for a microscope. *Int. J. Precis. Eng. Manuf.* **2015**, *16*, 2271–2279. [CrossRef]
15. Zhang, X.; Xu, Q. Design, fabrication and testing of a novel symmetrical 3-dof large-stroke parallel micro/nano-positioning stage. *Robot. Comput. Integr. Manuf.* **2017**, *54*, 162–172. [CrossRef]
16. Shinno, H.; Hashizume, H. High Speed Nanometer Positioning Using a Hybrid Linear Motor. *CIRP Ann.* **2001**, *50*, 243–246. [CrossRef]
17. Dong, L.; Chen, J.; Zhang, C.; Wu, D.; Yu, G. A Novel Voice Coil Motor Used in Nano-Positioning Device. In Proceedings of the 2015 18th International Conference on Electrical Machines and Systems (ICEMS), Pattaya, Thailand, 25–28 October 2015; IEEE: Piscataway, NJ, USA, 2016.
18. Youm, W.; Jung, J.; Lee, S.; Park, K. Control of voice coil motor nanoscanners for an atomic force microscopy system using a loop shaping technique. *Rev. Sci. Instrum.* **2008**, *79*, 013707. [CrossRef] [PubMed]

*actuators*

MDPI

*Article*

# A Disturbance Sliding Mode Observer Designed for Enhancing the LQR Current-Control Scheme of a Permanent Magnet Synchronous Motor

**Zhidong Zhang [1], Gongliu Yang [2,*], Jing Fan [3], Tao Li [4] and Qingzhong Cai [1,*]**

1 School of Instrumentation and Optoelectronic Engineering, Beihang University, Beijing 100191, China; zzdong@buaa.edu.cn
2 School of Mechanical Engineering, Zhejiang University, Hangzhou 310058, China
3 Systems Engineering Research Institute, China State Shipbuilding Corporation Limited, Beijing 100094, China; jingfan_807@163.com
4 Beijing Institute of Space Long March Vehicle, Beijing 100076, China; 15910638697@163.com
* Correspondence: yanggongliu@buaa.edu.cn (G.Y.); qingzhong_cai@buaa.edu.cn (Q.C.); Tel.: +86-8233-81112 (Q.C.)

**Abstract:** This paper introduces a current control method for permanent magnet synchronous motors (PMSMs) using a disturbance sliding mode observer (DSMO) in conjunction with a linear quadratic regulator (LQR). This approach enhances control performance, streamlines the tuning of controller parameters, and offers robust optimal control that is resistant to system disturbances. The LQR controller based on state feedback is advantageous for its simplicity in parameter adjustment and achieving an optimal control effect easily under specific performance indicators. It is suitable for the optimal control of strong linear systems that can be accurately modeled. However, most practical systems are difficult to model accurately, and the time-varying system parameters and existing nonlinearity limit the engineering application of LQR. In the PMSM current control loop, there is strong nonlinear disturbance manifesting as the nonlinearity of its dynamic model. Additionally, substantial noise and variations in system parameters within actual motor circuits hinder the linear quadratic regulator from attaining optimal performance. A disturbance sliding mode observer is proposed to enhance the LQR controller, enabling superior performance in nonlinear current loop control. Simulation and actual hardware experiments were conducted to verify the performance and robustness of the control scheme proposed in this paper. Compared with the widely used PI controller in engineering and sliding mode control (SMC) specialising in disturbance rejection, it offers the advantage of straightforward parameter tuning and can swiftly achieve the robust and optimal control performance that engineers prioritize.

**Keywords:** PMSM; LQR; sliding mode observer; disturbance suppression

## 1. Introduction

Thanks to the development of rare earth permanent magnet processing technology, humans can manufacture high magnetic energy product permanent magnets, which can make the motor smaller and lighter, while providing a stronger magnetic field and higher efficiency [1]. This has made permanent magnet synchronous motors the preferred solution for high-precision servo control applications in recent years.

Utilizing the decoupling of the flux and torque control scheme proposed by Park, an engineer from Siemens, and a space vector pulse width modulation (SVPWM) drive control scheme with smaller harmonic level, PMSMs can achieve excellent performance. Therefore, they are widely used in high-precision servo direct drive control [2] and precision measurement applications, for instance, in camera remote sensing stabilization platform systems [3]. In the axial servo control of the above systems, the motor current loop is the innermost loop, and the performance of its controller directly determines the performance level of the entire servo system [4–6]. Developing a robust and user-friendly optimal controller for the current loop is crucial for designing high-precision direct-drive servo systems with permanent magnet synchronous motors.

The challenges faced by PMSM current control in practical design mainly include the following points. The first issue is the inaccuracy of the motor model parameters, as engineers typically only obtain nominal motor parameters with significant error. The second challenge is that the motor parameters will change with the operating conditions. Even if accurate parameters of the motor are obtained through various efforts, they will only be accurate under certain operating condition. The motor parameters will change with the aging of its own components, changes in the surrounding working environment, and even after each startup. The third challenge is the limitation of computing resources and speed. In practical products, embedded current controllers usually operate at high control frequencies, usually above 10 kHz. Restricted computing resources limit the complexity of the control algorithm. Actual high-precision servo systems have higher requirements for control performance and long-term working accuracy stability of the motor actuator. However, most PMSM current loop control algorithms are in the laboratory validation stage. In other words, in a good working environment PMSMs can achieve considerable accuracy, but in other demanding working environments, their actual usage accuracy will significantly decrease. In order to design high-performance current controllers, many scholars have adopted various control schemes and algorithms, including modified classical PI control, current prediction control, data-driven neural network assisted control, and optimization control, such as LQR and MPC control, sliding mode control and so on. Each scheme presents a unique constellation of merits and demerits.

Many scholars have focused on studying how to model and calibrate the various parameters and errors of the motor in order to obtain as accurate prior information as possible [7]. There are also online estimators designed for a certain parameter to obtain real-time estimates. In references [8–11], offline or a combination of offline and online methods are used to identify the various parameters of the motor. The method is relatively complex or computationally intensive, and the calibrated parameters often change when the working environment changes. References [12–14] use online estimation methods to estimate certain parameters of the motor, such as coil resistance, inductance, etc. In order to improving the effectiveness and robustness of the PMSM controller, references [15,16] use a Kalman filter and extended Kalman filter to estimate the magnet flux ripple or demagnetization caused by the high operation temperature, or a stator winding current impulse and the nonlinearity of a voltage-source inverter, respectively. In reference [17], a method based on an unscented Kalman filter (UKF) algorithm is proposed to estimate the permanent magnet flux linkage by taking the inverter dead-time voltage error into consideration. Although the corresponding parameter changes can be estimated in real time, it is difficult to directly use the estimated parameters from the control law designed based on nominal parameter models. PI controllers that are widely used in practical engineering are like this. In the design stage, the initial values of the controller's relevant parameters can be determined based on accurate calibration parameters, and then adjusted through actual testing. The control law is fixed. When the motor parameters change or the working conditions change, it is difficult to ensure that the control performance does not decline. Consequently, to circumvent the degradation of the performance of a well-crafted control law in actual applications, the comprehensive estimation of various disparities, as well as internal and external disturbances within the control loop, can augment the system's robustness. References [18–20] used a Luenberger disturbance observer to cope with the divergences caused by motor parameter mismatch, magnetic saturation, cross-coupling, and other factors, such as internal unmodeled and external unknown interference during motor operation, to improve the dynamic performance and parameter robustness of deadbeat predictive current control. Aiming at the deterioration of control precision caused by unknown lumped disturbances that exist in a control system, a nonlinear disturbance observer is devised to estimate and further reject disturbance in reference [21]. Moreover, in order to improve the anti-interference capability of the system during operation, a second-order nonlinear disturbance observer model is developed, and the robustness of the system is improved by combining the model with linear sliding mode control in reference [22].

Various current prediction control schemes mainly use established models to predict the current, compensating for some uncertainties in the system through the error between the actual current measurement values and the predicted values [23]. However, in the algorithm design process, there is usually a lack of consideration for actual current sensor noise. The article [24] adopts predictive control to improve the deadbeat control. Compared with the traditional deadbeat controller, its predictive structure adds a numerical integration term to provide robustness to the parameter uncertainty and unmodeled dynamics. This integrator provides indirect disturbance rejection robustness and insufficient disturbance suppression speed. Reference [25] analyzed the impact of different parameter mismatches on system stability and current tracking error, and introduced corresponding compensation mechanisms through robust predictive current control strategies, effectively improving the system's robustness to parameter changes. Reference [26] directly uses the gradient descent method based on the current results to estimate the total disturbance. This method is simple and easy to implement, but it also does not consider the noise problem of actual current sensors. Moreover, the parameter adjustment of the PI controller used is very time-consuming and labor-intensive, making it difficult to achieve the optimal control goal.

With the popularity of artificial intelligence research, neural network (NN) control has once again been used in the design of PMSM current loop controllers, also known as data-driven. This approach can extract information about the controlled object using actual data. However, due to the difficulty of obtaining training data to cover actual operating conditions, the performance of the controller is unstable. In addition, network training and deployment on actual controllers are time-consuming and labor-intensive, which is not conducive to large-scale use in engineering. Reference [27] proposes an observer scheme based on NN to estimate the back electromotive force (BEMF) for current loop control without position sensors. Reference [28] develops a neural network-based vector controller to overcome the problem of inaccurate decoupling in traditional proportional integral (PI)-based vector control methods

Compared to black box solutions based on neural networks, the optimal control scheme has been favored by many researchers and engineers in recent years, including LQR and MPC [29]. These optimal control schemes are usually based on the design of the optimal control law based on the model, and the performance of the controller is limited by the accuracy of the model parameters. In reference [30], an internal input model was introduced to a control algorithm in order to eliminate the steady state speed error caused by the step reference speed as well as load torque variations. Reference [31] considers that there are always unmodeled dynamics in actual systems, so model-based control methods cannot guarantee accurate output results in all situations. A data-driven LQR control method based on Markov is proposed, which uses estimation of Markov parameters and an extended observability matrix to obtain the optimal feedback control gain without explicitly knowing the system's dynamic equations, which does not rely on an accurate mathematical model. However, the input and output data used in this method have difficulty covering the actual operating conditions of the system. Model predictive control is an online real-time optimization control scheme used to handle complex optimization control problems through rolling optimization and the ability to handle multiple constraints. In reference [32], an optimization approach to suppress torque disturbance caused by magnetic flux distortion in PMSM is adopted, but the impact of other parameter disturbance on torque spike fluctuation is not considered. In reference [33], a Finite Control Set Modulation Model Predictive Control (FCS-M2PC) is applied to improve the utilization of the PMSM DC bus voltage and reduce the voltage vector tracking error. In reference [34], an improved MPC algorithm was applied to PMSM current loop control. This algorithm selects the optimal voltage vector based on the current trajectory circle rather than the cost function, which can effectively reduce torque fluctuations and current harmonics in a PMSM drive. Reference [35] proposes an improved MPC based on the elimination of current prediction errors, overcoming the problem of parameter mismatch, and enhancing the control performance of the PMSM system. The MPC control algorithm is known for its online optimization and represents a very promising direction for the development of optimization control. However, currently, the MPC algorithm has a large computational load and is difficult to deploy and achieve high control frequencies on embedded controllers. In addition, its optimal solution may fail, which limits its application in high bandwidth current loop control schemes. Compared to traditional QR regulators, they can achieve optimal control in the context of the target, but without the aforementioned drawbacks of MPC, and are an optimal control method suitable for engineering applications. The main challenge faced by traditional QR regulators is the difficulty in obtaining an accurate mathematical model of the controlled object.

An LQR method based on a disturbance sliding mode observer is proposed in this paper to address the challenges faced by LQR in PMSM current loop design and the practical engineering application requirements. The uncertainty of various parameters, changes in motor state, and unmodeled parts of the PMSM motor are classified as total disturbance, and a system dynamics equation containing disturbances is derived. An SMO is used to estimate the real-time disturbance in the current loop in order to improve the control performance of the traditional current LQR used for PMSM.

The proposed control scheme base on a disturbance observer is shown in Figure 1.

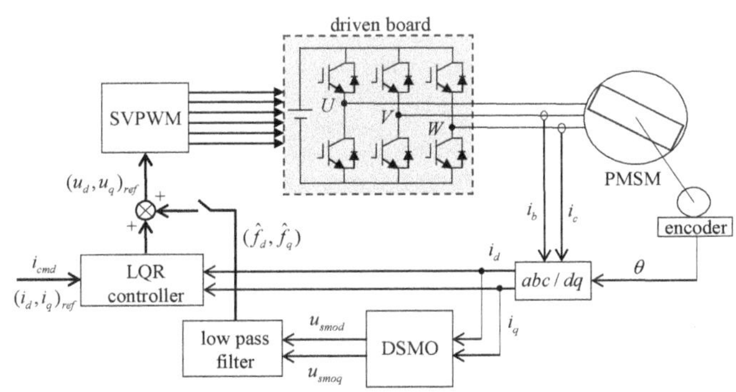


**Figure 1.** Current loop control scheme based on a disturbance sliding mode observer.

The remainder of this paper is organized as follows: In Section 2, the PMSM current loop equations with disturbance are established and the disturbance terms are analyzed. In Section 3, the disturbance and system state variable observer based on SMO is designed. Furthermore, the stability analysis of DSMO is conducted and the parameter selection criteria for convergence are derived. Subsequently, a current LQR tracking controller is designed based on DSMO in Section 4. And in Section 5, simulation and experimental verification are conducted to demonstrate the effectiveness of the proposed approach. The Section 6 provides a summary of the work in this paper.

## 2. Modeling of PMSM Current Loop with Disturbance

The magnetic linkage equation of PMSM in the $dq$-axis coordinate system in the FOC scheme is as follows:

$$\begin{aligned} \psi_d &= L_d i_d + \psi_m, \\ \psi_q &= L_q i_q, \end{aligned} \tag{1}$$

where $\psi_m$ is the permanent magnet flux linkage; $\psi_d$, $\psi_q$ represent the $dq$-axis flux linkage; $L_d$, $L_q$ is the $dq$-axis inductance; and $i_d$, $i_q$ is the current in the $dq$-axis, respectively.

The voltage equation under the $dq$-axis is

$$\begin{aligned} u_d &= R_s i_d + \frac{d\psi_d}{dt} - \omega_e \psi_q, \\ u_q &= R_s i_q + \frac{d\psi_q}{dt} + \omega_e \psi_d, \end{aligned} \tag{2}$$

where $u_d, u_q$, respectively, represent the voltage below the $dq$-axis; $R_s$ is the winding resistance; $\omega_e$ is the electrical angular velocity; and its relationship with the mechanical angular velocity is $\omega_e = p\omega$. The electromagnetic torque equation under the $dq$-axis is

$$T_m = p[\psi_m i_q + (L_d - L_q) i_d i_q], \tag{3}$$

where $T_m$ is the electromagnetic torque. By combining Equations (1) and (2), it can be concluded that the ideal current differential equation can be expressed as

$$\begin{aligned} \frac{di_d}{dt} &= -\frac{R_s}{L_d} i_d + \frac{1}{L_d} u_d + \frac{\omega_e \psi_q}{L_d}, \\ \frac{di_q}{dt} &= -\frac{R_s}{L_q} i_q + \frac{1}{L_q} u_q - \frac{\omega_e \psi_d}{L_q}. \end{aligned} \tag{4}$$

In an actual working environment, the hardware parameters of PMSMs, such as the resistance and inductance of the armature winding, as well as the magnetic flux of the permanent magnet, will inevitably vary with changes in the environment, each time the motor starts, the working state, and the operating time.

In this paper, $\Delta R_s, \Delta L_d, \Delta L_q$ and $\Delta\psi_m$ are used to represent the uncertain disturbances of the motor parameters relative to the calibration value.

Considering the uncertain disturbance $\Delta R_s$ caused by the resistance based on the calibration value, according to Equation (4), the corresponding disturbance in the current loop can be expressed as:

$$\begin{aligned} f_{d\_\Delta R_s} &= -\frac{\Delta R_s}{L_d} i_d, \\ f_{q\_\Delta R_s} &= -\frac{\Delta R_s}{L_q} i_q. \end{aligned} \tag{5}$$

Similarly, given the uncertain disturbance of the $dq$-axis inductance $\Delta L_d$, $\Delta L_q$, the new balance equation after introducing the inductance parameter disturbance is as follows:

$$\begin{aligned} \frac{di_d}{dt} &= -\frac{R_s}{L_d} i_d + \frac{1}{L_d} u_d + \frac{\omega_e \psi_q}{L_d} - \frac{\Delta L_d}{L_d}\frac{di_d}{dt}, \\ \frac{di_q}{dt} &= -\frac{R_s}{L_q} i_q + \frac{1}{L_q} u_q - \frac{\omega_e \psi_d}{L_q} - \frac{\Delta L_q}{L_q}\frac{di_d}{dt}. \end{aligned} \tag{6}$$

According to the circuit principle and Formulas (5) and (6), the following equation can be obtained:

$$\begin{aligned} \frac{di_d}{dt} &= -\frac{R_s}{L_d} i_d + \frac{1}{L_d} u_d + \frac{\omega_e \psi_q}{L_d} - \frac{\Delta L_d}{L_d}\frac{di_d}{dt} - \frac{\Delta R_s}{L_d} i_d, \\ \frac{di_q}{dt} &= -\frac{R_s}{L_q} i_q + \frac{1}{L_q} u_q - \frac{\omega_e \psi_d}{L_q} - \frac{\Delta L_q}{L_q}\frac{di_d}{dt} - \frac{\Delta R_s}{L_q} i_q. \end{aligned} \tag{7}$$

In addition, the uncertain perturbation of the permanent magnet flux $\Delta\psi_m$ and the inductance variation $\Delta L_d$, $\Delta L_q$ will also cause disturbance to the $dq$-axis flux in the current equation. The calculation model of the flux under the disturbance $dq$-axis is:

$$\begin{cases} \psi_d = \psi_m + \Delta\psi_{md} + (L_d + \Delta L_d) i_d, \\ \psi_q = \Delta\psi_{mq} + (L_q + \Delta L_q) i_q, \end{cases}$$
$$\begin{cases} \Delta\psi_{md} = \psi_m(\cos\alpha - 1) + \Delta\psi_m \cos\alpha, \\ \Delta\psi_{mq} = (\psi_m + \Delta\psi_m)\sin\alpha, \end{cases}$$

where $\Delta\psi_{md}$, $\Delta\psi_{mq}$ represent the disturbance of the permanent magnet flux in the axis direction and the distortion angle $\alpha$ in the direction of the permanent magnet flux, which is usually a small angle.

Finally, the PMSM current loop dynamic equation considering hardware parameter perturbation is established as

$$\begin{cases} \frac{di_d}{dt} = -\frac{R_s}{L_d} i_d + \frac{1}{L_d} u_d - \frac{1}{L_d} f_d, \\ \frac{di_q}{dt} = -\frac{R_s}{L_q} i_q + \frac{1}{L_q} u_q - \frac{1}{L_q} f_d, \\ f_d = \Delta R_s i_d + \Delta L_d \frac{di_d}{dt} - \omega_e \psi_q + \varepsilon_d, \\ f_q = \Delta R_s i_q + \Delta L_q \frac{di_q}{dt} + \omega_e \psi_d + \varepsilon_q, \end{cases} \tag{8}$$

where $f_d$ and $f_q$ represent the disturbance below the $dq$-axis, and $\varepsilon_d$ and $\varepsilon_q$ represent the unmodeled disturbance factors that have not been considered.

Equation (8) can be written in the form of a state space model as:

$$\begin{bmatrix} \dot{i}_d \\ \dot{i}_q \end{bmatrix} = \begin{bmatrix} -R_s/L_d & 0 \\ 0 & -R_s/L_q \end{bmatrix} \begin{bmatrix} i_d \\ i_q \end{bmatrix} + \begin{bmatrix} 1/L_d & 0 \\ 0 & 1/L_q \end{bmatrix} \begin{bmatrix} u_d \\ u_q \end{bmatrix} + \begin{bmatrix} -1/L_d & 0 \\ 0 & -1/L_q \end{bmatrix} \begin{bmatrix} f_d \\ f_q \end{bmatrix}, \tag{9}$$

which can be expressed as:

$$\dot{\boldsymbol{X}} = \boldsymbol{A}\boldsymbol{X} + \boldsymbol{B}\boldsymbol{u} - \boldsymbol{B}\boldsymbol{f}, \tag{10}$$

where

$$\boldsymbol{X} = \begin{bmatrix} i_d \\ i_q \end{bmatrix}, \quad \boldsymbol{A} = \begin{bmatrix} -R_s/L_d & 0 \\ 0 & -R_s/L_q \end{bmatrix}, \quad \boldsymbol{B} = \begin{bmatrix} 1/L_d & 0 \\ 0 & 1/L_q \end{bmatrix}, \quad \boldsymbol{u} = \begin{bmatrix} u_d \\ u_q \end{bmatrix}, \quad \boldsymbol{f} = \begin{bmatrix} f_d \\ f_q \end{bmatrix}.$$

The disturbance term in the above established model contains various influencing factors, such as the parameter uncertain perturbation, the changing back electromotive force, as well as unmodeled parts, including nonlinear and time-varying influencing factors, which makes it difficult to design a controller based on the classical linear model. Therefore, a disturbance estimator based on sliding mode control to estimate the total disturbance $\hat{f}$ in the system in real time is designed in this paper. And equivalent compensation is added to the control input $\boldsymbol{u}$ in the current loop to eliminate disturbance in a feedforward manner.

## 3. Design of Disturbance Sliding Mode Observer

### 3.1. Disturbance Observer Equation

To estimate the disturbance term in the actual current control loop, the system equation for constructing a sliding mode observer is as follows:

$$\begin{bmatrix} \dot{\hat{i}}_d \\ \dot{\hat{i}}_q \end{bmatrix} = \begin{bmatrix} -R_s/L_d & 0 \\ 0 & -R_s/L_q \end{bmatrix} \begin{bmatrix} \hat{i}_d \\ \hat{i}_q \end{bmatrix} + \begin{bmatrix} 1/L_d & 0 \\ 0 & 1/L_q \end{bmatrix} \begin{bmatrix} u_d \\ u_q \end{bmatrix} + \begin{bmatrix} -1/L_d & 0 \\ 0 & -1/L_q \end{bmatrix} \begin{bmatrix} u_{smod} \\ u_{smoq} \end{bmatrix}, \tag{11}$$

where $\hat{i}_d$ and $\hat{i}_q$ represent the estimated values of the PMSM's $dq$-axis currents by the sliding mode observer, $u_d$ and $u_q$, respectively, represent the actual input control effort of the observed motor, $u_{smod}$ and $u_{smoq}$, respectively, represent the error feedback correction of the $dq$-axis in the dynamic model of the sliding mode observer, and the other parameters are the nominal argumentations of the motor. And the sliding mode function used in SMO is selected as

$$S = \boldsymbol{\lambda} \boldsymbol{e} + \boldsymbol{k} \int \boldsymbol{e} dt, \tag{12}$$

where $\boldsymbol{e} = \begin{bmatrix} e_{id} \\ e_{iq} \end{bmatrix} = \begin{bmatrix} i_d \\ i_q \end{bmatrix} - \begin{bmatrix} \hat{i}_d \\ \hat{i}_q \end{bmatrix}, \boldsymbol{\lambda} = \begin{bmatrix} \lambda_1 & 0 \\ 0 & \lambda_2 \end{bmatrix}, \boldsymbol{k} = \begin{bmatrix} k_1 & 0 \\ 0 & k_2 \end{bmatrix}, \lambda_1 > 0, \lambda_2 > 0,$ $k_1 > 0, k_2 > 0$ satisfy the Hurwitz criterion. The feedback control law designed subsequently in the sliding mode observer can ensure that $S$ returns to zero, i.e., $\boldsymbol{e}$ and its integration get to zero. Upon achieving the aforementioned control objective, the convergence of the observer is realized.

So the derivative of the sliding function is:

$$\dot{S} = \boldsymbol{\lambda} \dot{\boldsymbol{e}} + \boldsymbol{k} \boldsymbol{e}. \tag{13}$$

By expanding the derivative function (13) of the sliding surface based on the dynamic Equation (11) of SMO, we have:

$$\begin{aligned} \dot{S} = & \begin{bmatrix} -\lambda_1 R_s / L_d & 0 \\ 0 & -\lambda_2 R_s / L_q \end{bmatrix} \begin{bmatrix} e_{id} \\ e_{iq} \end{bmatrix} + \begin{bmatrix} -\lambda_1 / L_d & 0 \\ 0 & -\lambda_2 / L_q \end{bmatrix} \begin{bmatrix} f_d \\ f_q \end{bmatrix} - \\ & \begin{bmatrix} -\lambda_1 / L_d & 0 \\ 0 & -\lambda_2 / L_q \end{bmatrix} \begin{bmatrix} u_{smod} \\ u_{smoq} \end{bmatrix} + \begin{bmatrix} k_1 e_{id} \\ k_2 e_{iq} \end{bmatrix}. \end{aligned} \tag{14}$$

The exponential convergence law is chosen in this paper to reach the sliding mode surface. It is shown as follows:

$$\dot{S} = \begin{bmatrix} -\varepsilon_1 \cdot \mathrm{sgn}(s_1) - p_1 s_1 \\ -\varepsilon_2 \cdot \mathrm{sgn}(s_2) - p_2 s_2 \end{bmatrix}, \tag{15}$$

where $s_1$ and $s_2$ are the components of $S = \begin{bmatrix} s_1 \\ s_2 \end{bmatrix}$, respectively. $\varepsilon_1$, $\varepsilon_2$ and $p_1$, $p_2$ are all greater than 0.

The feedback correction control law shown in Equation (16) for the sliding mode observer can be obtained by combining the two Equations (14) and (15). In other words, adjusting the observer in accordance with this derived control law enables the state to converge exponentially to the system's state variables.

$$\begin{aligned} u_{smod} &= R_s s_1 - L_d(\varepsilon_1 \cdot \mathrm{sgn}(s_1) + p_1 s_1)/\lambda_1 + f_d, \\ u_{smoq} &= R_s s_2 - L_q(\varepsilon_2 \cdot \mathrm{sgn}(s_2) + p_2 s_2)/\lambda_2 + f_q. \end{aligned} \tag{16}$$

Once the observer has converged, it implies that $s_1$ and $s_2$ approach zero, which means that the control efforts applied to the observer are equivalent to the disturbances within the system, thereby also achieving an estimation of the disturbances in the control loop.

### *3.2. Stability Analysis and Observer Tuning*

Due to the presence of unknown disturbances in the feedback control law of the sliding mode observer mentioned above, its upper and lower bounds can be estimated in practical applications based on its expression by Equation (8) using actual system parameters and operating conditions. Then, the convergence of the sliding mode observer can be ensured by designing the parameters $\varepsilon_1$, $\varepsilon_2$, $p_1$ and $p_2$ in the feedback control law. Assuming that the maximum disturbance values of the estimated $dq$-axis are $f_{mad}$ and $f_{maq}$, respectively, the feedback correction law of the sliding mode observer is:

$$\begin{aligned} u'_{smod} &= R_s s_1 - L_d(\varepsilon_1 \cdot \mathrm{sgn}(s_1) + p_1 s_1)/\lambda_1, \\ u'_{smoq} &= R_s s_2 - L_q(\varepsilon_2 \cdot \mathrm{sgn}(s_2) + p_2 s_2)/\lambda_2, \end{aligned} \tag{17}$$

where $u'_{smod}$ and $u'_{smoq}$ represent the actual control efforts used for the sliding mode observer because the disturbances cannot be accurately known.

Taking the $d$-axis as an example for analysis, in order to ensure the stability of the sliding mode observer, the Lyapunov function is taken as $V_1 = 1/2 * (s_1)^2$, which is a positive definite function. And its derivative is $\dot{V}_1 = s_1\dot{s}_1$.

By substituting $u'_{smod}$ into $u_{smod}$ in $\dot{s}_1$, which is the component of $\dot{S}$ shown in Equation (14), we can obtain the following equation:

$$\dot{s}_1 = -p_1 s_1 - \varepsilon_1 \cdot \mathrm{sgn}(s_1) - \lambda_1 f_d / L_d. \tag{18}$$

Substituting Equation (18) into Equation $\dot{V}_1 = s_1\dot{s}_1$, we obtain:

$$\begin{aligned} \dot{V}_1 &= s_1(-p_1 s_1 - \varepsilon_1 \cdot \mathrm{sgn}(s_1) - \lambda_1 f_d / L_d) \\ &= -p_1 (s_1)^2 - \varepsilon_1 |s_1| - \lambda_1 f_d s_1 / L_d. \end{aligned} \tag{19}$$

The first item on the right $-p_1(s_1)^2 < 0$, if only $\varepsilon_1 \geq |\lambda_1 f_d / L_d|_{\max}$ can be guaranteed, and then $\dot{V}_1 \leq 0$ can be achieved. Because the second term on the right-hand side $-\varepsilon_1|s_1|$ is always negative, once the aforementioned condition is met, its absolute value is always greater than that of the third term $-\lambda_1 f_d s_1 / L_d$. Therefore, regardless of the sign of the third term, it ensures that $\dot{V}_1$ is negative. According to Lyapunov stability theory, the observer is convergent. Similarly, for the $q$-axis, it is necessary to satisfy the condition to ensure the convergence of the sliding mode observer. In other words, the sliding mode observer will converge to the sliding surface. At this point, the observation error between the observer and the actual current is 0, and the feedback correction of the sliding mode observer is equal to the disturbance of each axis in a manner similar to PWM modulation. By using a low-pass filter, the disturbance of the corresponding axis can be obtained.

A flowchart of the disturbance sliding mode observer algorithm is shown as Algorithm 1:

**Algorithm 1** Disturbance Estimation Based on Sliding Mode Observer

1: Initialization: $\lambda_1$, $\lambda_2$, $k_1$, $k_2$, $\varepsilon_1$, $\varepsilon_2$ and $p_1$, $p_2$ according to $\varepsilon_1 \geq |\lambda_1 f_d / L_d|_{\max}$ and $\varepsilon_2 \geq |\lambda_2 f_q / L_q|_{\max}$.
2: **repeat**
3: Update the phase current measurement $i_b$, $i_c$ in Figure 1 from analog to digital converter of controller. And calculate $i_a$ using Kirchhoff's current law $i_a + i_b + i_c = 0$.
4: Update the $i_d$,$i_q$ by Clark and Park conversion using $i_a$, $i_b$ and $i_c$ from stator $abc$-axis to $dq$-axis synchronous coordinate system.
5: Calculate $\boldsymbol{e}$ using the updated $i_d$, $i_q$ and the last estimation $\hat{i}_d$, $\hat{i}_q$ from SMO , and update $\boldsymbol{S}$ according to Equation (12).
6: Update the $u'_{smod}$, $u'_{smoq}$ according to Equation (17).
7: Update the new estimation $\hat{i}_d$, $\hat{i}_q$ according to the SMO system Equation (11).
8: Update the disturbances estimation in control loop $\hat{f}_d$, $\hat{f}_q$ from the low pass filtering of $u'_{smod}$ and $u'_{smoq}$.
**Output:** $\hat{f}_d$, $\hat{f}_q$.
9: **until** Program exit.

## 4. Design of LQR

Real-time estimation and compensation can be achieved through a sliding mode disturbance estimator. When the disturbance observer converges quickly, feedforward control is introduced, which is shown as:

$$\mathbf{u} = \mathbf{u}_1 + \hat{\mathbf{f}}. \tag{20}$$

For $\mathbf{X} = \begin{bmatrix} i_d \\ i_q \end{bmatrix}$, $i_d = 0$, $i_q = i_{q\_ref}$ are always desired, i.e., tracking control command is the objective of the actual control loop.

New state variables are taken as:

$$\mathbf{X}^{new} = \begin{bmatrix} i_d \\ i_{q\_ref} - i_q \end{bmatrix} = \begin{bmatrix} i_d \\ i_{eq} \end{bmatrix}. \tag{21}$$

Because of the fact that the current loop is the innermost loop, its control frequency is usually more than 10 times that of the outer speed loop. It can be considered as a constant here, and after transformation, it is shown as follows:

$$\begin{aligned}
\dot{\mathbf{X}}^{new} &= \begin{bmatrix} \dot{i}_d \\ \dot{i}_{eq} \end{bmatrix} = \begin{bmatrix} \dot{i}_d \\ -\dot{i}_q \end{bmatrix} \\
&= \begin{bmatrix} -R_s/L_d & 0 \\ 0 & -R_s/L_q \end{bmatrix} \begin{bmatrix} i_d \\ -i_q \end{bmatrix} + \begin{bmatrix} 1/L_d & 0 \\ 0 & -1/L_q \end{bmatrix} \begin{bmatrix} u_{d1} \\ u_{q1} \end{bmatrix} \\
&= \begin{bmatrix} -R_s/L_d & 0 \\ 0 & -R_s/L_q \end{bmatrix} \begin{bmatrix} i_d \\ i_{eq} - q_ref \end{bmatrix} + \\
&\quad \begin{bmatrix} 1/L_d & 0 \\ 0 & -1/L_q \end{bmatrix} \begin{bmatrix} u_{d1} \\ u_{q1} \end{bmatrix} \\
&= \begin{bmatrix} -R_s/L_d & 0 \\ 0 & -R_s/L_q \end{bmatrix} \begin{bmatrix} i_d \\ i_{eq} \end{bmatrix} + \begin{bmatrix} 1/L_d & 0 \\ 0 & -1/L_q \end{bmatrix} \begin{bmatrix} u_{d1} \\ u_{q1} \end{bmatrix} + \\
&\quad \begin{bmatrix} 0 \\ R_s * q_ref/L_q \end{bmatrix}.
\end{aligned} \tag{22}$$

Substituting the system state:

$$\dot{\mathbf{X}}^{new} = \mathbf{A}\mathbf{X}^{new} + \mathbf{B}^{new}[\mathbf{u}_1 + \mathbf{u}_2], \tag{23}$$

where $\mathbf{B}^{new} = \begin{bmatrix} 1/L_d & 0 \\ 0 & -1/L_q \end{bmatrix}$, $\mathbf{u}_2 = \begin{bmatrix} 0 \\ -R_s * q\_ref \end{bmatrix}$.

The state feedback is shown as:

$$\begin{aligned} \mathbf{u}_1 &= \mathbf{u}_3 - \mathbf{u}_2, \\ \mathbf{u}_3 &= -\mathbf{K}\mathbf{X}^{new}. \end{aligned} \tag{24}$$

Substituting the above state feedback into the system control equation:

$$\begin{aligned} \dot{\mathbf{X}}^{new} &= \mathbf{A}\mathbf{X}^{new} + \mathbf{B}^{new}\mathbf{u}_3 \\ &= \mathbf{A}\mathbf{X}^{new} - \mathbf{B}^{new}\mathbf{K}\mathbf{X}^{new} \\ &= (\mathbf{A} - \mathbf{B}^{new}\mathbf{K})\mathbf{X}^{new} \\ &= \mathbf{A}_{cl}\mathbf{X}^{new}, \end{aligned} \tag{25}$$

where $\mathbf{A}_{cl} = (\mathbf{A} - \mathbf{B}^{new}\mathbf{K})$.

Select the cost function:

$$J = \frac{1}{2}\int_0^{\infty} (\mathbf{X}^{new})^T\mathbf{Q}\mathbf{X}^{new} + \mathbf{u}_3{}^T\mathbf{R}\mathbf{u}_3 dt. \tag{26}$$

Thus, the standard form of the LQR problem is obtained and the design of the controller is completed. By adjusting the weights of the matrices **Q** and **R**, it is relatively easy to adjust the performance that the controller focuses on.

## 5. Evaluation Results and Discussion

### 5.1. Simulation Protocol

In order to verify the effectiveness of the proposed algorithm, computer simulation experiments were conducted. In the Simulink environment, a simulation environment for PMSM was built using simscape multi-domain joint simulation. The parameters of the PMSM are shown in Table 1.

The cntrol simulation system sampling time is 0.0001 s. The amplitude of the current noise power spectral density (PSD) is set to $0.02 \times 0.02 \times 0.0001$, and the amplitude of the angular velocity noise PSD is set to $0.01 \times 0.01 \times 0.0001$.

**Table 1.** The true parameters of PMSM used in the simulation.

| Parameter | Value | Unit |
| --- | --- | --- |
| Number of pole pairs | 8 | |
| Flux Linkage | 0.6 | Wb |
| Ld | 0.00019 | H |
| Lq | 0.00025 | H |
| Rs | 0.013 | Ohm |
| Rotor inertia | 0.2 | $kg \times m^2$ |
| Rotor damping | 5.0 | $N \times m/(rad/s)$ |

The design of the sliding mode observer and LQR controller is carried out using parameters with an error of about 15% to 25% from the true values of the PMSM parameters in order to simulate the error that always exists in the nominal parameters. The nominal parameters used are shown in Table 2.

**Table 2.** The nominal parameters of PMSM used for algorithem design.

| Parameter | Value | Unit |
| --- | --- | --- |
| Number of pole pairs | 8 | |
| Ld | 0.00016 | H |
| Lq | 0.00019 | H |
| Rs | 0.015 | Ohm |

### *5.2. Simulation Results and Discussion*

PI control can usually achieve good results and high reliability, and is easy to use, so it is widely used in practical engineering. This paper conducted comparative experiments with the classical PI controller to verify the LQR control effect of the designed disturbance sliding mode observer. The parameter selection of the PI controller is aggressive, with the Kp parameter set to 2.0 and the Ki parameter set to 2000.0. The simulation experiment results are as follows:

Setting the $q$-axis command current of the current loop (iqRef in Figure 2) to 1.0 ampere and the $d$-axis command current to 0.0 ampere, the current tracking effect of the PI controller and the LQR based on DSMO are compared. The step response simulation test results are shown in Figures 2 and 3.

From Figures 2 and 3, it can be seen that when the current command suddenly changes, the current of both schemes can quickly rise to the reference value of the command, indicating that the Kp parameter setting of the PI controller is relatively aggressive. However, there is a significant control error in the subsequent PI control, which gradually stabilizes at the reference value over time. Although there is some tracking deviation with the LQR control scheme based on DSMO, it maintains a good control effect. Based on the theoretical analysis in the Section 2, it can be seen that the disturbance generated during the operation of the motor current loop is the reason for the control error. Although the integral effect in the PI controller has the ability to resist disturbances, the simulated Ki parameter set here is 2000.0, which is obviously more aggressive and can lead to instability in actual hardware motor control. However, it still cannot effectively suppress disturbances in the circuit. Compared to this, the disturbance feedforward scheme based on a sliding mode observer proposed in this paper shows better control performance.

Figure 3 is a comparison of the current tracking error results between the scheme proposed by this paper and the classic PI scheme.

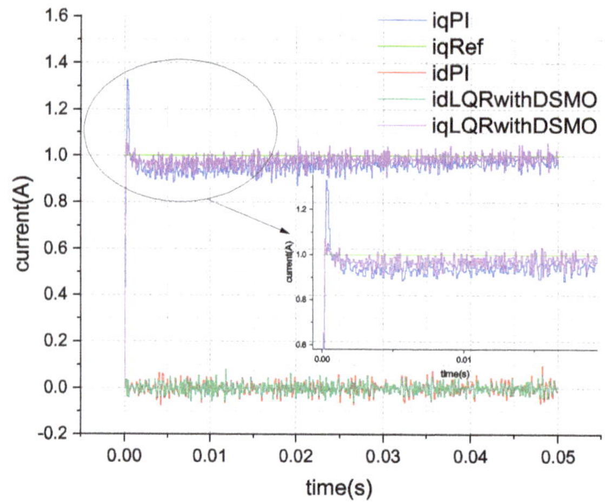


**Figure 2.** Comparison diagram of step response control effects of the two schemes.

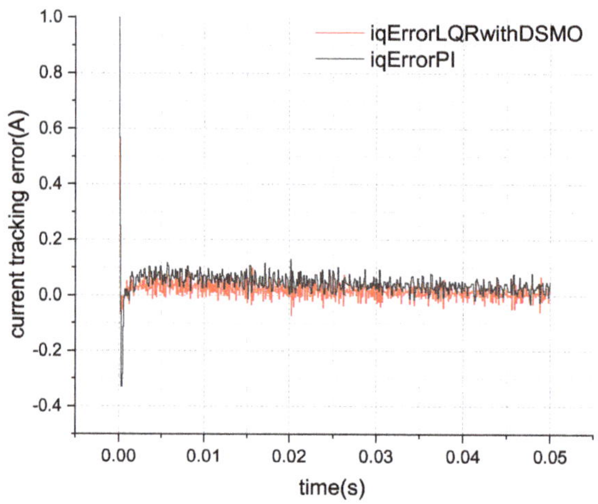


**Figure 3.** Comparison diagram of current tracking error of the two schemes.

By calculating the root mean square error (RMSE) of the current, the RMSE of the classical PI controller is 0.0695 A, and the root mean square error of the LQR current control based on a disturbance sliding mode observer is 0.0489 A, resulting in an improvement in the control accuracy of approximately 29.64%. Clearly, the control scheme proposed in this paper has achieved the expected results.

The current tracking effect of the disturbance sliding mode observer is a very intuitive criterion for evaluating the convergence of the observer. The results of the current value estimated by a sliding mode observer and the actual current value are shown in Figure 4.

In Figure 4, the red line represents the actual $q$-axis current value, and the green line represents the $q$-axis current value of the sliding mode observer. It can be seen that the sliding mode observer can quickly converge to the tracked $q$-axis current, and the same is true for the $d$-axis. This also verifies the correctness of the convergence analysis of the observer in the Section 3.

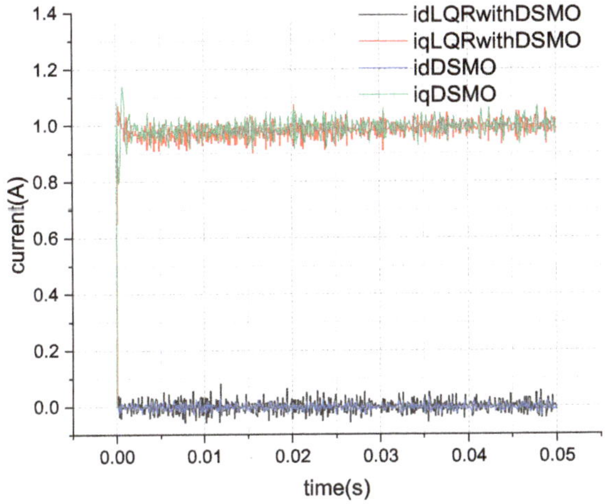


**Figure 4.** Estimation and convergence result of the sliding mode observer.

The estimation of disturbances in the circuit by the sliding mode observer is shown in Figure 5.

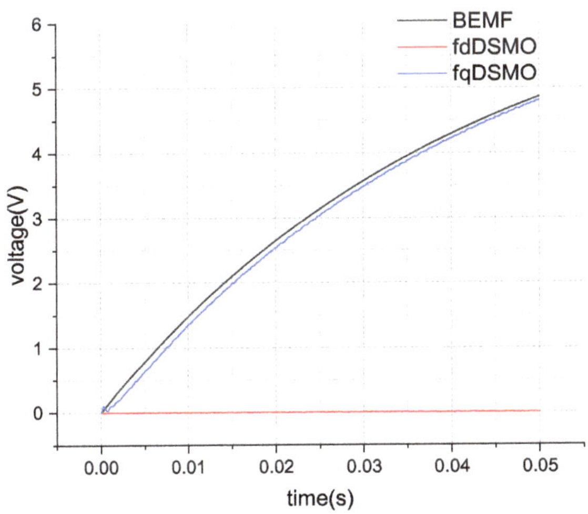


**Figure 5.** Estimation result of the disturbance in the $dq$-axis by the sliding mode observer.

Under simulation conditions, the back electromotive force term in the circuit can be accurately obtained, as shown in the black curve in the figure. According to the theoretical analysis of the disturbance model Equation (8), it can be seen that this is a larger part of the $q$-axis disturbance. From the simulation results, it can also be seen that the variation trend of the $q$-axis disturbance (blue curve) is the same as its main part of the back electromotive force, indicating the effectiveness of the disturbance estimator design.

In order to further verify the performance of the designed control scheme, current loop sweep frequency experiments were conducted under simulation conditions compared with the classical PI controller. The reference value of the current loop adopts a chirp signal with an amplitude of 1.0 A, and the frequency rises from 100 Hz to 1000 Hz within 0.05 s. The result is shown in Figure 6.

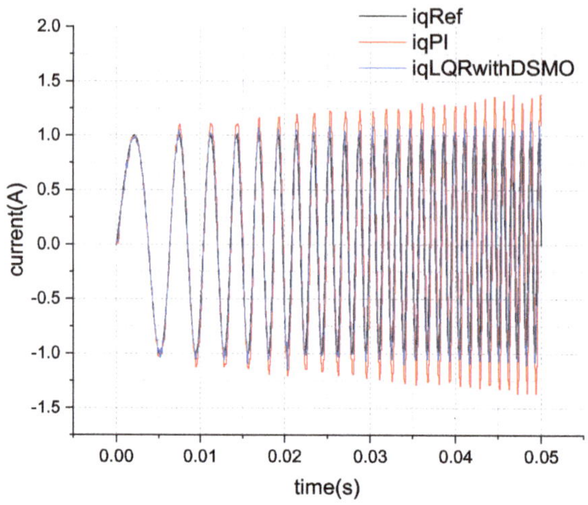


**Figure 6.** Current tracking result when the frequency varies from 100 Hz to 1000 Hz.

The comparison results of the current tracking error are shown in Figure 7. From Figure 7, it can be seen that as the frequency of current changes increases, the tracking error of the classic PI controller significantly increases, with a current tracking RMSE of 0.4289. The current tracking RMSE of the control scheme proposed in this paper is 0.1569, which improves the performance by about 63.41%.

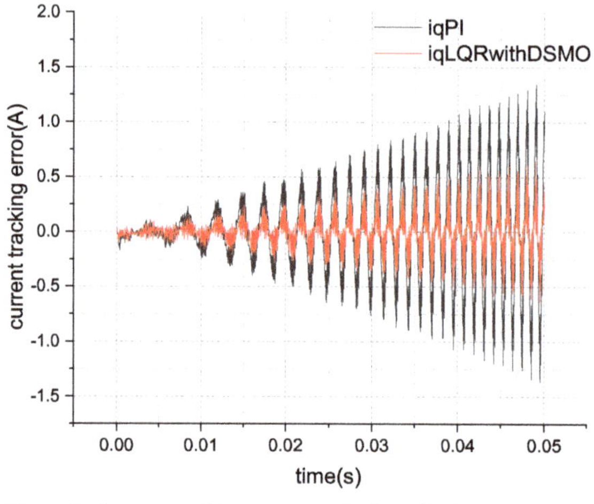


**Figure 7.** Current tracking error result when the frequency varies from 100 Hz to 1000 Hz.

The current tracking variation near the frequency sweep signal of about 1000 Hz is shown in Figure 8.

In Figure 8, the green curve represents the reference value of the $q$-axis current, the blue curve represents the current tracking result of the control scheme proposed in this paper, and the red curve represents the current tracking result of the classical PI controller. In comparison, the classical PI controller shows a significant phase lag and tracking error.

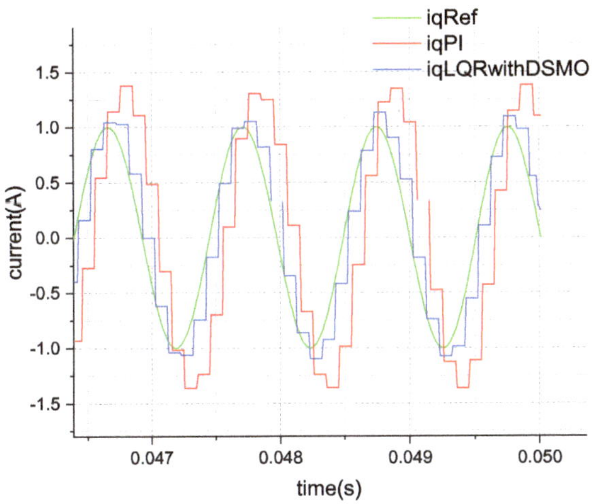


**Figure 8.** Current tracking variation at frequency 1000 Hz.

The primary objective of the disturbance observer proposed in this paper is to suppress disturbances within the control loop, which aligns with the goal of direct sliding mode control. To further validate the performance of the method introduced in this paper, a comparative analysis was conducted with an improved sliding mode control scheme. The improved sliding mode control strategy employs a saturation function in lieu of the signum function to mitigate the chattering phenomenon typically associated with traditional sliding mode control, which is a commonly adopted approach. The comparative results between the method proposed in this paper and the improved sliding mode control scheme, as demonstrated through computer simulations, are illustrated in Figure 9.

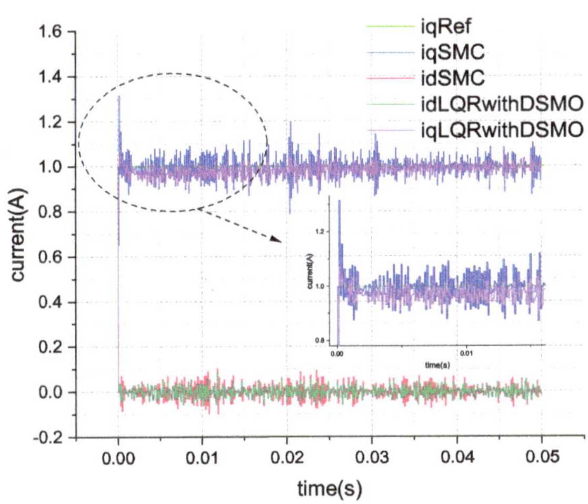


**Figure 9.** Comparative results of current step tracking performance between the proposed method and the improved SMC scheme.

The step response current tracking RMSE for the proposed scheme, the improved sliding mode control scheme, and the classical PI control scheme are presented in Table 3.

**Table 3.** The step response current tracking RMSE of different schemes.

| Scheme | Current Tracking RMSE | Unit |
|---|---|---|
| PI | 0.0695 | A |
| SMC | 0.0647 | A |
| LQR + DSMO | 0.0489 | A |

From the rise phase of the current in Figure 9, it can be observed that the current under sliding mode control can rapidly reach the reference current and fluctuate around it, indicating that sliding mode control possesses a strong disturbance rejection capability. However, the dynamic error of these fluctuations also becomes significantly larger. As shown in Table 3, the calculation results of the RMSE for current tracking demonstrate that the performance of the improved sliding mode control in tracking current is superior to that of the PI controller in terms of root mean square error, but it is not as effective as the method proposed in this paper. This is attributed to the fact that the disturbance rejection and rapid response of sliding mode control are achieved by swiftly changing the direction of the control effort, thereby forcing the system state to oscillate around the sliding surface. Consequently, despite modifications to the switching function in the sliding mode control, the inherent chattering phenomenon remains challenging to eliminate. The comparison of control efforts during the current tracking process is depicted in Figure 10.

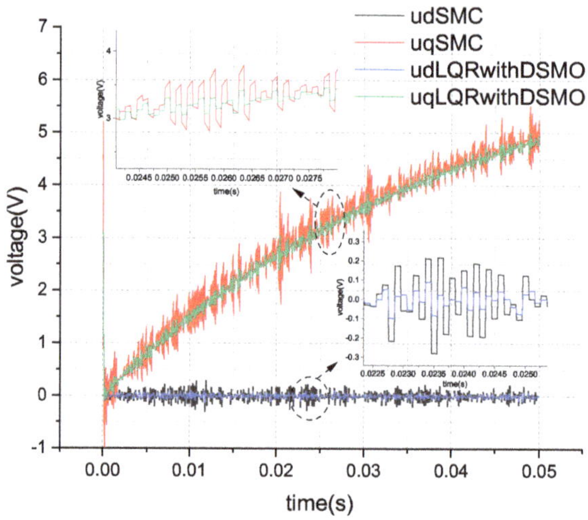


**Figure 10.** Comparative results of control effort between the proposed method and the improved SMC scheme.

The simulation results depicted in Figure 10 reveal that the red curve, labeled 'uqSMC', represents the control effort of the improved sliding mode control for the $q$-axis voltage, while the green curve, marked 'uqLQRwithDSMO', corresponds to the control effort of the method proposed in this paper. It is evident that, in order to suppress disturbances, the sliding mode control exhibits a larger fluctuation in its control effort, which can induce vibrations throughout the system and potentially reduce the service life of the mechanical structure.

The sweep frequency simulation results for the current loop are illustrated in Figures 11 and 12. The sweep frequency current tracking RMSE of the three control schemes is presented in Table 4.

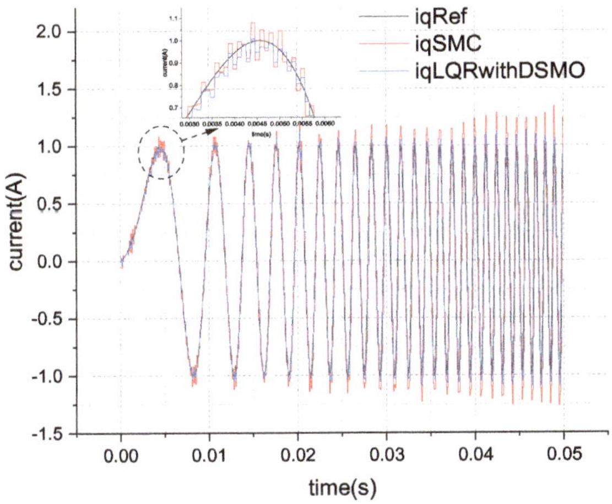


**Figure 11.** Current tracking comparative result when the frequency varies from 100 Hz to 1000 Hz between the proposed method and the improved SMC scheme.

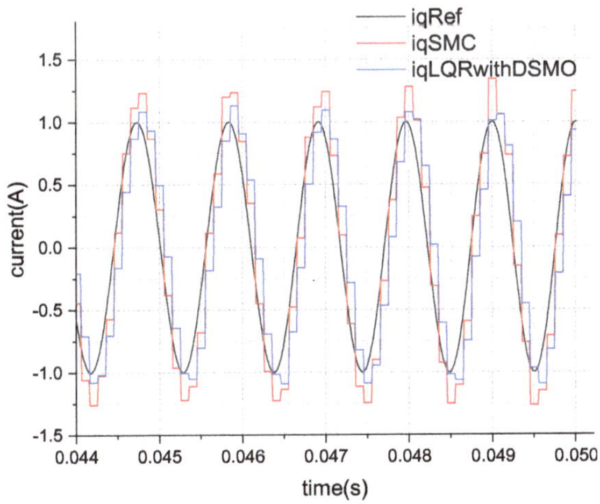


**Figure 12.** Current tracking comparative result between the proposed method and the improved SMC scheme at frequency around 1000 Hz.

**Table 4.** The sweep frequency response current tracking RMSE of different schemes.

| Scheme | Current Tracking RMSE | Unit |
|---|---|---|
| PI | 0.4289 | A |
| SMC | 0.9831 | A |
| LQR + DSMO | 0.1569 | A |

From Figures 11 and 12, it can be observed that the tracking current under sliding mode control fluctuates rapidly around the reference current, yet with considerable error, as indicated in Table 4, which is still indicative of chattering. It is evident that to achieve good disturbance rejection performance, even with improved sliding mode control, the inherent chattering phenomenon is difficult to eliminate. In contrast, the control method proposed in this paper does not have the aforementioned issues while resisting disturbance.

### 5.3. Experiment Protocol

Validation experiments are conducted to compare the classical PI control method used widely in engineering with the current loop control scheme proposed in this paper. The hardware experimental environment used is shown in Figure 13.

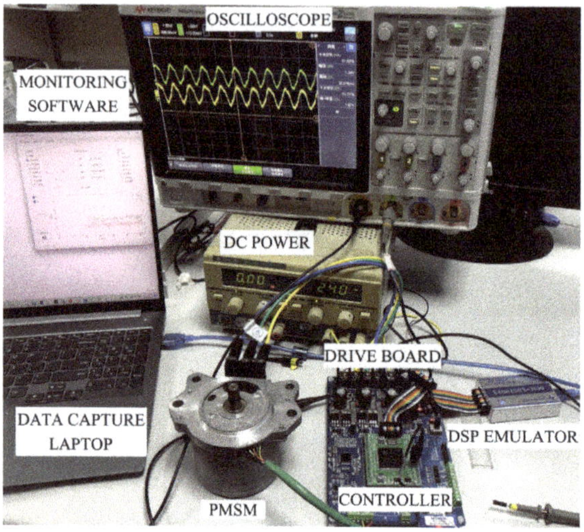


**Figure 13.** Experimental setup.

The PMSM tested in the validation experiment is the MITSUBA-PA66-GF30, and the real-time controller 32-bit floating point DSP (TMS320F28335, manufactured by Texas Instruments in Dallas, TX, USA) is used, which can run at 150 MHz system frequency, in order to process the measurement information, the FOC control algorithm, the PI control law, the proposed disturbance observer, and the output SVPWM waveform driving power board.

Two current Hall sensors (ACS758LCB-050B, manufactured by Allegro MicroSystems in Manchester, NH, USA) are used to measure the phase current of the PMSM, which are converted into a digital signal through the 12 bit ADC of DSP. The current of the third phase is calculated using Kirchhoff's current law for the conversion from a stator *abc*-axis to a *dq*-axis synchronous coordinate system. A resolver scheme is adopted to measure the shaft rotation by the angle signal acquisition chip (AD2S1205, manufactured by Analog Devices in Norwood, MA, USA), which can achieve 12-bit resolution digital conversion. Six metal oxide semiconductor field effect transistors (HY3610, manufactured by Hooyi Semiconductor, Xi'an, China) are used to construct the motor-driven circuit board in the form of three half-bridge circuits, achieving three-phase command voltage through an inverter supplied by DC 24 V power to drive the PMSM.

The nominal parameters of the PMSM used to validate the proposed observer scheme are given in Table 5.

**Table 5.** The parameters of PMSM used in the validation experiment.

| Parameter | Value | Unit |
|---|---|---|
| Number of pole pairs | 3 | |
| Flux Linkage | 0.00612 | Wb |
| Ld | 0.000313 | H |
| Lq | 0.000325 | H |
| Rs | 0.015 | Ohm |
| Rotor inertia | 0.00025 | $kg \times m^2$ |

### 5.4. Experiment Results and Discussion

After setting the current loop $q$-axis command to 1.0 A and the $d$-axis command to 0.0 A, and starting the current loop control circuit, the comparison results obtained between the proposed control scheme and the classical PI controller current response are shown in Figure 14.

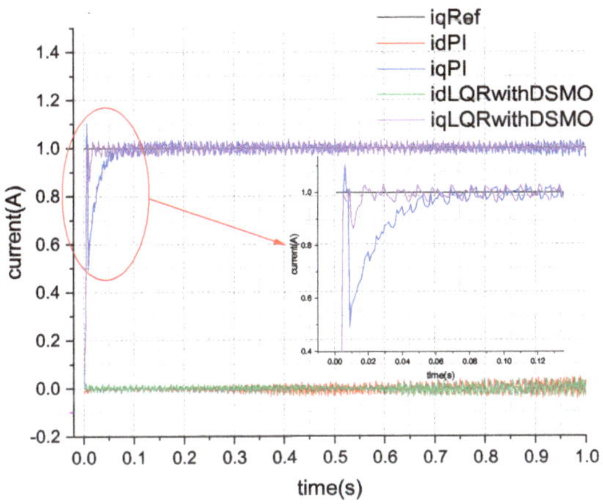


**Figure 14.** Comparison diagram of step response control effects of the two schemes.

From the actual comparison results of current tracking, it can be seen that the current of the classic PI controller cannot stabilize at the reference value after rapidly rising above it. This is due to the internal disturbance generated by the circuit after operation, and the control error gradually decreases under the effect of integration. Compared to other methods, the LQR control scheme based on DSMO has a good anti-interference effect and can quickly and stably track the current reference value. The current tracking RMSE of classical PI control is 0.0716 A, while the current tracking RMSE of the LQR scheme based on DSMO is 0.0560 A, resulting in a 21.79% improvement in control performance.

The current state convergence results of DSMO are shown in Figure 15. From Figure 15, it can be seen that DSMO can quickly and accurately track the changes in actual current, indicating that the observer has achieved fast convergence. The estimation of disturbances in the $dq$-axis by the sliding mode observer is shown in Figure 16.

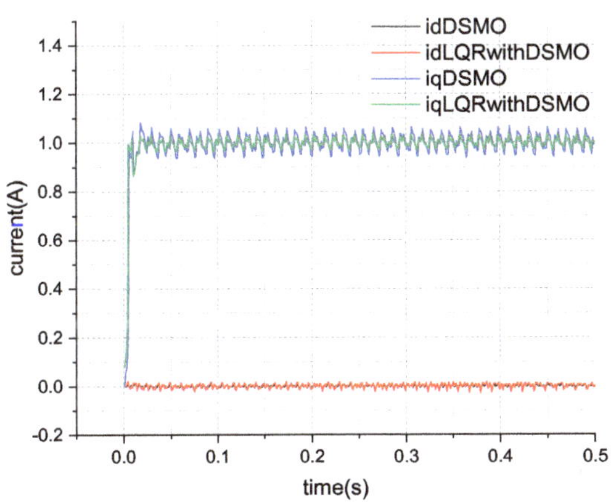


**Figure 15.** Estimation and convergence experimental result of the sliding mode observer.

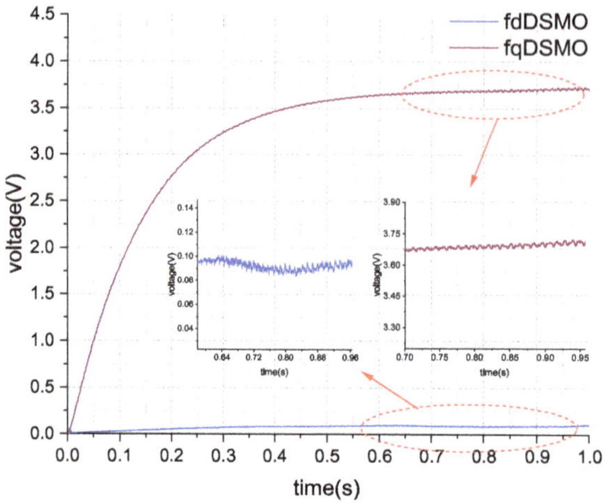


**Figure 16.** Estimation result of the disturbance in the $dq$-axis in the experiment.

In order to fully verify the performance of the proposed control scheme, a sweep frequency test with an amplitude of 1.0 A and frequency change from 1.0 Hz to 101.0 Hz was conducted in actual PMSM current loop testing experiments, with a frequency increase time of 4.0 s. The results are shown in Figure 17.

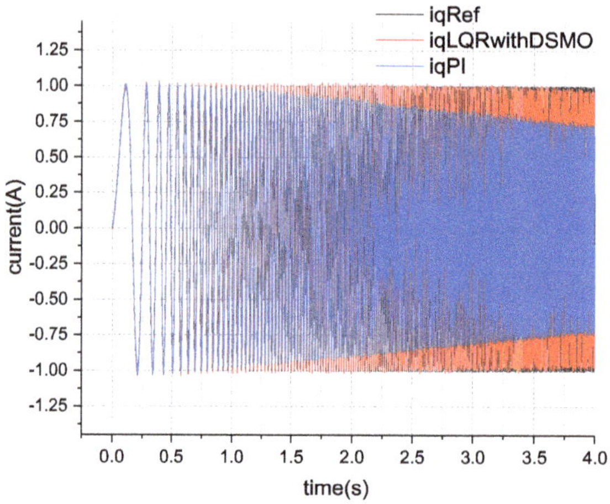


**Figure 17.** Comparison result of linear sweep frequency experiment from 1.0 Hz to 101.0 Hz.

An enlarged diagram of the current tracking during the frequency sweep test is shown in Figure 18. A comparison of current tracking errors in the frequency sweep test of the two schemes is shown in Figure 19.

From the comparison results in Figures 17 and 18 tests, it can be seen that classical PI control can track the current changes well in the low-frequency range. When the current changes quickly, the tracking effect of LQR control based on DSMO is significantly better than that of classical PI control. From the current amplification diagram during the frequency sweep process, it can be seen that for the current command curve (black curve, iqRef), there is a significant phase delay and tracking error in the classical PI control (blue curve, iqPI) compared to the DSMO-based LQR (red curve, iqLQRwithDSMO). After calculation, the current tracking RMSE of the classical PI control is 0.5621 A, while the current tracking RMSE of the proposed control scheme is 0.2582 A. Under the frequency sweep test conditions, the control accuracy was improved by about 54.07% using the scheme proposed in this paper.

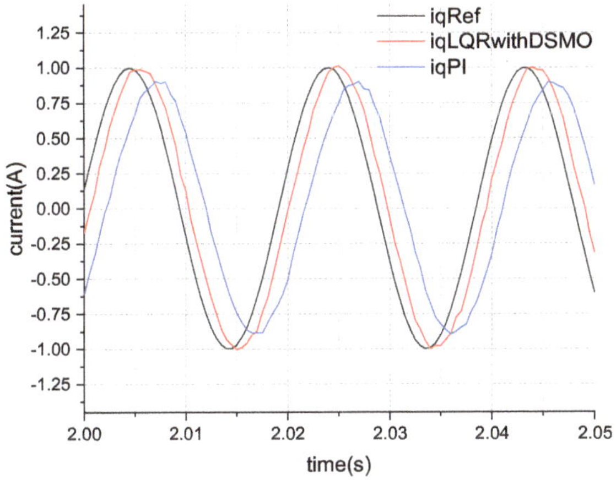


**Figure 18.** Enlarged comparison results of linear sweep frequency experiment.

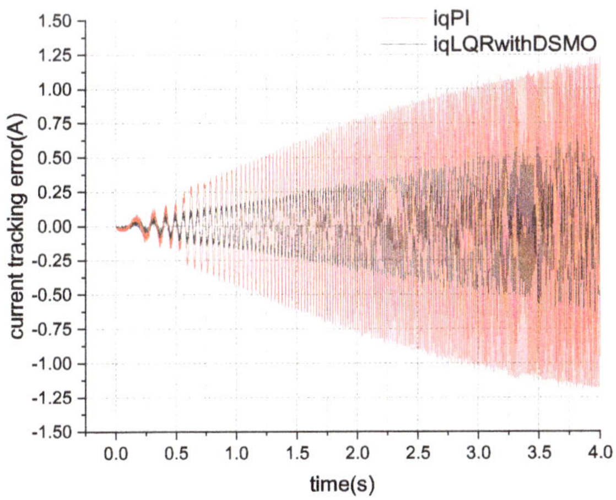


**Figure 19.** Comparison tracking error result of linear sweep frequency experiment from 1.0 Hz to 101.0 Hz.

The comparative experimental results between the method proposed in this paper and the improved SMC scheme are shown in Figure 20. The step response current tracking RMSE for the three control schemes is depicted in Table 6.

The hardware expriment results in Figure 20 indicate that the tracking current of the improved sliding mode control exhibits significant fluctuations around the reference value, and the chattering phenomenon is difficult to completely eliminate. The RMSE calculation results for current tracking in Table 6 show that the tracking performance of the improved sliding mode control is superior to that of the classical PI controller, but it is not as effective as the method proposed in this paper, which is in basic agreement with the simulation results and discussion conclusions. The actual control efforts during the experimental process are compared as shown in Figure 21.

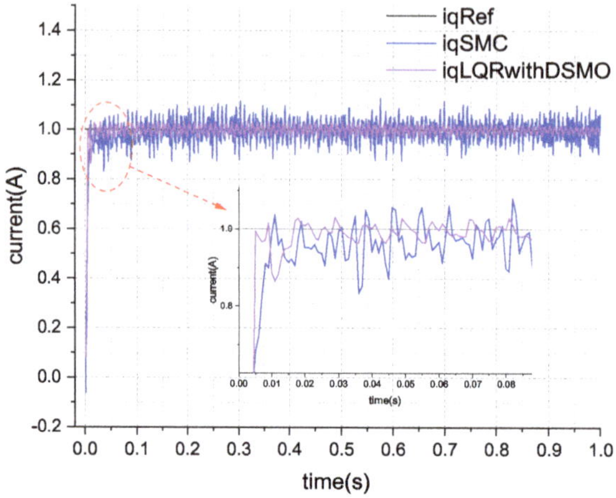


**Figure 20.** The comparative experimental results of current step tracking performance between the proposed method and the improved SMC scheme.

**Table 6.** Experimental step response current tracking RMSE of different schemes.

| Scheme | Current Tracking RMSE | Unit |
|---|---|---|
| PI | 0.0716 | A |
| SMC | 0.0701 | A |
| LQR + DSMO | 0.0560 | A |

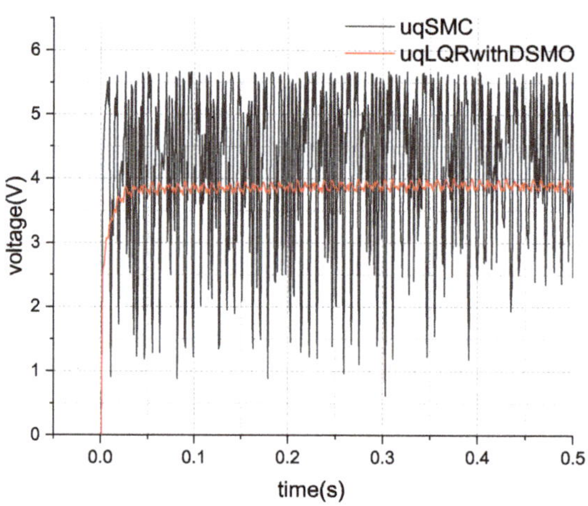


**Figure 21.** Experimental comparative results of control effort between the proposed method and the improved SMC scheme.

The experimental validation results depicted in Figure 21 reveal that the black curve labeled 'uqSMC' corresponds to the control effort of the $q$-axis voltage in the improved sliding mode control, while the red curve marked "uqLQRwithDSMO' represents the actual control effort of the method proposed in this paper. It is evident that, in order to suppress disturbances, the control effort in sliding mode control fluctuates significantly, which may induce vibrations throughout the system and potentially reduce the service life of the mechanical structure. This finding is consistent with the conclusions drawn from the simulation analysis.

The comparative experimental results of the sweep frequency test for the current loop are illustrated in the subsequent Figures 22 and 23. The RMSE of the sweep frequency experimental current tracking for the three control schemes is presented in Table 7.

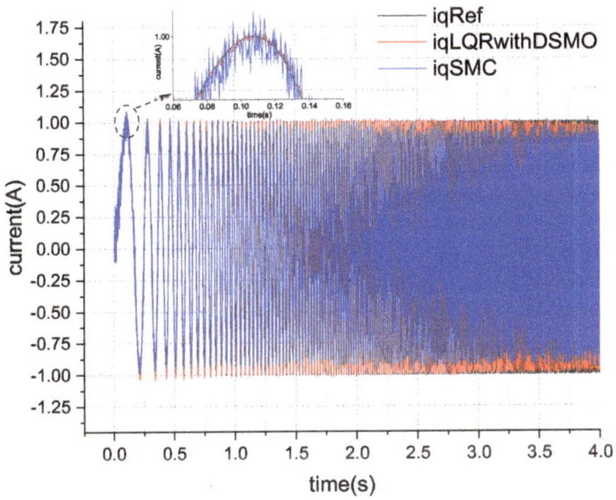


**Figure 22.** Current tracking comparative experimental result when the frequency varies from 1.0 Hz to 101.0 Hz between the proposed method and the improved SMC scheme.

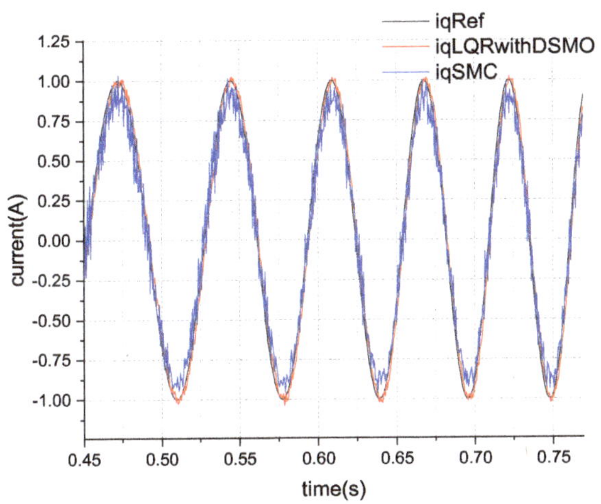


**Figure 23.** Enlarged comparison results of linear sweep frequency experiment between the proposed method and the improved SMC scheme.

**Table 7.** Sweep frequency experimental response current tracking RMSE of different schemes.

| Scheme | Current Tracking RMSE | Unit |
|---|---|---|
| PI | 0.5621 | A |
| SMC | 0.8161 | A |
| LQR + DSMO | 0.2582 | A |

From Figures 22 and 23, it is observed that the actual current under sliding mode control fluctuates rapidly around the reference current, albeit with a significant error, as demonstrated in Table 7, which is indicative of persistent chattering. To achieve robust disturbance rejection, the improved sliding mode control still struggles to eliminate its inherent chattering issues. In contrast, the control method proposed in this paper effectively resists disturbance without the chattering problem. The actual experimental validation results and analytical conclusions are fundamentally consistent with the simulation findings.

## 6. Conclusions

In this paper, a current loop control model with disturbance is first established, and subsequently an LQR controller based on a disturbance sliding mode observer is proposed to improve current tracking performance. There are inevitable disturbances in the $dq$-axis decoupling control loop of permanent magnet synchronous motors, resulting in current tracking errors within the controller. The integration effect in the classic PI controller widely used in engineering demonstrates a certain disturbance suppression capability. Because of the time required for integration, the disturbance suppression speed is obviously slow. Moreover, increasing the integration coefficient of the PI controller is not a good solution, as excessive integration coefficients can cause system instability. Furthermore, through comparative simulations and experimental results with the improved sliding mode control scheme, the disturbance suppression scheme based on the sliding mode observer proposed in this paper possesses the capability to resist disturbances without the encumbrance of chattering issues. In summary, the disturbance sliding mode observer proposed in this paper can achieve rapid real-time estimation and compensation of the disturbances in the control loop. Consequently, the linearized feedforward LQR controller by DSMO can improve the precision of current loop control.

In conclusion, this paper represents a straightforward and effective current control scheme suitable for servo-engineering applications of surface-mounted PMSMs.

**Author Contributions:** This paper was accomplished by all the authors. Z.Z., J.F. and T.L. proposed the idea, performed the analysis, and designed the simulation; Z.Z. and J.F. carried out the simulations and validation experiments; and Z.Z., J.F., G.Y. and Q.C. co-wrote the manuscript. All authors have read and agreed to the published version of the manuscript.

**Funding:** This work was supported by the National Natural Science Foundation of China (61803015).

**Data Availability Statement:** The data are not publicly available due to privacy.

**Conflicts of Interest:** Author Jing Fan was employed by the company China State Shipbuilding Corporation Limited. The remaining authors declare that the research was conducted in the absence of any commercial or financial relationships that could be construed as a potential conflict of interest.

## Abbreviations

The following abbreviations are used in this manuscript:

| | |
|---|---|
| PMSM | Permanent magnet synchronous motor |
| DSMO | Disturbance sliding mode observer |
| SMO | Sliding mode observer |
| SMC | Sliding mode control |
| LQR | Linear quadratic regulator |
| SVPWM | Space vector pulse width modulation |
| FCS-M2PC | Finite control set modulation model predictive control |
| NN | Neural network |
| PI | Proportional integral |
| MPC | Model predictive control |

## References

1. Rafaq, M.S.; Jung, J.W. A Comprehensive Review of State-of-the-Art Parameter Estimation Techniques for Permanent Magnet Synchronous Motors in Wide Speed Range. *IEEE Trans. Ind. Inform.* **2020**, *16*, 4747–4758. [CrossRef]
2. Inoue, T.; Inoue, Y.; Morimoto, S.; Sanada, M. Maximum Torque Per Ampere Control of a Direct Torque-Controlled PMSM in a Stator Flux Linkage Synchronous Frame. *IEEE Trans. Ind. Appl.* **2016**, *52*, 2360–2367. [CrossRef]
3. Königseder, F.; Kemmetmüller, W.; Kugi, A. Attitude control strategy for a camera stabilization platform. *Mechatronics* **2017**, *46*, 60–69. [CrossRef]

4. Liu, H.; Li, S. Speed Control for PMSM Servo System Using Predictive Functional Control and Extended State Observer. *IEEE Trans. Ind. Electron.* **2012**, *59*, 1171–1183. [CrossRef]
5. Yin, Y.; Liu, L.; Vazquez, S.; Xu, R.; Dong, Z.; Liu, J.; Leon, J.I.; Wu, L.; Franquelo, L.G. Disturbance and Uncertainty Attenuation for Speed Regulation of PMSM Servo System Using Adaptive Optimal Control Strategy. *IEEE Trans. Transp. Electrif.* **2023**, *9*, 3410–3420. [CrossRef]
6. Bu, F.; Yang, Z.; Gao, Y.; Pan, Z.; Pu, T.; Degano, M.; Gerada, C. Speed Ripple Reduction of Direct-Drive PMSM Servo System at Low-Speed Operation Using Virtual Cogging Torque Control Method. *IEEE Trans. Ind. Electron.* **2021**, *68*, 160–174. [CrossRef]
7. Yuan, T.; Chang, J.; Zhang, Y. Parameter Identification of Permanent Magnet Synchronous Motor with Dynamic Forgetting Factor Based on H-infinite Filtering Algorithm. *Actuators* **2023**, *12*, 453. [CrossRef]
8. Candelo-Zuluaga, C.; Riba, J.R.; Garcia, A. PMSM Parameter Estimation for Sensorless FOC Based on Differential Power Factor. *IEEE Trans. Instrum. Meas.* **2021**, *70*, 1–12. [CrossRef]
9. Sandre-Hernandez, O.; Morales-Caporal, R.; Rangel-Magdaleno, J.; Peregrina-Barreto, H.; Hernandez-Perez, J.N. Parameter Identification of PMSMs Using Experimental Measurements and a PSO Algorithm. *IEEE Trans. Instrum. Meas.* **2015**, *64*, 2146–2154. [CrossRef]
10. Wang, Q.; Wang, G.; Zhao, N.; Zhang, G.; Cui, Q.; Xu, D. An Impedance Model-Based Multiparameter Identification Method of PMSM for Both Offline and Online Conditions. *IEEE Trans. Power Electron.* **2021**, *36*, 727–738. [CrossRef]
11. Wallscheid, O.; Bocker, J. Global Identification of a Low-Order Lumped-Parameter Thermal Network for Permanent Magnet Synchronous Motors. *IEEE Trans. Energy Convers.* **2016**, *31*, 354–365. [CrossRef]
12. Pulvirenti, M.; Scarcella, G.; Scelba, G.; Testa, A.; Harbaugh, M.M. On-Line Stator Resistance and Permanent Magnet Flux Linkage Identification on Open-End Winding PMSM Drives. *IEEE Trans. Ind. Appl.* **2019**, *55*, 504–515. [CrossRef]
13. Raja, R.; Sebastian, T.; Wang, M.; Chowdhury, M. Stator inductance estimation for permanent magnet motors using the PWM excitation. In Proceedings of the 2017 IEEE Energy Conversion Congress and Exposition (ECCE), Cincinnati, OH, USA, 1–5 October 2017; IEEE: Piscataway, NJ, USA, 2017. [CrossRef]
14. Rafaq, M.S.; Mwasilu, F.; Kim, J.; Choi, H.H.; Jung, J.W. Online Parameter Identification for Model-Based Sensorless Control of Interior Permanent Magnet Synchronous Machine. *IEEE Trans. Power Electron.* **2017**, *32*, 4631–4643. [CrossRef]
15. Li, X.; Kennel, R. General Formulation of Kalman-Filter-Based Online Parameter Identification Methods for VSI-Fed PMSM. *IEEE Trans. Ind. Electron.* **2021**, *68*, 2856–2864. [CrossRef]
16. Xiao, X.; Chen, C.; Zhang, M. Dynamic Permanent Magnet Flux Estimation of Permanent Magnet Synchronous Machines. *IEEE Trans. Appl. Supercond.* **2010**, *20*, 1085–1088. [CrossRef]
17. Li, T.; Liang, C.; Su, D. UKF-based Offline Estimation of PMSM Magnet Flux Linkage Considering Inverter Dead-time Voltage Error. In Proceedings of the 2022 IEEE Applied Power Electronics Conference and Exposition (APEC), Houston, TX, USA, 20–24 March 2022; IEEE: Piscataway, NJ, USA, 2022. [CrossRef]
18. Yang, N.; Zhang, S.; Li, X.; Li, X. A New Model-Free Deadbeat Predictive Current Control for PMSM Using Parameter-Free Luenberger Disturbance Observer. *IEEE J. Emerg. Sel. Top. Power Electron.* **2023**, *11*, 407–417. [CrossRef]
19. He, L.; Wang, F.; Wang, J.; Rodriguez, J. Zynq Implemented Luenberger Disturbance Observer Based Predictive Control Scheme for PMSM Drives. *IEEE Trans. Power Electron.* **2020**, *35*, 1770–1778. [CrossRef]
20. Wang, W.; Liu, C.; Song, Z.; Dong, Z. Harmonic Current Suppression for Dual Three-Phase PMSM Based on Deadbeat Control and Disturbance Observer. *IEEE Trans. Ind. Electron.* **2023**, *70*, 3482–3492. [CrossRef]
21. Zhao, Y.; Liu, X.; Yu, H.; Yu, J. Model-free adaptive discrete-time integral terminal sliding mode control for PMSM drive system with disturbance observer. *IET Electr. Power Appl.* **2020**, *14*, 1756–1765. [CrossRef]
22. Yang, C.; Hua, T.; Dai, Y.; Huang, X.; Zhang, D. Second-Order Nonlinear Disturbance Observer Based Adaptive Disturbance Rejection Control for PMSM in Electric Vehicles. *J. Electr. Eng. Technol.* **2022**, *18*, 1919–1930. [CrossRef]
23. Ahn, H.; Kim, S.; Park, J.; Chung, Y.; Hu, M.; You, K. Adaptive Quick Sliding Mode Reaching Law and Disturbance Observer for Robust PMSM Control Systems. *Actuators* **2024**, *13*, 136. [CrossRef]
24. Turker, T.; Buyukkeles, U.; Bakan, A.F. A Robust Predictive Current Controller for PMSM Drives. *IEEE Trans. Ind. Electron.* **2016**, *63*, 3906–3914. [CrossRef]
25. Li, Y.; Li, Y.; Wang, Q. Robust Predictive Current Control With Parallel Compensation Terms Against Multi-Parameter Mismatches for PMSMs. *IEEE Trans. Energy Convers.* **2020**, *35*, 2222–2230. [CrossRef]
26. Mohamed, Y.R. Design and Implementation of a Robust Current-Control Scheme for a PMSM Vector Drive With a Simple Adaptive Disturbance Observer. *IEEE Trans. Ind. Electron.* **2007**, *54*, 1981–1988. [CrossRef]
27. Tan, L.N.; Cong, T.P.; Cong, D.P. Neural Network Observers and Sensorless Robust Optimal Control for Partially Unknown PMSM With Disturbances and Saturating Voltages. *IEEE Trans. Power Electron.* **2021**, *36*, 12045–12056. [CrossRef]
28. Li, S.; Won, H.; Fu, X.; Fairbank, M.; Wunsch, D.C.; Alonso, E. Neural-Network Vector Controller for Permanent-Magnet Synchronous Motor Drives: Simulated and Hardware-Validated Results. *IEEE Trans. Cybern.* **2020**, *50*, 3218–3230. [CrossRef]
29. Meng, Q.; Bao, G. A Novel Low-Complexity Cascaded Model Predictive Control Method for PMSM. *Actuators* **2023**, *12*, 349. [CrossRef]
30. Grzesiak, L.M.; Tarczewski, T. Permanent magnet synchronous motor discrete linear quadratic speed controller. In Proceedings of the 2011 IEEE International Symposium on Industrial Electronics, Gdansk, Poland, 27–30 June 2011; IEEE: Piscataway, NJ, USA, 2011. [CrossRef]

31. Suleimenov, K.; Do, T.D. Data-Driven LQR for Permanent Magnet Synchronous Machines. In Proceedings of the 2019 IEEE Vehicle Power and Propulsion Conference (VPPC), Hanoi, Vietnam, 14–17 October 2019; IEEE: Piscataway, NJ, USA, 2019. [CrossRef]
32. Mora, A.; Orellana, A.; Juliet, J.; Cardenas, R. Model Predictive Torque Control for Torque Ripple Compensation in Variable-Speed PMSMs. *IEEE Trans. Ind. Electron.* **2016**, *63*, 4584–4592. [CrossRef]
33. Yu, F.; Li, K.; Zhu, Z.; Liu, X. An Over-Modulated Model Predictive Current Control for Permanent Magnet Synchronous Motors. *IEEE Access* **2022**, *10*, 40391–40401. [CrossRef]
34. Sun, X.; Wu, M.; Lei, G.; Guo, Y.; Zhu, J. An Improved Model Predictive Current Control for PMSM Drives Based on Current Track Circle. *IEEE Trans. Ind. Electron.* **2021**, *68*, 3782–3793. [CrossRef]
35. Zhang, X.; Cao, Y.; Zhang, C.; Niu, S. Model Predictive Control for PMSM Based on the Elimination of Current Prediction Errors. *IEEE J. Emerg. Sel. Top. Power Electron.* **2024**, *12*, 2651–2660. [CrossRef]

*actuators* MDPI

*Article*

# Fast Parameter Identification of the Fractional-Order Creep Model

**Shabnam Tashakori [1,2,*], Andres San-Millan [2], Vahid Vaziri [2] and Sumeet S. Aphale [2]**

1 Department of Mechanical Engineering, Shiraz University of Technology, Shiraz 71557-13876, Iran
2 Centre for Applied Dynamics Research, School of Engineering, University of Aberdeen, Aberdeen AB24 3UE, UK; andres.san-millanrodriguez@abdn.ac.uk (A.S.-M.); vahid.vaziri@abdn.ac.uk (V.V.); s.aphale@abdn.ac.uk (S.S.A.)
* Correspondence: tashakori@sutech.ac.ir or shabnam.tashakori@abdn.ac.uk

**Abstract:** In this study, a parameter identification approach for the fractional-order piezoelectric creep model is proposed. Indeed, creep is a wide-impacting phenomenon leading to time-dependent deformation in spite of constant persistent input. The creep behavior results in performance debasement, especially in applications with low-frequency responses. Fractional-Order (FO) modeling for creep dynamics has been proposed in recent years, which has demonstrated improved modeling precision compared to integer-order models. Still, parameter uncertainty in creep models is a challenge for real-time control. Aiming at a faster identification process, the proposed approach in this paper identifies the model parameters in two layers, i.e., one layer for the fractional-order exponent, corresponding to creep, and the other for the integer-order polynomial coefficients, corresponding to mechanical resonance. The proposed identification strategy is validated by utilizing experimental data from a piezoelectric actuator used in a nanopositioner and a piezoelectric sensor.

**Keywords:** system identification; fractional-order model; nanopositioning; creep

## 1. Introduction

Creep describes the gradual changes in displacement that occur after the voltage has been adjusted. This phenomenon mainly impacts slow-pace or start-from-previous-stop processes [1]. The creep behavior has been seen, e.g., in piezoelectric actuators or sensors [2], polymers, dielectric elastomers, concrete [3], etc. In piezoelectric actuators, increasing the operation speed while reducing the operation time can help diminish creep [4]. Nevertheless, in many applications, like scanning probe microscopes, this level of accuracy is not sufficient, because even small motion makes the measured image distorted [5]. To reach a higher accuracy, feedback methods can be employed, whereas those are not applicable in many cases because they rely on mounting several sensors over the system [4]. This highlights the necessity of a precise creep model, especially in applications like nanopositioning [6,7] and soft actuators [8,9].

Most papers in the literature focusing on the mathematical modeling and control of the creep dynamics lie within the span of integer-order calculus, e.g., the logarithmic model [10] and the Linear Time-Invariant (LTI) model [11]. In recent years, however, fractional calculus and fractional-order models have been explored to accommodate the dynamic behavior of systems that cannot be adequately encapsulated via integer-order models [12–16]. Fractional-order control is also being investigated, which usually achieves more precise tracking performance, with similar robustness and fewer controller gains to tune compared to integer-order controllers [17–23]. Fractional-order calculus has also gained attention as a viable candidate for creep modeling and control [24]. Generally speaking, the superiority of fractional order methods has been revealed for non-local behaviors and memory effects compared to classical methods [25]. In [4], a fractional-order model is proposed for the creep phenomenon, where the piezoelectric actuator is considered as a resistocaptance. In [26], a fractional-order Maxwell resistive capacitor approach is employed to model the creep phenomenon. Another simplified fractional-order creep model is proposed in [27]. A comparison between integer-order models (logarithmic and LTI models) and a fractional-order model is presented in [28].

Frequency-domain and time-domain system identification for fractional-order models have been first introduced in [29]. Fractional-order system identification mainly lies in two groups: (i) equation-error-based models and (ii) output-error-based models [30–32]. The main shortcoming of equation-error-based models is their need for prior knowledge of fractional orders, while output-error-based models can identify fractional orders and coefficients, simultaneously [4]. Nevertheless, identification of piezoelectric actuator parameters is tedious by using elementary output-error-based algorithms since, firstly, the sampling frequency should be high in order to capture mechanical resonant dynamics, secondly, the total time frame has to be long in order to capture creep, and thirdly, the fractional memory effect normally causes time-consuming computation [4]. Note that although methods based on optimization algorithms such as Particle Swarm Optimization (PSO) [33,34] and Genetic Algorithm (GA) [35] are applicable for the identification of a fractional-order system, they are not suitable for online identification as their computation times are normally high [36].

In recent years, several tools have been developed for fractional-order system analysis, modeling, and controller synthesis. Among these tools are MATLAB toolboxes CRONE [37], NINTEGER [38], and FOMCON [39]. The FOMCON toolbox is an extension to the mini toolbox introduced in [17,40], which is comprised of the following modules: (i) Main module (fractional system analysis), (ii) Identification module (system identification in time and frequency domains), and (iii) Control module (fractional PID controller design, tuning and optimization tools, as well as some additional features).

In this paper, the parameter identification of the fractional-order piezoelectric creep model is studied. In fact, the fractional-order piezoelectric model consists of two parts with different time scales: (i) a fractional-order part, describing the creep phenomenon, and (ii) an integer-order part, modeling the mechanical resonant. Accordingly, a two-layer identification approach is proposed in this paper in order to make the identification process faster. In this idea, fractional exponents and integer-order coefficients are estimated in a separate manner in the FOMCON toolbox (v1.50.1) with different sampling frequencies and time frames, resulting in a faster identification. This also remedies the FOMCON limitation in simultaneous identification of exponents and coefficients when the overall description of the model is fixed. Furthermore, a discussion on how to utilize the proposed method for online identification is provided.

The paper is organized as follows. Section 2 reviews some preliminaries of fractional calculus relevant to this paper. The fractional-order piezoelectric model and parameter identification algorithm are presented in Section 4. The experimental setups used to validate the presented identification approach are introduced in Section 5. Simulation and experimental results are shown and discussed in Section 6. Furthermore, Section 7 concludes this paper.

## 2. Preliminaries

Fractional calculus is a more general form of differentiating and integrating with fractional (i.e., non-integer) order. A general fractional-order transfer function has the following form:

$$G(s) = \frac{b_m s^{\beta_m} + b_{m-1} s^{\beta_{m-1}} + \ldots + b_0 s^{\beta_0}}{a_n s^{\alpha_n} + a_{n-1} s^{\alpha_{n-1}} + \ldots + a_0 s^{\alpha_0}}, \tag{1}$$

where either the exponents $\alpha_i, \beta_i \in \mathbb{R}$ or the coefficients $a_i, b_i \in \mathbb{R}$ can be fractional. In time domain, the fractional operator $\mathcal{D}$ is defined as follows [39]:

$$_{t_0}\mathcal{D}_t^{\alpha} = \begin{cases} \frac{d^{\alpha}}{dt^{\alpha}} & \mathcal{R}(\alpha) > 0 \\ 1 & \mathcal{R}(\alpha) = 0 \\ \int_{t_0}^{t} (dt)^{-\alpha} & \mathcal{R}(\alpha) < 0, \end{cases} \tag{2}$$

with $\alpha$ a non-integer number, and $\mathcal{R}(\alpha)$ its real part. Different definitions have been proposed for the fractional differentiation/integration, among which the Riemann–Liouville [41,42], Grunwald–Letnikov [43], and Caputo [44] are most exploited in the literature.

The Grunwald–Letnikov (GL) definition of the fractional differintegral introduced in (2), is as follows [39,45]:

$$ {}_{t_0}\mathcal{D}_t^{\alpha} f(t) = \lim_{h \to 0} \frac{1}{h^{\alpha}} \sum_{j=0}^{\left[\frac{t-t_0}{h}\right]} (-1)^j \binom{\alpha}{j} f(t - jh), \tag{3} $$

where $h$ is the time step and [.] denotes the integer part. Note that the GL definition is also valid for non-zero initial condition with taking the interval $(-\infty, 0)$ into account as presented in [46] in detail. Interested readers may refer to [47,48] for more information on fractional calculus.

## 3. Problem Formulation

Delving into the complexities of piezo-actuated smart mechatronic systems, nonlinearities caused by piezoelectric actuators are classified into (i) electrical nonlinearity issues, i.e., hysteresis and creep, and (ii) mechanical nonlinearities, which are primarily due to vibrations [2]. Creep becomes more significant at lower speeds, affecting the accuracy of piezoelectric actuators in open-loop operations. Vibrations, on the other hand, are related to the dynamic behavior of the system, characterized by high stiffness and low structural damping [49]. To address these challenges, researchers typically follow three steps: modeling, system identification, and control.

There are various types of piezoelectric actuators, such as stacks [50], tubes [51], and benders [52], which are utilized in different configurations in nanopositioning systems, with or without a sensing resistor or capacitor. Compared to tube piezos, stack piezos are (i) more cost-effective, (ii) yield significantly larger forces and displacements, and (iii) exhibit less coupling in different positioning directions. However, the nonlinear effects previously mentioned are much more marked in stack piezos [50].

Modeling piezo-actuated nanopositioners involves various variables, including [51]: (i) piezoelectric voltage (the voltage across the actuator, which equals the driving voltage if no sensing element is present) [53], (ii) induced voltage [54], (iii) sensing voltage [55], (iv) electric charge, and (v) displacement (position). Models that map any combination of these variables (or other relevant ones) to each other are referred to as models of the piezoelectric actuator. Regardless of the selected model, it involves unknown parameters that require an appropriate parameter identification algorithm for determination.

This study employs a model of the piezoelectric actuator, using driving voltage and displacement as the primary input and output while focusing on both electrical and mechanical nonlinearities, specifically creep and vibration. Although a more extensive data set would lead to a more accurate and reliable data-driven model, this paper opts for a different approach. Instead of a complex and nonlinear global model, which cannot be directly used in control system design, a fractional transfer function with time-varying parameters is continuously identified as a locally valid model. This approach is more practical for applications such as model-based adaptive control systems. Additionally, the identification method is designed to be fast, making it suitable for online applications.

## 4. Creep Model Identification

Piezoelectricity is accompanied by the memory effect and hence can be effectively described by a fractional-order system [56]. Therein, the creep phenomenon is modeled by a fractional-order integrator with an order between 0 and 1. A piezoelectric system is hence described by the following fractional-order transfer function [4]:

$$ G(s) := \frac{Y(s)}{U(s)} = \frac{b}{s^{\alpha}(1 + a_1 s + a_2 s^2)}, \tag{4} $$

where $Y(s)$ and $U(s)$ are, respectively, the output displacement and the driving voltage in frequency domain. The fractional-order parameter $0 < \alpha < 1$ specifies the creep rate, integer-order parameters $a_1, a_2$ depend on mechanical components, and parameter $b$ is the gain [4].

After a certain time $t_c$, the mechanical response can be disregarded when the input voltage is constant, which melts down the transfer function in Equation (4) to the following fractional-order integrator:

$$G(s) = \frac{b}{s^{\alpha}}, \quad t \geq t_c. \tag{5}$$

Subsequently, an approximate displacement for a unit step input is given by [57]:

$$y(t) = \mathscr{L}^{-1}\{G(s)U(s)\} = \frac{bt^{\alpha}}{\alpha\Gamma(\alpha)}, \quad t \geq t_c, \tag{6}$$

where $\Gamma(\alpha)$ is the gamma function. The so-called "double-logarithmic" creep model is hence described by:

$$log(y(t)) = \alpha log(t) + log(\frac{b}{\alpha\Gamma(\alpha)}) \quad t \geq t_c. \tag{7}$$

Four parameters are involved in the full piezoelectric model, Equation (4), i.e., fractional-order parameter $\alpha$, and integer-order parameters $a_1, a_2$, and $b$. These parameters should then be identified. The output-error approach treats the identification as a least square problem. More detailedly, the model in Equation (4) can be characterized by the following parameters vector [4]:

$$v = [a_1 \quad a_2 \quad \alpha \quad b]^T. \tag{8}$$

Note that the above set of parameters is unique. Firstly, the integer-order parameters characterize the mechanical components of the structure (mass, damping, stiffness), which are fixed for a given setup. Secondly, the creep rate is also fixed, depending on the piezoelectric material.

Suppose that the input $\mathbf{u}(t)$ and measured data $\mathbf{y}^*(t)$ are acquired at sample times $t_1, t_2, \ldots, t_k$. The parameter vector is then estimated as $\hat{v}$ such that it minimizes the quadratic norm of the error, given by:

$$J(\hat{v}) = \mathbf{e}^T\mathbf{e}, \tag{9}$$

where $\mathbf{e}$ is the output estimation error vector, with $\hat{\mathbf{y}}$ the estimated output, as follows:

$$\mathbf{e}(t, \hat{v}) = \mathbf{y}^*(t) - \hat{\mathbf{y}}(t, \hat{v}). \tag{10}$$

To solve the minimization problem $min(J(\hat{v}))$, two popular methods are (i) Levenberg Marquardt [58,59], and (ii) Trust region reflective [60,61], which are introduced below.

### 4.1. Trust Region Reflective Algorithm

In this method, the cost function $J$ is approximated with a quadratic function $\psi(x)$ by its Taylor series expansion around a point $x_i$ (in a trust region $\mathbf{N}$ of $x_i$). The algorithm then seeks a point $x_{i+1}$ in $\mathbf{N}$ that leads to a lower cost function compared to $x_i$. The minimization problem is hence defined by $\min_{p \in \mathbf{N}} \psi_i(p)$, where $p_i = x_{i+1} - x_i$ is the iteration step. The function $\psi_i(p)$ is given by:

$$\psi_i(p) = g(x_i)^T p + \frac{1}{2} p^T H(x_i) p, \tag{11}$$

where $g(x)$, and $H(x)$ are the gradient and Hessian of the cost function $J$, respectively [61].

This algorithm is time-consuming due to the huge amount of collected data. More precisely, firstly, the sampling time has to be relatively small (less than 0.1 ms), since the mechanical resonant frequency of piezoelectric actuator/sensor is high (of the order of kHz), and secondly, the time frame has to be relatively long (more than tens of seconds) as creep is exhibited in the long term. Moreover, this algorithm (employed in the FOMCON package [39,45]) is not directly applicable to our piezoelectric model in Equation (4). Herein, three identification modes can be considered, (i) identification of coefficients with fix exponents, (ii) identification of exponents with fix coefficients, and (iii) free identification of all coefficients and exponents, none of which are solely suitable for our case. Because the model in Equation (4) has one exponent and three coefficients to be identified. Note that the free identification mode can not be used since the exponents of the polynomial in the denominator of Equation (4) are fixed. Therefore, the two-layer trust region reflective algorithm is introduced in the following.

### 4.2. Two-Layer Trust Region Reflective Algorithm

A two-layer trust region reflective algorithm is proposed for the parameter identification of the fractional-order piezoelectric model in Equation (4), which has four parameters to identify. This approach identifies the fractional-order exponent $\alpha$ in a separate layer than the integer-order polynomial coefficients $a_1, a_2$, and $b$. With this idea, the creep phenomenon and the mechanical resonant can be identified using different sampling times, time frames, and identification modes.

**Remark 1.** *It is observed that, when using the FOMCON package, the initial identified creep rate is not usually precise enough. This exponent is hence re-tuned based on the following strategy:*

$$\alpha = \alpha_0 \times \frac{y_f^* - y_c^*}{\hat{y}_f - \hat{y}_c}, \tag{12}$$

*where $y^*$ and $\hat{y}$ are, respectively, the measured and estimated outputs with subscript $f$ the final value and $c$ corresponding to the time $t_c$, the time after which the mechanical response can be disregarded.*

The strategy is hence as follows, as also shown schematically in Figure 1:

- **Layer 1:** The fractional-order exponent $\alpha$ is initially identified based on the creep model in Equation (5) with its numerator fixed as 1 and considering free identification mode. In this layer, a relatively greater sampling time can be used. Note that a value is also identified for $\frac{1}{b}$, which is disregarded as this parameter will be re-identified in the next layer. The estimated value for $\alpha$ is then re-tuned using Equation (12) to arrive at a more accurate creep rate.
- **Layer 2:** integer-order coefficients $a_1, a_2$, and $b$ are identified by using the determined $\alpha$ in layer 1. Here, the sampling time should be smaller while the time frame can be selected as $0 < t < t_c$, with $t_c$ defined in Section 4. In this layer, the fix exponent mode is considered. Note that to alleviate the computational burden, the numerator can still be fixed at 1. As shown in Figure 1, with fixed exponents and numerator, the coefficients of the denominator $\frac{a_2}{b}$, $\frac{a_1}{b}$, and $\frac{1}{b}$ are identified. As the coefficient of $s^\alpha$ should be 1 (see Equation (4)), all three parameters $a_2, a_1$, and $b$ are obtained.

**Remark 2.** *As stated above, one parameter is identified in layer 1, and three other parameters in layer 2. For online identification, one can first perform layer 2 in the time frame $0 < t < t_c$ with a guess on $\alpha$. Then, for $t > t_c$, $\alpha$ can be updated by performing layer 1. The initial guess on $\alpha$ does not need to be precisely accurate since the creep is mostly relevant when the input is constant or with a slow changing rate, i.e., it is mostly the case in $t > t_c$ and not $0 < t < t_c$.*

As shown in Figure 2, for online identification, the idea of a two-layer trust region reflective algorithm can be performed as follows. First, the coefficients $a_2, a_1, b$ are identified by using an initial guess for $\alpha$ in $0 < t < t_c$. Then, for $t > t_c$, using the identified coefficients from Layer 1, the creep rate $\alpha$ is found. In the end, this value can be retuned using Equation (12) for more accuracy.

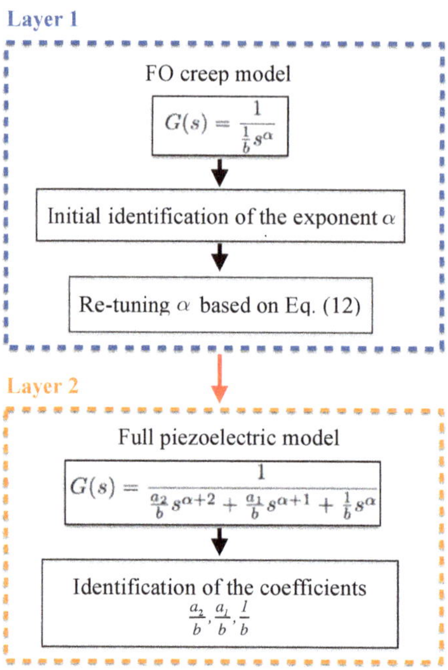


**Figure 1.** The general idea of the two-layer trust region reflective identification algorithm for the fractional-order piezoelectric model.

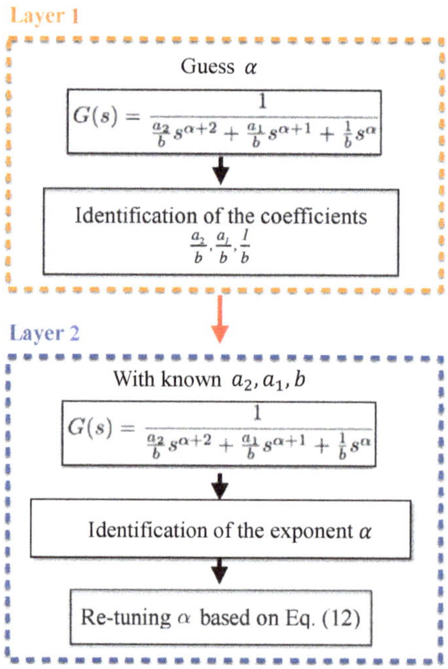


**Figure 2.** Two-layer trust region reflective algorithm for online identification of the fractional-order piezoelectric model.

**Remark 3.** *The readers interested in online identification of the FO creep model should note that GL approximation is recursive, i.e., the output of the fractional integrator is used to compute the new output. Therefore, one should be careful when the value of α changes as new samples arrive at the estimator. If, in that case, the computations need to be performed all over, this will require a batch implementation over a fixed time window. However, we believe that small changes in α may not interrupt the identification procedure, and it can proceed consecutively with the most recent estimated α.*

## 5. Experimental Setups

Two experimental setups are utilized to validate the identification approach proposed in this paper, i.e.:

- Setup 1: A stack of piezo-actuated serial kinematic nanopositioning stage, designed by the EasyLab, University of Nevada, Reno, USA [62], is shown in Figure 3. A voltage amplifier supplies the actuation voltage in the range of 0–200 V, and the output displacement (±20 μm) is measured by a high-resolution capacitive sensor in real-time [63]. Note that the experiment was conducted with an input of 0.7455 V, and the output was recorded every 0.005 s. As shown in Figure 4, the first resonant mode of the nanopositioner utilized in this paper is almost at 700 Hz. Therefore, the estimation time must be in the order of 1 ms, since, as stated in Section 4, the estimation time should be less than a period of the resonant mode frequency targeted, which provides the condition to tune the controller gains quickly enough. This figure also reveals some unmodeled dynamics that the identification approach needs to be robust against. As shown in Figure 4, there are several high-frequency resonant modes in the recorded frequency response, and the frequencies of the second and third modes are very close. Therefore, the third mode will influence the output signal and may lead to error in the system parameters identification as the system is modeled as a second-order resonant system.

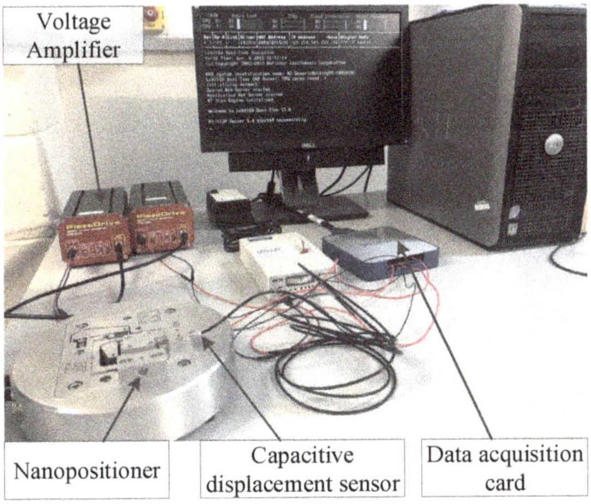


**Figure 3.** Experimental setup 1: two-axis piezo-actuated serial kinematic nanopositioning stage [63].

- Setup 2: Kistler 9272 four component dynamometer mounted on a vertical drill-string assembly, housed in the Drill-string Laboratory at the Centre for Applied Dynamics research (CADR), University of Aberdeen, Aberdeen, Scotland, UK [64]. A schematic of setup 2 is shown in Figure 5. This setup uses a stacked sensor configuration. The load-cell generates electric charge proportional to the measured forces and torques, relying on the principle of piezoelectricity. To convert the generated charges to corresponding voltage levels, a charge amplifier is required, for which a Kistler 5073A-type charge amplifier is utilized. Note that the experiment was conducted with different scenarios, i.e., one-step 50 N and 150 N loads, two-step 200 N input, and four-step 400 N input, and the output was recorded every 0.005 s.

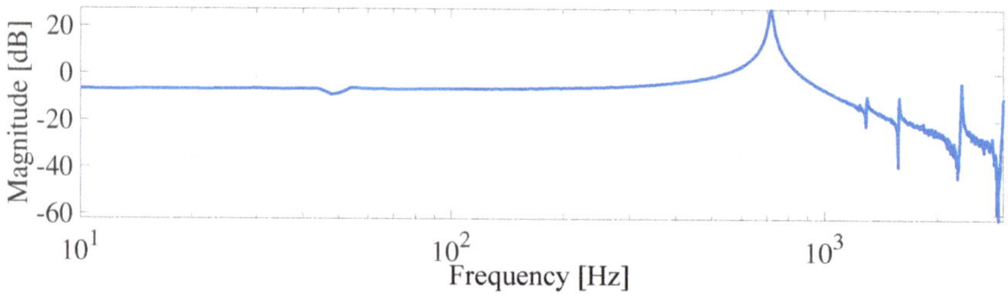


**Figure 4.** Frequency response function of the x−axis and y−axis in the experimental setup 1.

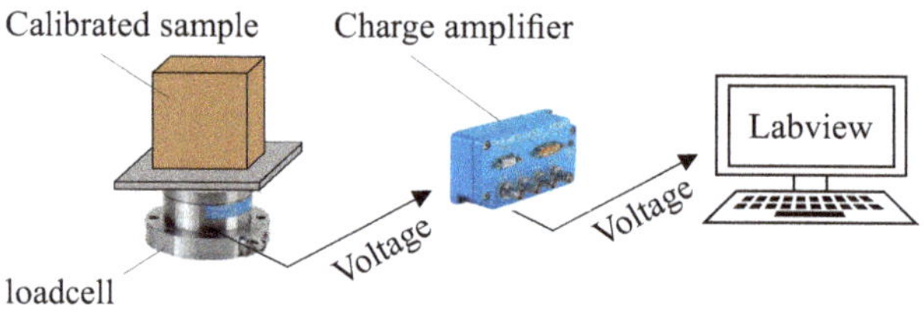


**Figure 5.** Experimental setup 1, including Kistler 9272 load-cell, a four component dynamometer, and ICAM5073A charge amplifier.

## 6. Results and Discussion

In this section, the fractional-order piezoelectric model (Equation (4)) is identified based on the time-domain experimental step response, which demonstrates slow creep phenomenon. The validation of the two-layer trust region reflective algorithm, as proposed in Section 4, is shown by employing two experimental setups, presented in Section 5.

To obtain the time-domain response, the revised Grunwald–Letnikov method is utilized in this paper based on (3), as follows [17,39]:

$$y(t) = \frac{1}{\sum_{i=0}^{2} \frac{a_i}{h^{\alpha_i}}} \left[u(t) - \sum_{i=0}^{2} \left(\frac{a_i}{h^{\alpha_i}} \sum_{j=1}^{\frac{t-t_0}{h}} \omega_j^{(\alpha_i)} y(t-jh)\right)\right], \tag{13}$$

with $a_0 = 1$, $a_1$ and $a_2$ as identified, $\alpha_0 = \alpha, \alpha_1 = \alpha + 1$, and $\alpha_2 = \alpha + 2$ with $\alpha$ identified. Note that the time step h should not be too large, as this may reduce the accuracy of the simulation. Conversely, it should not be too small, as this would result in unnecessarily long computation times. In Equation (13), $\omega_j^{(\alpha)}$ is recursively computed, as follows:

$$\omega_j^{(\alpha)} = (1 - \frac{\alpha+1}{j})\omega_{j-1}^{(\alpha)}, \quad j = 1, 2, \ldots, \frac{t-t_0}{h} \tag{14}$$

with $\omega_0^{(\alpha)} = 1$.

As discussed earlier, the time-domain identification is considered a least squares problem that seeks the minimum of the error norm by searching for a set of parameters (Equation (8)) where the error is defined in Equation (10) as the difference between the identified and measured output. Note that since the natural frequency of the mechanical part is high, the output signal is prone to noise in the transient regime. However, generally speaking, the computation time of the identification increases dramatically when the experimental data have too many high-frequency contents. Therefore, the acquired output vector could be filtered using a low-pass filter that guarantees zero phase distortion. It is fortunate that this high-frequency dynamics can be disregarded in many applications [4]. Another idea that can alleviate the computational burden is that the time step can be considered bigger after the transient response, as the dynamics in this regime are slower, and hence not much information would be lost. Moreover, note that the output signal is shifted such that $y(t = 0)$ is zero.

Applying the two-layer trust region reflective algorithm method, the governing piezoelectric model for setup 1 is identified as follows:

$$G(s) = \frac{0.1500}{s^{0.0133}(1 - 0.1200 \times 10^{-5}s + 0.3298 \times 10^{-5}s^2)}. \tag{15}$$

Figure 6 illustrates the experimental vs. the identified outputs, as well as the output identification error. As shown, the identified model fits 71.69 percent of the experimental data with almost 0.5 $\mu$m error norm, which demonstrates the efficacy of the proposed model. In this figure, two zoomed areas are demonstrated: one compares the experimental and identified outputs in the transient time, and the other illustrates the creep over time.

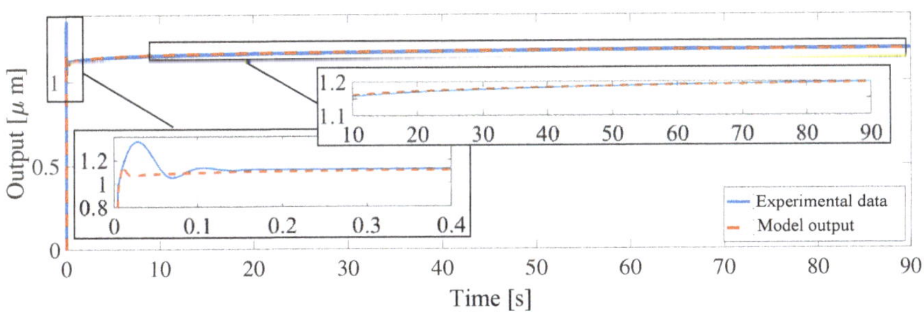


(**a**) The experimental output vs. the identified output.

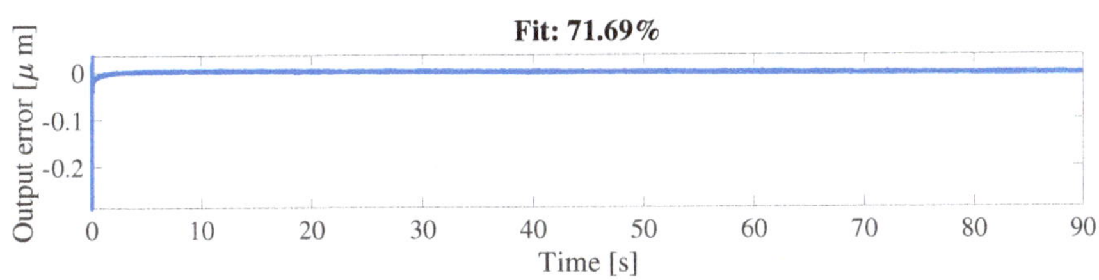


(**b**) The output identification error.

**Figure 6.** Identification of the piezoelectric model for setup 1.

**Remark 4.** *The time required to run the code for the identification approach proposed in this paper depends on the size of the dataset utilized. In other words, the smaller the time step of the acquired data, the greater the convergence time, indicated by "Elapsed time" in the FOMCON toolbox. Specifically, for time steps of 0.005, 0.01, 0.02, and 0.03 s, the elapsed time of the creep model identification (Layer 1) is 29, 7, 2, and 1 s, respectively. For time steps greater than 0.04 s, the elapsed time is less than a second, which can be regarded as "fast identification", with minimal loss of accuracy, as illustrated in Figure 7.*

Figure 7 presents a zoomed-in view of the identified creep model output after performing Layer 1 with varying time steps for the input data. As shown, the plots for 0.005-s and 0.04-s time steps are relatively close. This indicates that for the identification of the creep rate in Layer 1, the proposed approach can be effectively utilized with larger time steps, meaning it does not necessarily require an extensive dataset.

Figure 8 shows a zoomed-in view of the identified full model output after performing both Layer 1 and Layer 2. Although the time step influences the accuracy of the identification in the transient regime, as stated before, the time frame in Layer 2 can be selected as $0 < t < t_c$. Therefore, even with small time step, the size of the dataset utilized in Layer 2 is not large as the time frame is small.

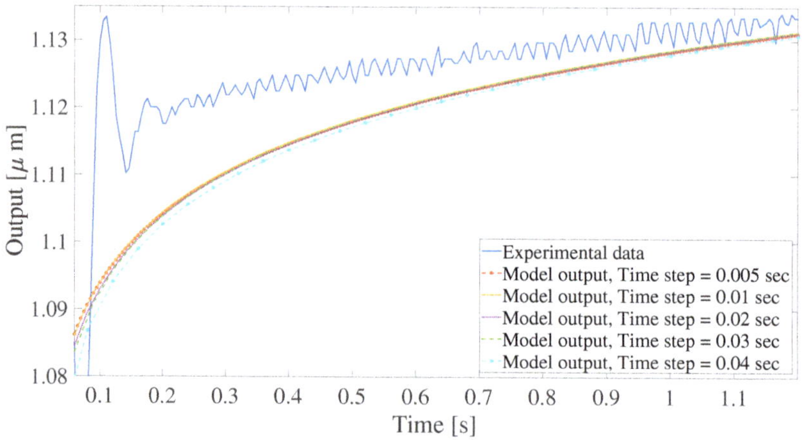


**Figure 7.** Comparison of the identified creep model outputs (Layer 1) using different time steps for the input data.

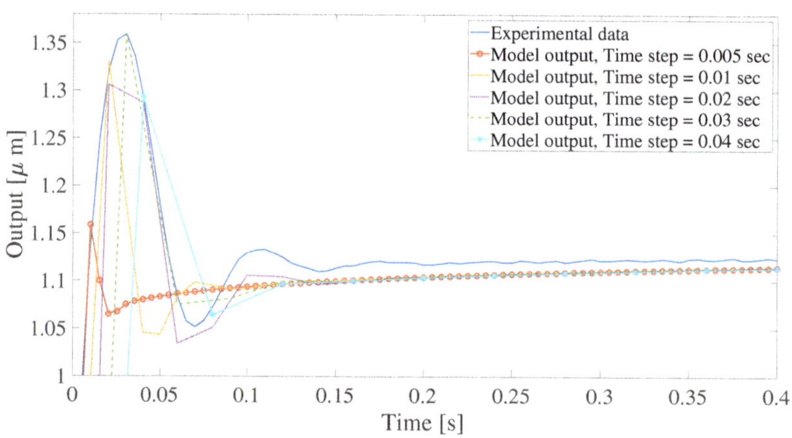


**Figure 8.** Comparison of the identified full model outputs (Layers 1 and 2) using different time steps for the input data.

The validation of the proposed two-layer trust region reflective identification is investigated by utilizing the experimental data of setup 2. With 50 N input force, the piezoelectric model is identified as follows:

$$G(s) = \frac{0.9922}{s^{0.085}(1 - 0.0363s + 0.0013s^2)}, \tag{16}$$

which is compared with the experimental data in Figure 9. As illustrated, the identified output is 49.15 percent fit to the measured output.

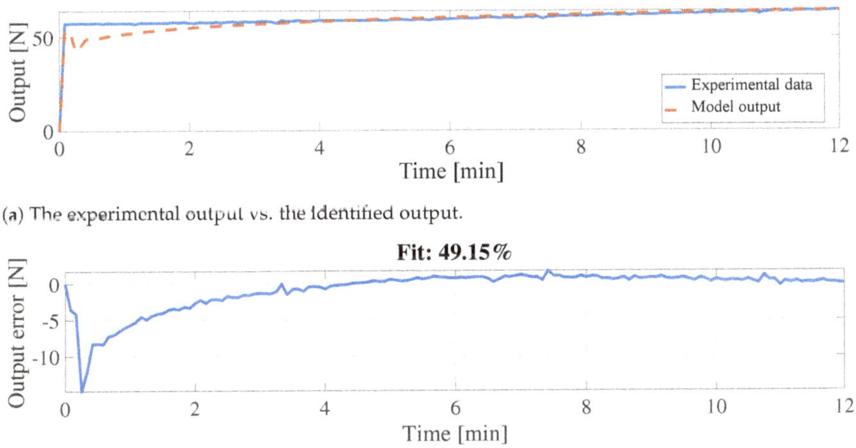


(a) The experimental output vs. the identified output.



(b) The output identification error.

**Figure 9.** Identification of the piezoelectric model for setup 2 with 50 N input.

Experimental setup 2 is again utilized to produce a new set of data with an input force of 150 N. In this case, the model is identified as follows:

$$G(s) = \frac{0.9792}{s^{0.037}(1 - 0.0172s + 0.0006s^2)}, \tag{17}$$

which is 79.15 percent fit to the experimental data, as shown in Figure 10.

**Remark 5.** *In the full fractional-order piezoelectric model in Equation* (4), $\alpha$ *specifies the creep rate,* $b$ *adjusts the error at the final time, and* $a_2, a_1$ *characterize the output in the transient condition, e.g., the overshoot and the settling time. Therefore, having reference data, for which the governing model is identified, can lead to intuitively approximating the parameters of an unknown model governing another set of data without running the identification procedure all over again. This is, of course, applicable when high modeling accuracy is not essential.*

Figures 11 and 12 illustrate the identified piezoelectric model for a two-step and four-step input. Note that to obtain these figures the identification algorithm is performed only once. The resulting identified transfer functions are as follows:

$$G(s) = \frac{0.9861}{s^{0.0173}(1 + 0.0032s + 0.8805 \times 10^{-4}s^2)}, \tag{18}$$

for the two-step input scenario, and

$$G(s) = \frac{0.9954}{s^{0.0094}(1 - 0.0017s + 0.2991 \times 10^{-3}s^2)}, \tag{19}$$

for the four-step scenario. These figures show the validity of the proposed identification approach for other input functions than a simple step.

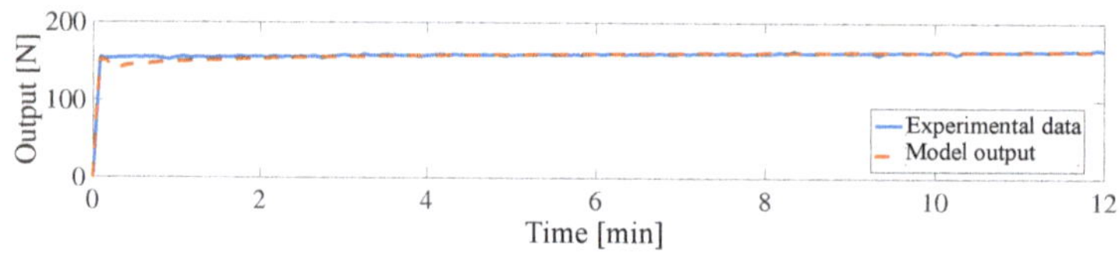


(**a**) The experimental output vs. the identified output.

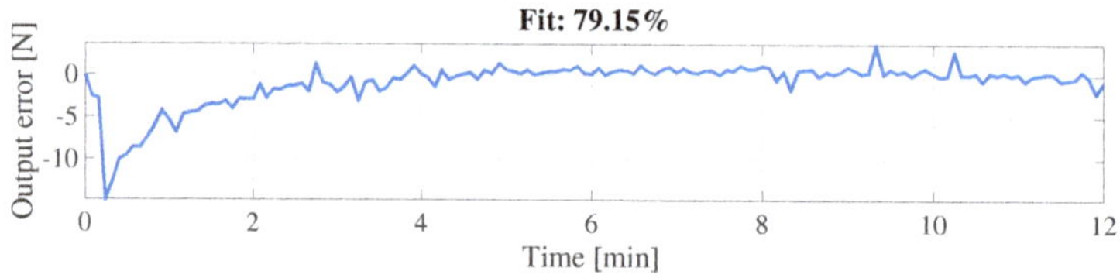


(**b**) The output identification error.

**Figure 10.** Identification of the piezoelectric model for setup 2 with 150 N input.

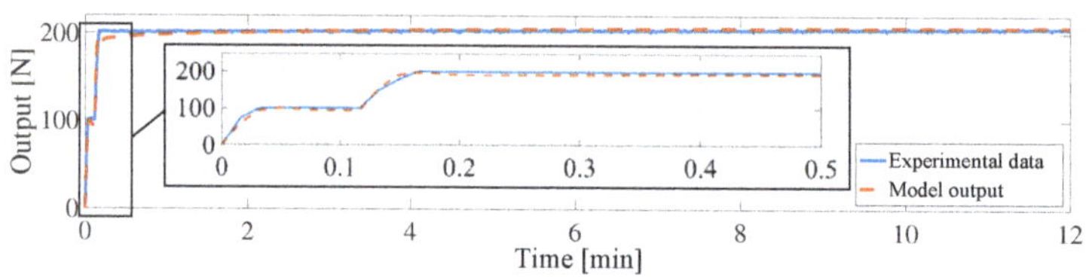


(**a**) The experimental output vs. the identified output.

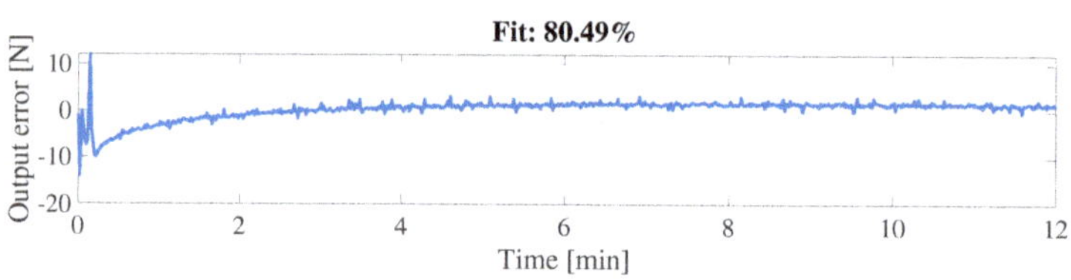


(**b**) The output identification error.

**Figure 11.** Identification of the piezoelectric model for setup 2 with a two-step input.

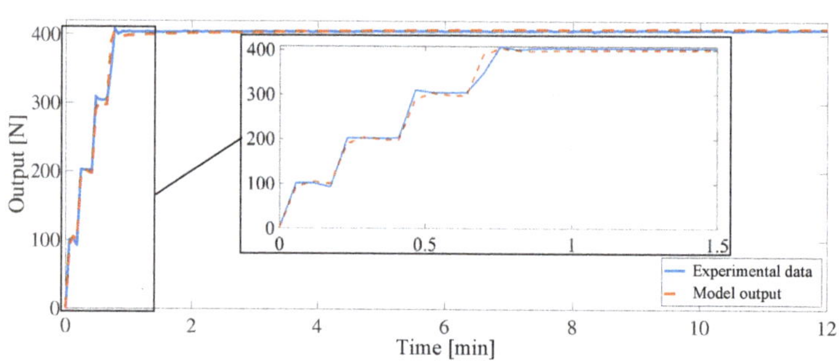


(**a**) The experimental output vs. the identified output.

**Figure 12.** *Cont.*

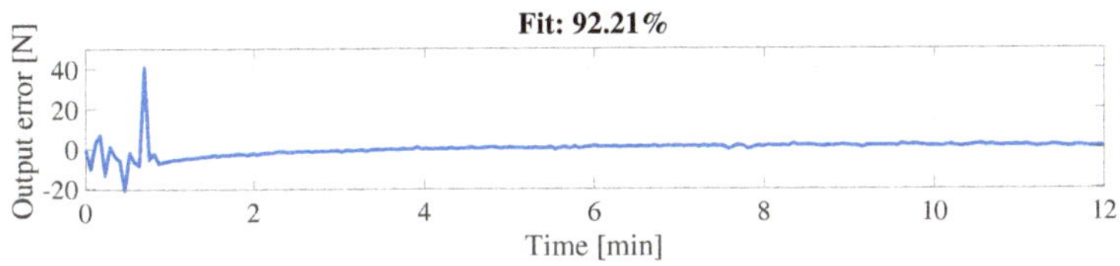


(**b**) The output identification error.

**Figure 12.** Identification of the piezoelectric model for setup 2 with a four-step input.

In brief, the results of two experimental set-ups validate the efficacy of the fractional-order modeling and the proposed two-layer identification for piezoelectric actuators/sensors model with consideration of the creep phenomenon.

## 7. Conclusions

Precise modeling and identification are increasingly demanded for micro/nano-scale positioning in low frequencies, which performance is deteriorated due to the creep phenomenon. To improve modeling precision, fractional-order models have been presented in recent years. In this paper, a piezoelectric fractional-order model is employed according to the properties of the creep effect. A parameter identification approach, named the two-layer trust region reflective method, is proposed, aiming at a faster identification. In this method, the fractional-order exponent, describing the creep phenomenon, is identified in a separate layer as the integer-order coefficients, describing the mechanical resonant, with different time frames and sampling times. This idea is hence appropriate for applications that need online identification and control.

The effectiveness and validation of the proposed identification technique are experimentally shown by using the data acquired from two different experimental setups. The first setup is a piezo-actuated serial kinematic nanopositioner, and the second setup is a dynamometer, relying on the principle of piezoelectricity, mounted on a vertical drill-string assembly. Step responses of both setups show slow creep phenomenon. The time-domain identification is then performed, seeking a set of parameters that minimizes the difference between the identified and measured output. The measured output vector is first filtered using a low-pass filter that guarantees zero phase distortion since the signal is prone to noise in the transient regime due to the high natural frequency of the mechanical part. To obtain the time-domain response of the identified model, the revised Grunwald–Letnikov method is utilized. The validity of the proposed identification approach is also demonstrated for other input functions than a simple step, i.e., two-step and four-step scenarios.

The proposed fractional-order identification of the creep phenomenon is a stepping stone toward improved control schemes, aiming at positioning precision, which is under investigation by the authors in their future work.

**Author Contributions:** Conceptualization, S.S.A., V.V. and S.T.; methodology, S.S.A., S.T. and V.V.; software, S.T. and S.S.A.; validation, S.S.A., V.V. and S.T.; formal analysis, S.T., S.S.A. and V.V.; investigation, S.T., S.S.A. and V.V.; resources, S.S.A. and V.V.; data curation, S.T., S.S.A. and V.V.; writing—original draft preparation, S.T.; writing—review and editing, S.S.A., V.V. and A.S.-M.; visualization, S.T.; supervision, S.S.A., V.V. and A.S.-M.; project administration, S.S.A. and V.V. All authors have read and agreed to the published version of the manuscript.

**Funding:** This research received no external funding.

**Data Availability Statement:** The raw data supporting the conclusions of this article will be made available by the authors on request.

**Conflicts of Interest:** The authors declare no conflicts of interest.

## References


1. Sabarianand, D.; Karthikeyan, P.; Muthuramalingam, T. A review on control strategies for compensation of hysteresis and creep on piezoelectric actuators based micro systems. *Mech. Syst. Signal Process.* **2020**, *140*, 106634. [CrossRef]
2. Yang, C.; Youcef-Toumi, K. Principle, implementation, and applications of charge control for piezo-actuated nanopositioners: A comprehensive review. *Mech. Syst. Signal Process.* **2022**, *171*, 108885. [CrossRef]
3. Chen, J.; Gong, L.; Meng, R. Application of Fractional Calculus in Predicting the Temperature-Dependent Creep Behavior of Concrete. *Fractal Fract.* **2024**, *8*, 482. [CrossRef]
4. Liu, Y.; Shan, J.; Qi, N. Creep modeling and identification for piezoelectric actuators based on fractional-order system. *Mechatronics* **2013**, *23*, 840–847. [CrossRef]
5. El-Rifai, O.M.; Youcef-Toumi, K. Creep in piezoelectric scanners of atomic force microscopes. In Proceedings of the 2002 American Control Conference (IEEE Cat. No. CH37301), IEEE, Anchorage, AK, USA, 8–10 May 2002; Volume 5, pp. 3777–3782.
6. Ge, R.; Wang, X.; Long, J.; Chen, Z.; Zhang, X. Creep modeling and control methods of piezoelectric actuators based on fractional order theory. In Proceedings of the Sixth International Conference on Electromechanical Control Technology and Transportation (ICECTT 2021), Chongqing, China, 14–16 May 2021; SPIE: San Francisco, CA, USA, 2022; Volume 12081, pp. 71–80.
7. Voda, A.; Charef, A.; Idiou, D.; Machado, M.M.P. Creep modeling for piezoelectric actuators using fractional order system of commensurate order. In Proceedings of the 2017 21st International Conference on System Theory, Control and Computing (ICSTCC), IEEE, Sinaia, Romania, 19–21 October 2017; pp. 120–125.
8. Tominaga, K.; Hamai, K.; Gupta, B.; Kudoh, Y.; Takashima, W.; Prakash, R.; Kaneto, K. Suppression of electrochemical creep by cross-link in polypyrrole soft actuators. *Phys. Procedia* **2011**, *14*, 143–146. [CrossRef]
9. Gu, G.Y.; Gupta, U.; Zhu, J.; Zhu, L.M.; Zhu, X. Modeling of viscoelastic electromechanical behavior in a soft dielectric elastomer actuator. *IEEE Trans. Robot.* **2017**, *33*, 1263–1271. [CrossRef]
10. Pesotski, D.; Janocha, H.; Kuhnen, K. Adaptive compensation of hysteretic and creep non-linearities in solid-state actuators. *J. Intell. Mater. Syst. Struct.* **2010**, *21*, 1437–1446. [CrossRef]
11. Yang, Q.; Jagannathan, S. Creep and hysteresis compensation for nanomanipulation using atomic force microscope. *Asian J. Control* **2009**, *11*, 182–187. [CrossRef]
12. Panja, P. Dynamics of a fractional order predator-prey model with intraguild predation. *Int. J. Model. Simul.* **2019**, *39*, 256–268. [CrossRef]
13. Deressa, C.T.; Duressa, G.F. Investigation of the dynamics of COVID-19 with SEIHR nonsingular and nonlocal kernel fractional model. *Int. J. Model. Simul.* **2022**, *42*, 1030–1048. [CrossRef]
14. Qureshi, S.; Abro, K.A.; Gómez-Aguilar, J. On the numerical study of fractional and non-fractional model of nonlinear Duffing oscillator: A comparison of integer and non-integer order approaches. *Int. J. Model. Simul.* **2022**, *43*, 362–375. [CrossRef]
15. Xu, Y.; Luo, Y.; Luo, X.; Chen, Y.; Liu, W. Fractional-Order Modeling of Piezoelectric Actuators with Coupled Hysteresis and Creep Effects. *Fractal Fract.* **2023**, *8*, 3. [CrossRef]
16. Li, J.; Hu, B.; Sheng, J.; Huang, L. A Fractional-Order Creep-Damage Model for Carbonaceous Shale Describing Coupled Damage Caused by Rainfall and Blasting. *Fractal Fract.* **2024**, *8*, 459. [CrossRef]
17. Chen, Y.; Petras, I.; Xue, D. Fractional order control-a tutorial. In Proceedings of the 2009 American Control Conference, IEEE, St. Louis, MO, USA, 10–12 June 2009; pp. 1397–1411.
18. Shahri, E.S.A.; Alfi, A.; Tenreiro Machado, J. Stabilization of fractional-order systems subject to saturation element using fractional dynamic output feedback sliding mode control. *J. Comput. Nonlinear Dyn.* **2017**, *12*, 031014. [CrossRef]
19. Delavari, H.; Heydarinejad, H. Fractional-order backstepping sliding-mode control based on fractional-order nonlinear disturbance observer. *J. Comput. Nonlinear Dyn.* **2018**, *13*, 111009. [CrossRef]
20. Muñoz-Vázquez, A.J.; Parra-Vega, V.; Sánchez-Orta, A. Fractional-Order Nonlinear Disturbance Observer Based Control of Fractional-Order Systems. *J. Comput. Nonlinear Dyn.* **2018**, *13*, 071007. [CrossRef]
21. Muñoz-Vázquez, A.J.; Martínez-Reyes, F. Output feedback fractional integral sliding mode control of robotic manipulators. *J. Comput. Nonlinear Dyn.* **2019**, *14*, 054502. [CrossRef]
22. Soukkou, A.; Soukkou, Y.; Haddad, S.; Benghanem, M.; Rabhi, A. Finite-Time Synchronization of Fractional-Order Energy Resources Demand-Supply Hyperchaotic Systems Via Fractional-Order Prediction-Based Feedback Control Strategy with Bio-Inspired Multiobjective Optimization. *J. Comput. Nonlinear Dyn.* **2023**, *18*, 031003. [CrossRef]
23. Muñoz-Vázquez, A.J.; Treesatayapun, C. Discrete-Time Adaptive Fractional Nonlinear Control Using Fuzzy Rules Emulating Networks. *J. Comput. Nonlinear Dyn.* **2023**, *18*, 071002. [CrossRef]
24. San-Millan, A.; Feliu-Batlle, V.; Aphale, S.S. Fractional order implementation of Integral Resonant Control–A nanopositioning application. *ISA Trans.* **2018**, *82*, 223–231. [CrossRef]
25. Conejero, J.A.; Franceschi, J.; Picó-Marco, E. Fractional vs. ordinary control systems: What does the fractional derivative provide? *Mathematics* **2022**, *10*, 2719. [CrossRef]
26. Liu, Y.; Shan, J.; Gabbert, U.; Qi, N. Hysteresis and creep modeling and compensation for a piezoelectric actuator using a fractional-order Maxwell resistive capacitor approach. *Smart Mater. Struct.* **2013**, *22*, 115020. [CrossRef]
27. Liu, L.; Yun, H.; Li, Q.; Ma, X.; Chen, S.L.; Shen, J. Fractional order based modeling and identification of coupled creep and hysteresis effects in piezoelectric actuators. *IEEE/ASME Trans. Mechatron.* **2020**, *25*, 1036–1044. [CrossRef]

28. Tashakori, S.; Vaziri, V.; Aphale, S.S. A comparative quantification of existing creep models for piezoactuators. In Proceedings of the 10th International Conference on Wave Mechanics and Vibrations, WMVC2022, Lisbon, Portugal, 4–6 July 2022; Springer: Cham, Switzerland, 2022.
29. Le Lay, L. Identification Fréquentielle et Temporelle par Modèle non Entier. Ph.D. Thesis, University of Bordeaux, Bordeaux, France, 1998. Available online: http://www.theses.fr/1998BOR10605 (accessed on 27 October 2024).
30. Malti, R.; Aoun, M.; Sabatier, J.; Oustaloup, A. Tutorial on system identification using fractional differentiation models. *IFAC Proc. Vol.* **2006**, *39*, 606–611. [CrossRef]
31. Malti, R.; Victor, S.p.; Nicolas, O.; Oustaloup, A. System identification using fractional models: State of the art. In Proceedings of the International Design Engineering Technical Conferences and Computers and Information in Engineering Conference, Las Vegas, NV, USA, 4–7 September 2007; Volume 4806, pp. 295–304.
32. Malti, R.; Victor, S.; Oustaloup, A. Advances in system identification using fractional models. *J. Comput. Nonlinear Dyn.* **2008**, *3*. [CrossRef]
33. Gupta, R.; Gairola, S.; Diwiedi, S. Fractional order system identification and controller design using PSO. In Proceedings of the 2014 Innovative Applications of Computational Intelligence on Power, Energy and Controls with their impact on Humanity (CIPECH), IEEE, Ghaziabad, India, 28–29 November 2014; pp. 149–153.
34. Liu, L.; Shan, L.; Jiang, C.; Dai, Y.; Liu, C.; Qi, Z. Parameter identification of the fractional-order systems based on a modified PSO algorithm. *J. Southeast Univ. (Engl. Ed.)* **2018**, *34*, 6–14.
35. Zhou, S.; Cao, J.; Chen, Y. Genetic algorithm-based identification of fractional-order systems. *Entropy* **2013**, *15*, 1624–1642. [CrossRef]
36. Zhang, L.; Chen, Y.; Sun, R.; Jing, S.; Yang, B. A task scheduling algorithm based on PSO for grid computing. *Int. J. Comput. Intell. Res.* **2008**, *4*, 37–43. [CrossRef]
37. Oustaloup, A.; Melchior, P.; Lanusse, P.; Cois, O.; Dancla, F. The CRONE toolbox for Matlab. In Proceedings of the CACSD. Conference Proceedings-IEEE International Symposium on Computer-Aided Control System Design (Cat. No. 00TH8537), IEEE, Anchorage, AK, USA, 25–27 September 2000; pp. 190–195.
38. Valerio, D.; Da Costa, J.S. Ninteger: A non-integer control toolbox for MatLab. In *Proceedings of the Fractional Differentiation and Its Applications*; ENSEIRB: Bordeaux, France, 2004.
39. Tepljakov, A.; Petlenkov, E.; Belikov, J. FOMCON: A MATLAB toolbox for fractional-order system identification and control. *Int. J. Microelectron. Comput. Sci.* **2011**, *2*, 51–62.
40. Xue, D.; Chen, Y.; Atherton, D.P. *Linear Feedback Control: Analysis and Design with MATLAB*; SIAM: Philadelphia, PA, USA, 2007.
41. Jumarie, G. Modified Riemann–Liouville derivative and fractional Taylor series of nondifferentiable functions further results. *Comput. Math. Appl.* **2006**, *51*, 1367–1376. [CrossRef]
42. Atangana, A.; Gómez-Aguilar, J. Numerical approximation of Riemann–Liouville definition of fractional derivative: From Riemann–Liouville to Atangana-Baleanu. *Numer. Methods Partial. Differ. Equ.* **2018**, *34*, 1502–1523. [CrossRef]
43. Scherer, R.; Kalla, S.L.; Tang, Y.; Huang, J. The Grünwald–Letnikov method for fractional differential equations. *Comput. Math. Appl.* **2011**, *62*, 902–917. [CrossRef]
44. Ortigueira, M.D.; Machado, J.T. What is a fractional derivative? *J. Comput. Phys.* **2015**, *293*, 4–13. [CrossRef]
45. Tepljakov, A. FOMCON: Fractional-order modeling and control toolbox. In *Fractional-Order Modeling and Control of Dynamic Systems*; Springer: Cham, Switzerland, 2017; pp. 107–129.
46. Sabatier, J.; Farges, C. Initial value problems should not be associated with fractional model descriptions whatever the derivative definition used. *AIMS Math* **2021**, *6*, 11318–11329. [CrossRef]
47. Kilbas, A.A.; Srivastava, H.M.; Trujillo, J.J. *Theory and Applications of Fractional Differential Equations*; Springer: Cham, Switzerland, 2006; Volume 204.
48. Monje, C.A.; Chen, Y.; Vinagre, B.M.; Xue, D.; Feliu-Batlle, V. *Fractional-Order Systems and Controls: Fundamentals and Applications*; Springer Science & Business Media: Cham, Switzerland, 2010.
49. Yuan, Z.; Zhou, S.; Zhang, Z.; Xiao, Z.; Hong, C.; Chen, X.; Zeng, L.; Li, X. Piezo-actuated smart mechatronic systems: Nonlinear modeling, identification, and control. *Mech. Syst. Signal Process.* **2024**, *221*, 111715. [CrossRef]
50. Salapaka, S.; Sebastian, A.; Cleveland, J.P.; Salapaka, M.V. High bandwidth nano-positioner: A robust control approach. *Rev. Sci. Instrum.* **2002**, *73*, 3232–3241. [CrossRef]
51. Mohammadzaheri, M.; Grainger, S.; Bazghaleh, M. A system identification approach to the characterization and control of a piezoelectric tube actuator. *Smart Mater. Struct.* **2013**, *22*, 105022. [CrossRef]
52. Moore, S.I.; Yong, Y.K.; Omidbeike, M.; Fleming, A.J. Serial-kinematic monolithic nanopositioner with in-plane bender actuators. *Mechatronics* **2021**, *75*, 102541. [CrossRef]
53. Aphale, S.; Fleming, A.J.; Moheimani, S. High speed nano-scale positioning using a piezoelectric tube actuator with active shunt control. *Micro Nano Lett.* **2007**, *2*, 9–12. [CrossRef]
54. Kuiper, S.; Schitter, G. Active damping of a piezoelectric tube scanner using self-sensing piezo actuation. *Mechatronics* **2010**, *20*, 656–665. [CrossRef]
55. Fleming, A.J.; Moheimani, S.R. Sensorless vibration suppression and scan compensation for piezoelectric tube nanopositioners. *IEEE Trans. Control Syst. Technol.* **2005**, *14*, 33–44. [CrossRef]
56. Das, S. *Functional Fractional Calculus*; Springer: Berlin/Heidelberg, Germany, 2011; Volume 1.

57. Chen, Y.; Petras, I.; Vinagre, B. A List of Laplace and Inverse Laplace Transforms Related to Fractional Order Calculus. 2001. Available online: http://people.tuke.sk/ivo.petras/foclaplace.pdf (accessed on 30 November 2001).
58. Bellavia, S.; Gratton, S.; Riccietti, E. A Levenberg–Marquardt method for large nonlinear least-squares problems with dynamic accuracy in functions and gradients. *Numer. Math.* **2018**, *140*, 791–825. [CrossRef]
59. Gavin, H.P. *The Levenberg–Marquardt Algorithm for Nonlinear Least Squares Curve-Fitting Problems*; Department of Civil and Environmental Engineering, Duke University: Durham, NC, USA, 2019; Volume 19.
60. Le, T.M.; Fatahi, B.; Khabbaz, H.; Sun, W. Numerical optimization applying trust-region reflective least squares algorithm with constraints to optimize the non-linear creep parameters of soft soil. *Appl. Math. Model.* **2017**, *41*, 236–256. [CrossRef]
61. Ahsan, M.; Choudhry, M.A. System identification of an airship using trust region reflective least squares algorithm. *Int. J. Control Autom. Syst.* **2017**, *15*, 1384–1393. [CrossRef]
62. Fairbairn, M.W.; Moheimani, S.R. Control techniques for increasing the scan speed and minimizing image artifacts in tapping-mode atomic force microscopy: Toward video-rate nanoscale imaging. *IEEE Control Syst. Mag.* **2013**, *33*, 46–67.
63. Babarinde, A.K.; Li, L.; Zhu, L.; Aphale, S.S. Experimental validation of the simultaneous damping and tracking controller design strategy for high-bandwidth nanopositioning—A PAVPF approach. *IET Control Theory Appl.* **2020**, *14*, 3506–3514. [CrossRef]
64. Vaziri, V.; Oladunjoye, I.O.; Kapitaniak, M.; Aphale, S.S.; Wiercigroch, M. Parametric analysis of a sliding-mode controller to suppress drill-string stick-slip vibration. *Meccanica* **2020**, *55*, 2475–2492. [CrossRef]

*actuators*

MDPI

*Article*

# Study on the Vibration Characteristics of Separated Armature Assembly in an Electro-Hydraulic Servo Valve Under Interference Fit

**Tong Li [1], Jinghui Peng [1], Songjing Li [1,*], Juan Zhang [2,*] and Aiying Zhang [3]**

[1] Department of Fluid Control and Automation, Harbin Institute of Technology, Harbin 150001, China; 18b908121@stu.hit.edu.cn (T.L.); jpeng@hit.edu.cn (J.P.)
[2] Training Base of Army Engineering University, Xuzhou 221001, China
[3] Qingan Group Co., Ltd., Xi'an 710082, China; zhangxy_heyg@163.com
* Correspondence: lisongjing@hit.edu.cn (S.L.); zjuan426@163.com (J.Z.); Tel.: +86-451-8641-8318 (S.L.)

**Abstract:** The electro-hydraulic servo valve is a critical component that transforms electrical signals into hydraulic signals, thereby controlling the hydraulic system. It finds extensive application in precision control systems. The stability of the electro-hydraulic servo valve is primarily influenced by the armature assembly. Unlike integral armature assembly, the separated armature assembly, comprising the armature, spring tube, flapper, and feedback spring, is joined through an interference fit, which introduces prestress within the assembly. The existence of prestress may affect the operational mode of the armature assembly. Consequently, this paper investigates the vibration characteristics of the separated armature assembly under interference fit conditions. Comparative analysis reveals that interference fit indeed generates prestress, which cannot be overlooked. To further validate the reliability of the simulation results, the natural frequency of the separated armature assembly is determined by applying a sweeping frequency signal to the torque motor using an electric drive, thereby verifying the feasibility of the simulation analysis. Additionally, the impact of interference on the vibration characteristics of the separated armature assembly is examined, confirming the accuracy of the simulation analysis method based on the interference fit. The research on vibration characteristics of a separated armature assembly provides technical support for the structural optimization design of the electro-hydraulic servo valve, thereby enhancing its performance.

**Keywords:** servo valve; separated armature assembly; interference fit; vibration characteristic

## 1. Introduction

The nozzle flapper servo valve has the advantages of fast response, high precision, and high sensitivity, and is widely used in hydraulic servo control systems. The occasional whistling during the operation of the servo valve [1], causes the spring tube rupture, resulting in servo valve failure, and even affects the function of the hydraulic system. The whistling is caused by the phenomenon of self-excited oscillations. The reported research shows that the self-excited oscillations are not only related to the pressure pulsation of the pilot-stage flow field [2,3] but also to the vibration characteristics of the armature assembly [4].

Mode is the inherent vibration characteristic of the structure, so the mode of the armature assembly is studied in this paper. As shown in Figure 1, the separated armature assembly is composed of four parts: armature, spring tube, flapper, and feedback spring.

The characteristic of the separated armature assembly is that the flapper and feedback spring are two independent parts, while in the integral armature assembly, the flapper and feedback spring are integrated. Many researchers have studied the modal analysis of components. Zhang [5] used a modified transfer matrix method to calculate the modal calculation of the rotor in the turbomolecular pump, and the experiment proved that the method could calculate the modal of the rotor assembly with high accuracy. Kim [6] carried out natural frequency and Campbell diagram analysis of the turbine blades according to their assembly conditions, and verified the modal characteristics and resonance stability. Zhai [7] calculated the natural frequency and mode shape of the rotor assembly without a thrust disk through the transfer matrix method according to the actual assembly relationship, and verified the results by using the finite element method and a new modal experiment method, the interference fit and clearance fit between the parts were also considered in FEM. Daouk [8] studied the relationship between the number of bolts, the position of the bolts, and the pre-loaded torque value on the modal damping of the assembly. Yang [9] conducted modal analysis experiments using the force hammer excitation method and obtained the natural frequency and modal modes of stacked chip components.

The modal analysis of components cannot be simplified in some cases, and the influence of external conditions on the stiffness needs to be considered [10–12]. Orlowska [13] studied the effect of prestress on the natural frequency and dynamic characteristics of eccentric prestressed glass fiber reinforced polymer composite beams. The results show that prestress increases the first bending frequency and decreases the second bending frequency, and a comparison of the numerical and experimental data confirmed this effect. Hu [14] studied the effect of prestress on the dynamic characteristics of rail transit gearboxes and compared it with the model without prestress; the results showed that prestress has a significant effect on the modal characteristics of the gearbox. Through finite element analysis, Wang [15] constructed a prestressed modal module to obtain the strength and modal characteristics of the steering device. Li [16] studied the sealing interface of the ultra-high pressure diaphragm compressor cylinder head, and found that some modal frequencies changed transition with bolt loosening. Liu [17] analyzed the dynamic behavior of the printed circuit board assembly (PCBA) and the reliability of the board-level solder joints under thermal–vibration combined load. The results showed that when the temperature changes from 22 °C to 60 °C, the natural frequencies of PCBA decreases, resulting in a decrease in the fatigue life of solder joints. Therefore, in some cases, external conditions have an effect on the dynamic characteristics of components.

Different from the general components, the separated armature assembly is composed of four parts connected by interference fit. Interference fit usually occurs under the condition of shaft and hole fit, and the magnitude of interference has an effect on the modes of the components [18,19]. Fan [20] used Ansys Workbench software to study and analyze the modes of the rotor assembly connected by the interference fit between the elastic shaft and the rigid thick disk and obtained the natural frequency of the component.

In this paper, in order to explore the vibration characteristics of the separated armature assembly and provide technical support for the structural design and optimization of the servo valve, the influence of bonded, friction contact type, and contact surface nodes coupling or not on the modes of armature assembly is studied, and the influence of interference on the modes of armature assembly is further studied.

## 2. Armature Assembly

### 2.1. Geometry Model of Armature Assembly

There are two types of armature assembly: integral and separated, the separated armature assembly is composed of four parts: armature, spring tube, flapper, and feedback spring, and the four parts are assembled together by interference fit. As shown in Figure 1, armature and spring tube interference assembly, spring tube and flapper interference assembly, flapper, and feedback spring interference assembly. Therefore, there is prestress in the armature assembly under normal working conditions. The research shows that the existence of prestress has an effect on the modes of the structure [13–15].

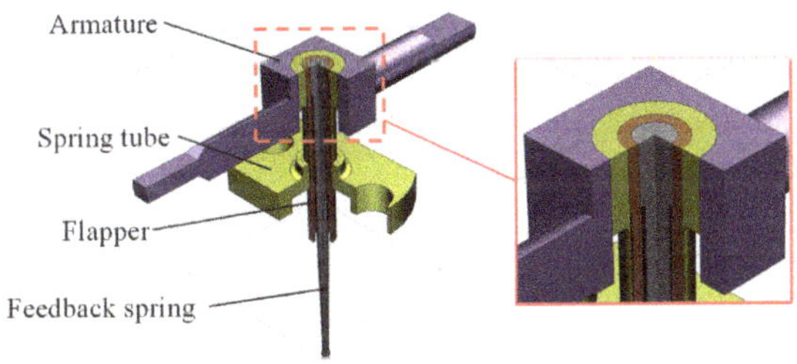


**Figure 1.** Separated armature assembly.

*2.2. Dynamic Model of Armature Assembly Under Prestressed Condition*

For armature assembly, the free vibration differential equation without damping can be described as follows [21]:

$$[M]\{\ddot{u}\} + [K]\{u\} = \{0\} \tag{1}$$

where $[M]$ is the mass matrix of the armature assembly, $[K]$ is the stiffness matrix of the armature assembly, $\{\ddot{u}\}$ is the vibration acceleration matrix, and $\{u\}$ is the displacement matrix.

Under certain initial conditions, armature assembly move at the same frequency according to simple harmonic vibration.

$$\{u\} = \{\phi\}\sin(\omega t + \theta) \tag{2}$$

Then, the equation can be simplified as

$$([K] - \omega_i^2[M])\{\phi_i\} = 0 \tag{3}$$

$$\omega_i^2 = \frac{\{\phi_i\}^{\mathrm{T}}[K]\{\phi_i\}}{\{\phi_i\}^{\mathrm{T}}[M]\{\phi_i\}} \tag{4}$$

where $\{\phi_i\}$ and $\omega_i$ are modal shape and natural frequency of the $i$th mode, respectively.

Due to the interference fit of the armature assembly, the assembly prestress $[\sigma]$ is generated, which generates a stress stiffness matrix $[S]$. The stress stiffness effect can be realized by superimposing the stress stiffness matrix into the elastic stiffness matrix of the structure.

$$[K_s] = [K] + [S] \tag{5}$$

where $[S]$ represents the stress stiffness matrix that accounts for assembly prestress, $[K_s]$ represents the stiffness matrix of the armature assembly considering the assembly prestress. This result is derived by substituting Equation (5) into Equation (3).

$$([K_s] - \omega_{is}^2[M])\{\phi_i\} = 0 \tag{6}$$

$$\omega_{is}^2 = \frac{\{\phi_i\}^{\mathrm{T}}[K_s]\{\phi_i\}}{\{\phi_i\}^{\mathrm{T}}[M]\{\phi_i\}} \tag{7}$$

where $\omega_{is}$ is the natural frequency of the $i$th mode considering the assembly prestress.

Thus, the $\omega_{is}$ can be described as

$$\omega_s^2 = \omega_i^2 + \frac{\{\phi_i\}^{\mathrm{T}}[S]\{\phi_i\}}{\{\phi_i\}^{\mathrm{T}}[M]\{\phi_i\}} \tag{8}$$

## 3. Modal Analysis of Interference Connected Assembly

In general, in the modal analysis of the assembly, the connection relationship between the two parts is simplified. For the interference fit assembly, there is prestress in the contact surface, which cannot be simplified according to the above method. For the modal analysis of armature assembly, previous studies [22,23] have used the direct-bonded method for the integrated armature assembly, but this method ignores the existence of prestress, therefore this type of contact is not suitable for separated armature assemblies. In order to simulate more closely to the real state, four schemes are used for comparative analysis and research.

Scheme A: Bonded contact; Scheme B: Friction contact; Scheme C: Bonded contact under node coupling condition; Scheme D: Frictional contact under node coupling conditions. The modal simulation analysis is completed by Ansys Workbench 2021 software, the geometric model is drawn by NX10.0 software, the size is referred to the real structure, the interference between each part is 0, and the interference is completed by the contact setting in the software. Constraints are set based on actual application scenarios, the bottom surface of the spring tube is set as a fixed constraint. The armature assembly is securely connected to the valve body through two holes in the spring tube base. The ambient temperature is configured to the default setting as specified by the software. Each part and the corresponding material parameters are shown in Table 1.

**Table 1.** Part name and corresponding material parameters.

| Part | Armature | Spring Tube | Flapper | Feedback Spring |
|---|---|---|---|---|
| Materials | 1J50 | QBe1.9 | 3J1 | 3J1 |
| Density (kg/m$^3$) | 8200 | 8230 | 8000 | 8000 |
| Young's Modulus (GPa) | 157 | 125 | 190 | 190 |
| Poisson's Ratio | 0.3 | 0.35 | 0.3 | 0.3 |
| Yield strength (MPa) | 685 | 1035 | 882 | 882 |

### 3.1. Scheme A: Setting of Bonded Contact

The contact modes between the armature and spring tube, spring tube and flapper, flapper and feedback spring are all set as bonded contact. Tetrahedral mesh is used to draw the whole components, in which the mesh size of the armature is 0.4 mm, the spring tube is 0.3 mm, the flapper is 0.2 mm, and the feedback spring is 0.2 mm. The thin wall of the spring tube is sealed. From the perspective of directions E and F in Figure 2, the armature assembly grid is shown in Figure 2a, and as shown in the figure, the nodes on each contact surface do not coincide. The results of the first-order mode to the fourth-order mode are analyzed.

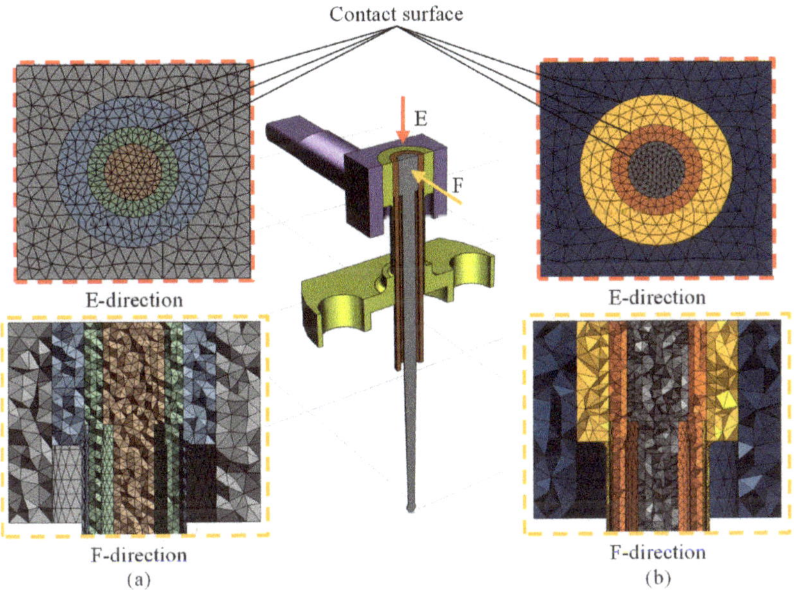


**Figure 2.** Meshing. (**a**) Node coupling is not considered in meshing; (**b**) node coupling is considered in meshing.

### 3.2. Scheme B: Setting of Friction Contact

When friction contact is set, the flexible body is generally selected as the contact surface. In interference fit, the hole has a radial deviation and produces interference. Therefore, when the friction contact is set in this paper, the hole is the contact surface and the shaft is the target surface. Friction contact is made between the armature and spring tube. The surface belonging to the spring tube in the contact pair is the target surface, as shown in Figure 3. While the surface belonging to the armature is the contact surface. The friction coefficient is 0.1 and the value of the interference is 0.012 mm. Tetrahedral mesh is used in the whole assembly during mesh division, in which the mesh size of the armature is 0.4 mm, the spring tube is 0.3 mm, the flapper is 0.2 mm, and the feedback spring is 0.2 mm, and the thin wall of the spring tube is sealed. The frictional contact of interference fit is a nonlinear problem. During the simulation analysis, the error limits of grid control are set to aggressive mechanical. Shape inspection can ensure the prediction of unit distortion in the process of large strain analysis, so as to improve the quality of the unit, element order is set to quadratic when the cells are divided, and the grid is divided by quadratic higher-order cells. For large deformation of the model, the stiffness matrix needs to be adjusted in multi-step iterations to adapt to the influence of stress hardening, and the large deflection in solver controls is set to on during the analysis setting, which illustrates the difference between prestressed modal analysis and conventional modal analysis. Augmented lagrange contact equation is used at the contact pair to perform forced contact coordination. In the simulation analysis, the static stress analysis is carried out first, then the modal analysis is carried out, and the results of the first-order mode to the fourth-order mode are analyzed.

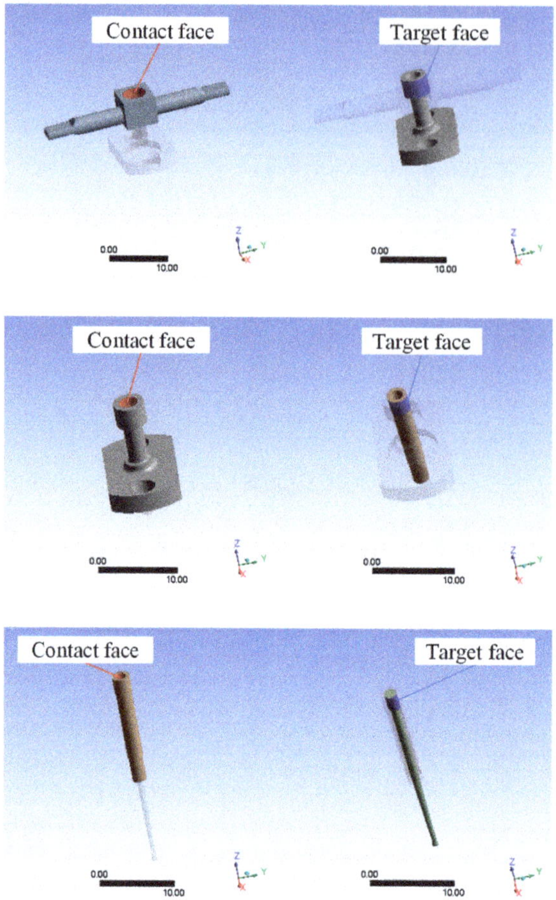


**Figure 3.** Contact setting.

### 3.3. Scheme C: Setting of Bonded Contacts with the Condition of Node Coupling

Node coupling refers to the sharing of grid nodes between different parts, ensuring a continuous grid. The shared topology function in Workbench allows for common node functionality across different parts but does not allow for contact between parts to be set. Therefore, node coupling is achieved through HyperMesh software 2021, and Scheme C is carried out through co-simulation of HyperMesh and Workbench software 2021R1.

(1) Modeling: The model is created using NX10.0 software and imported into HyperMesh software 2021 in stp format.
(2) Unit type and material properties definition: The unit type is set as solid186, and the material properties are set according to Table 1.
(3) Operation of armature assembly geometry model: Boolean operations are performed between the armature and spring tube, the spring tube and flapper, as well as the flapper and feedback spring. These operations serve a similar purpose as the shared topology function in Workbench.
(4) Tetrahedral mesh division is used throughout the component. The mesh size for the armature is set as 0.4 mm; spring tube mesh size is 0.3 mm with thin wall encryption applied; flapper mesh size is 0.2 mm; feedback spring mesh size is also 0.2 mm.

(5) Shared nodes separation: Using the detach function, boolean parts are separated into two so that shared nodes belong to two different parts while maintaining their original positions. This allows for node sharing while keeping both parts independent. If the contact surface nodes are not separated, they default to being part of a whole entity which prevents proper contact setting from being achieved, the meshed result is shown in Figure 2b.
(6) Format conversion: Mesh generated by HyperMesh software 2021 is exported then converted into Mechanical APDL Product format before importing it back into Workbench software 2021R1.
(7) The solver continues to utilize the Workbench software 2021R1. Bonded contact is employed for the interaction between the armature, spring tube, flapper, and feedback spring. A fixed constraint is applied to the bottom surface of the spring tube, and results of first-order mode to fourth-order mode are analyzed.

### 3.4. *Scheme D: Frictional Contact Under Node Coupling Conditions*

The node coupling treatment is the same as in (1)–(6) in Section 3.3. The frictional contact settings are the same as in Section 3.2, except for meshing.

## 4. Modal Analysis Results and Comparative Analysis

### 4.1. *Modal Analysis Results of Different Schemes*

The simulation analysis is carried out according to the four ABCD schemes, respectively, and the results of the first-order mode to the fourth-order mode are analyzed. The deformation direction of the same mode of different schemes is the same, but the natural frequencies are different.

Figure 4 shows the natural frequencies of the first to fourth modes of the armature assembly obtained by simulation analysis of different schemes. The figure shows that the results obtained by the second scheme are basically consistent. The results of the second mode are quite different. The natural frequencies of Schemes A and C are close to each other, and those of Schemes B and D are close to each other. The choice of contact type has a significant influence on the natural frequency of the armature assembly, while node coupling has little influence. Therefore, it is necessary to consider the interference fit when conducting a modal analysis of the interference fit armature assembly.

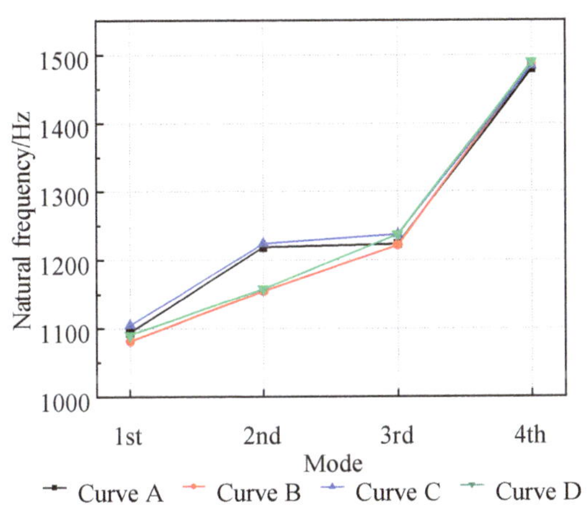


**Figure 4.** Calculation results.

## 4.2. Static Structural Analysis of Armature Assembly

In order to further explore the impact of the interference fit on modes of the armature assembly, static structure analysis was carried out according to Scheme B, and the specific parameters were set by referring to Section 3.2. The strain diagram and stress diagram of the armature assembly under interference fit conditions were obtained and shown in Figures 5 and 6, respectively. The figures show that under the action of the interference fit, there is prestress in the armature assembly, which leads to a certain amount of deformation, and the maximum value of deformation occurs at both ends of the armature, as shown in Figure 5. As shown in Figure 6, the stress of the armature assembly is mainly concentrated in the contact area of the armature assembly parts, the stress value is around 500 MPa, and the yield strength of the armature is 685 Mpa, indicating that the stress formed by the interference fit is relatively large. In addition, the maximum stress value is 1324.9 MPa. Although the value is relatively large, it appears at the corners and has little impact on the overall component, so it can be ignored.

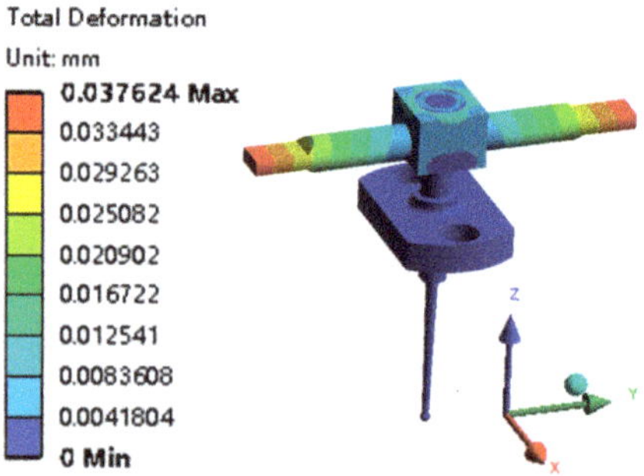


Figure 5. Deformation nephogram.

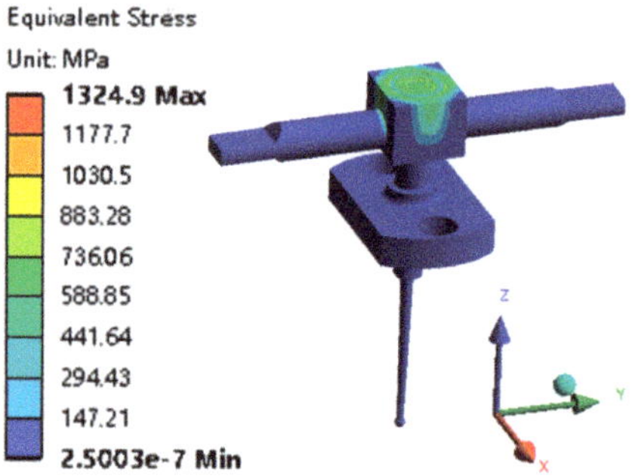


Figure 6. Stress nephogram.

## 4.3. Modal Analysis of Armature Assembly

After the structural statics analysis, the modes of the prestressed armature assembly were analyzed. The first four modes of the armature assembly were obtained, referring to Section 3.2. for parameter setting, as shown in Figure 7. The natural frequency of the first mode is 1081.7 Hz, and the deformation form is the swing of the armature and the feedback spring in the ZOY plane. The natural frequency of the second mode is 1154.8 Hz, and the deformation form is the rotation of the armature around the Z axis. The natural frequency of the third mode is 1222 Hz, and the deformation form is the swing of the feedback spring in the ZOX plane. The natural frequency of the fourth mode is 1486.1 Hz, and the deformation form is the swing of the feedback spring in the ZOY plane.

The natural frequency of the fifth mode is 2342 Hz, and the deformation form is the swing of the armature and the feedback spring in the ZOX plane. The natural frequency of the sixth mode is 3809.3 Hz, and the deformation form is a translation of the armature in the ZOY plane and the swing of the feedback spring in the ZOY plane.

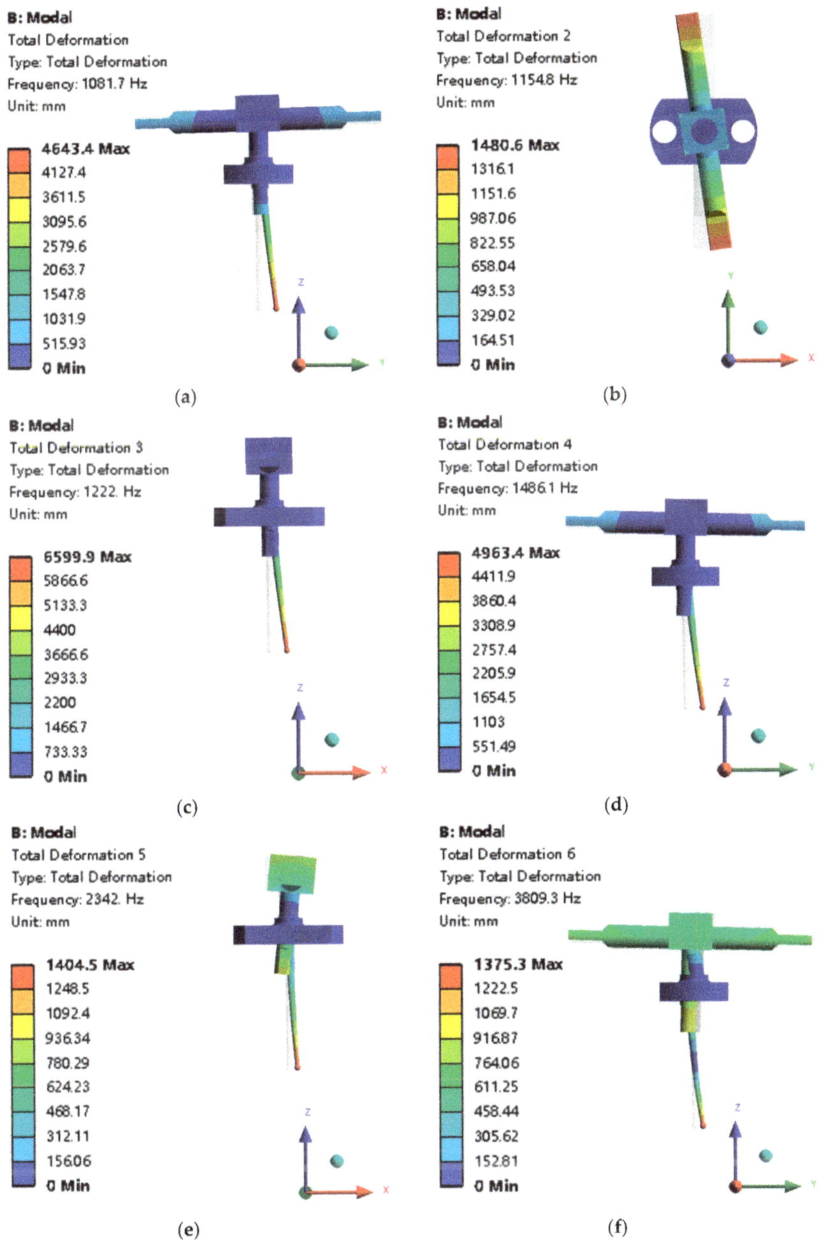


**Figure 7.** Modal analysis results under interference fit. (**a**) First-order vibration mode. (**b**) Second-order vibration mode. (**c**) Third-order vibration mode. (**d**) Fourth-order vibration mode. (**e**) Fifth-order vibration mode. (**f**) Sixth-order vibration mode.

### 4.4. Verification of Modal Experiments

The mode test of the conventional parts is carried out using the hammering method [9], but the size of the armature assembly is small and precise, so the hammering method cannot be used, and a softer electric driving method is used in combination with its actual working characteristics. In order to verify the results of the simulation analysis, the armature assembly is installed in the torque motor, and the modes of the armature assembly are tested. The principle of the test system is shown in Figure 8a. The signal generator provides the input signal of the sweeping frequency to the torque motor. Driven by the input signal, the armature assembly has a bias, and the displacement of the bias will be detected by the laser displacement sensor. In order to detect the amplitude of the armature deflection, a plastic sheet is pasted on the end of the armature. When the frequency of the sweep signal is close to the natural frequency of the armature assembly, the armature assembly has a resonance phenomenon, and the vibration amplitude of the armature assembly increases significantly. Therefore, the natural frequency of the armature assembly can be identified by detecting the displacement of the armature assembly under the sweep signal.

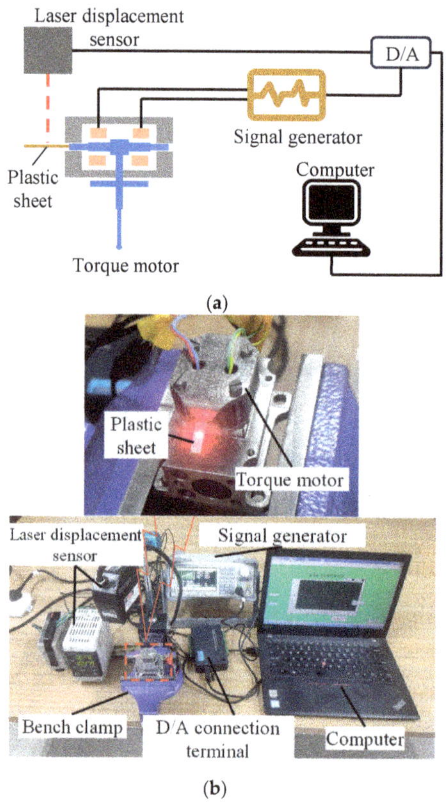


**Figure 8.** Modal test of armature assembly. (**a**) Schematic diagram of the experimental system. (**b**) Data acquisition system.

The measured armature assembly is a separate armature assembly. The interference between the armature and the spring tube is 0.012 mm, the interference between the spring tube and the flapper is 0.015 mm, and the interference between the flapper and the feedback spring is 0.01 mm. The actual interference is the same as that in the simulation model. As shown in Figure 8b, the two coils in the torque motor are powered in parallel, the resistance of the coils in parallel is 25 Ω, and the rated current is 40 mA. The input signal is the frequency sweep signal provided by the signal generator. The voltage of the frequency sweep signal is 0 to 1 V, the frequency sweep time is 10 s, and the frequency sweep range is 1 to 4000 Hz. The model of the signal generator is SIGLENT, SDG 1022X, manufactured by Shenzhen Siglent Technology Co., Ltd., China, the sampling frequency is 150 MSa/s, and the maximum frequency of the signal generator is 25 MHz. The torque motor is installed on the valve body, and the valve body is fixed by a vise. The probe of the laser displacement sensor is adsorbed on the vice through the magnetic base. The model of the laser displacement sensor is KEYENCE LK-G150, made in Japan, the sampling rate is 100 µs, and the repetition accuracy is 0.5 µm. The D/A connection terminal is mainly used to collect data and transfer it to the computer. The model of the D/A connection terminal is ADVANTECH USB-4716, and the sampling frequency is 200 kS/s. A computer is used to observe, record, and analyze data.

The voltage data detected by the laser displacement sensor in the test is converted into displacement data for the Fourier transform as the longitudinal coordinate, and the frequency as the horizontal coordinate, as shown in Figure 9. The extreme values appear when the frequency is 991 Hz and 1572.4 Hz, indicating that the resonance phenomenon occurs in the armature assembly under the action of this frequency, and the frequency value can be considered as the natural frequency of the mode. As shown in Figure 8, the displacement detected by the laser displacement sensor is the deflection of the armature end in the ZOY plane, corresponding to the deflection direction of the armature in Figure 7a,d. The signal frequency ranges from 0 to 4000 Hz, and the natural frequency corresponding to the first six modes of the armature assembly is 3809.3 Hz, the highest natural frequency of the sixth mode. Thus, the two extremes correspond to the natural frequencies of the first and fourth modes, respectively.

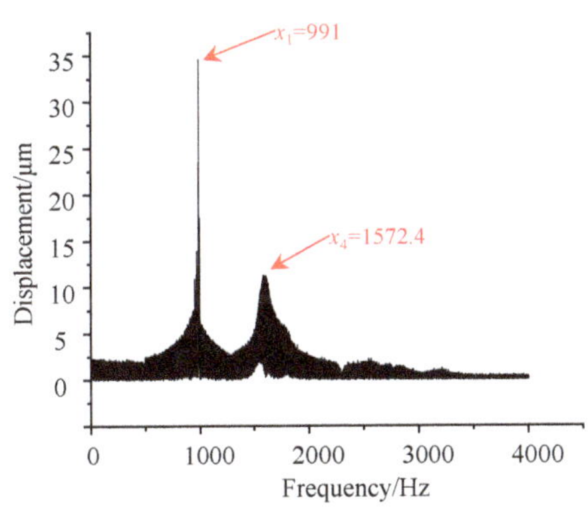


**Figure 9.** Frequency domain analysis of armature end displacement under sweep signal.

The natural frequency value obtained from the test is compared with the results obtained from the simulation, as shown in Tables 2 and 3. As shown in Table 2, the difference between the measured natural frequency of the first mode of the armature assembly and the simulation results is about 10%, among which the difference in Scheme B is the smallest, which is 9.15%. As shown in Table 3, the difference between the measured natural frequency of the fourth mode of the armature assembly and the simulation results is about 5.5%, in which the difference is the smallest in Scheme D, which is 5.28%. From the overall data, the error between the simulation results and the experimental results is less than 12%, indicating that the simulation results have a certain reliability, and the error may be caused by the plastic sheet at the top of the armature which is easy to measure. On the other hand, because the simulation results are close, the experimental results cannot judge which scheme is best, so further study is required.

**Table 2.** Simulation and experimental results analysis of first-order modes.

| Simulation Plan | Plan A | Plan B | Plan C | Plan D |
|---|---|---|---|---|
| Calculated f1 (Hz) | 1093.8 | 1081.7 | 1104.7 | 1090.4 |
| Measured f1 (Hz) | 991 | 991 | 991 | 991 |
| Error (%) | 10.37% | 9.15% | 11.47% | 10.03% |

**Table 3.** Simulation and experimental results analysis of fourth-order modes.

| Simulation Plan | Plan A | Plan B | Plan C | Plan D |
|---|---|---|---|---|
| Calculated f4 (Hz) | 1479.2 | 1486.1 | 1483.6 | 1489.4 |
| Measured f4 (Hz) | 1572.4 | 1572.4 | 1572.4 | 1572.4 |
| Error (%) | 5.93% | 5.49% | 5.65% | 5.28% |

### *4.5. Influence of Interference on Modal Analysis Results*

It can be seen from Figure 1 that the armature assembly is composed of armature, spring tube, flapper and feedback spring through interference fit. In order to explore the influence of interference on the modes of armature assembly, the modal analysis of armature assembly is carried out according to scheme B, with interference as a variable, as shown in Figure 10. The influence of interference between the armature and the spring tube on the modes of the armature assembly is shown in Figure 10a. The influence of interference between the spring tube and the flapper on the modes of the armature assembly is Figure 10b. The influence of the interference between the flapper and the feedback rod on the modes of the armature assembly is shown in Figure 10c. As can be seen from the figure, among the first four modes, the corresponding natural frequencies of the first, third and fourth modes all decrease with the increase in interference, but the change is little; while the corresponding natural frequencies of the second mode decrease with the increase in interference, and the change is more obvious. Therefore, the change in interference has an effect on the modes of armature assembly, and the bonded contact setting in scheme A and scheme C is unscientific.

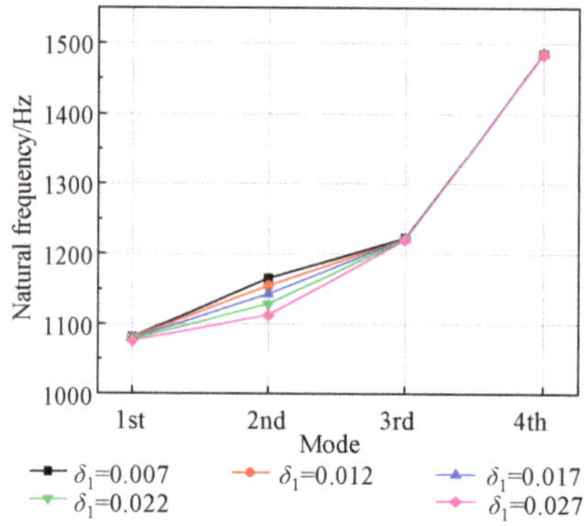


(**a**)

**Figure 10.** *Cont.*

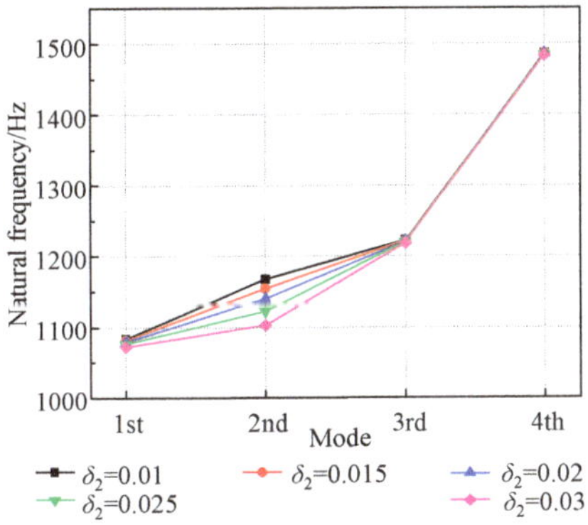


(**b**)

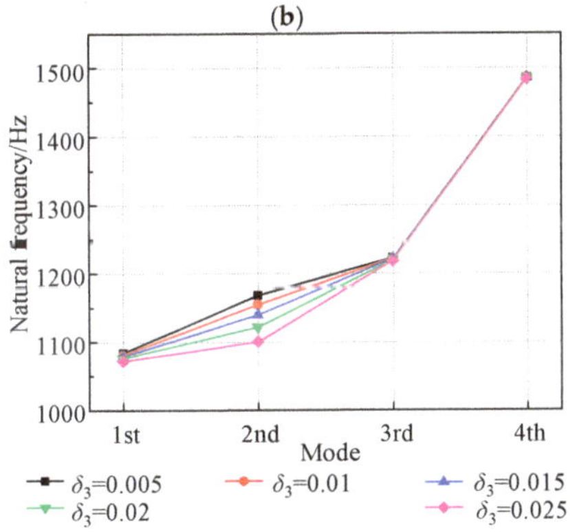


(**c**)

**Figure 10.** The influence of interference on the modes of armature assembly. (**a**) The influence of interference between the armature and the spring tube on the modes. (**b**) The influence of interference between the spring tube and the flapper on the modes. (**c**) The influence of interference between the flapper and the feedback spring on the modes.

In addition, Figure 4 shows that the simulation results of Scheme B and Scheme D are very close. Although common nodes on the contact surface can make the simulation results more accurate, the mesh needs to be divided by HyperMesh software 2021 and the operation is very complicated. The deviation of the calculation results between the two is negligible, so Scheme B is the optimal scheme for the modal analysis of armature assembly.

## 5. Precautions for the Design of Servo Valves

In the case of either integral or separated armature assembly, the mechanism of self-excited oscillation is similar. The natural frequency of the armature assembly is coupled with the pressure pulsation in the pilot stage flow field. The problem of internal stress caused by the interference fit of the separated armature assembly is relatively prominent, which cannot be simplified as the integral armature assembly.

The stiffness of the armature assembly is critical to the function and performance of the servo valve and must be prioritized as the primary consideration. In terms of vibration characteristics, the vibration source is mainly the pressure pulsation of the pilot valve, it is recommended to adopt the following schemes to reduce the probability of self-excited oscillation of the servo valve.

(1) For a separated armature assembly, while ensuring compliance with the assembly's stiffness requirements, the natural frequency of the armature assembly can be altered by modifying the structure or adjusting the interference fit.
(2) While ensuring the fundamental functionality of the pilot stage of the servo valve, it is possible to modify the flow field structure to alter the frequency and amplitude of pressure pulsations. This adjustment helps to prevent the coupling of the pressure pulsation frequency with the natural frequency of the armature assembly, thereby reducing the likelihood of self-excited oscillations in the servo valve and enhancing its overall performance.

## 6. Conclusions

In this paper, the vibration characteristics of the separated armature assembly based on the interference fit are studied. Four different simulation methods are used to analyze the modes of the armature assembly, and the feasibility of the simulation analysis is verified by experiments. The following conclusions were obtained:

(1) Due to the interference connection between the parts of the separated armature assembly, prestress is present and cannot be ignored. The method of bonded contact simplification in modal simulation analysis is unscientific. Whether the contact surface has common nodes or not has little influence on the modal simulation results.
(2) The electric drive method is used to provide sweeping frequency signals to the armature assembly, and the natural frequencies of first-order and fourth-order modes of the armature assembly are obtained by the resonance principle, which verifies the feasibility of the simulation.
(3) Interference has an effect on the modes of the armature assembly. With the increase in interference, the natural frequency corresponding to the first, third, and fourth modes of the armature assembly, all decrease with increasing interference, but the change is little, while the natural frequency corresponding to the second mode gradually decreases.
(4) In the design of a servo valve with a separable armature assembly, while ensuring the fundamental functionality of the servo valve, it is possible to enhance its performance through two approaches. On the one hand, the natural frequency of the armature assembly can be modified by altering its structure or adjusting the interference fit. On the other hand, the frequency and amplitude of pressure pulsations can be changed by modifying the flow field structure. These adjustments help to prevent the coupling of the pressure pulsation frequency with the natural frequency of the armature assembly, thereby reducing the likelihood of self-excited oscillations in the servo valve and improving its overall performance.

**Author Contributions:** Conceptualization, T.L., S.L., J.Z. and A.Z.; methodology, software, writing—original draft preparation, T.L. and S.L.; formal analysis, J.P.; investigation, T.L.; resources, A.Z.; writing—review and editing, J.P. and J.Z.; supervision, J.P. All authors have read and agreed to the published version of the manuscript.

**Funding:** This research was funded by NSFC, grant number 51675119.

**Data Availability Statement:** The data presented in this study are available on request from the corresponding author.

**Conflicts of Interest:** Aiying Zhang is employed by the company Qingan Group Co., Ltd. The remaining authors declare that the research was conducted in the absence of any commercial or financial relationships that could be construed as a potential conflict of interest.

## References


1. Aung, N.Z.; Li, S. A numerical study of cavitation phenomenon in a flapper-nozzle pilot stage of an electrohydraulic servo-valve with an innovative flapper shape. *Energy Convers. Manag.* **2014**, *77*, 31–39. [CrossRef]
2. Zhang, S.; Aung, N.Z.; Li, S. Reduction of undesired lateral forces acting on the flapper of a flapper–nozzle pilot valve by using an innovative flapper shape. *Energy Convers. Manag.* **2015**, *106*, 835–848. [CrossRef]
3. Glaun, A. Avoiding flow-induced Sympathetic vibration in control valves. *Power* **2012**, *156*, 80–83.
4. Peng, J.; Li, S.; Han, H. Damping properties for vibration suppression in electrohydraulic servo-valve torque motor using magnetic fluid. *Appl. Phys. Lett.* **2014**, *104*, 171905. [CrossRef]
5. Zhang, Y.; Tang, J.; Wen, T. A Modified Transfer Matrix Method for Modal Analysis of Stepped Rotor Assembly Applied in the Turbomolecular Pump. *Shock. Vib.* **2022**, *2022*, 3692081. [CrossRef]
6. Kim, K.K.; Lee, Y.S. Modal characteristics according to the tip shape and assembly condition of the turbine blade. *J. Mech. Sci. Technol.* **2013**, *27*, 3409–3417. [CrossRef]
7. Zhai, L.; Han, B.; Zhang, Y.; Ye, X.; Sun, M. A Modal Analysis Method for Turbomolecular Pump Rotor Assembly with Separable Thrust Disk. In Proceedings of the 2018 21st International Conference on Electrical Machines and Systems (ICEMS), Jeju, Republic of Korea, 7–10 October 2018; pp. 2561–2564. [CrossRef]
8. Daouk, S.; Louf, F.; Cluzel, C.; Dorival, O.; Champaney, L. Variability of a bolted assembly through an experimental modal analysis. In *Model Validation and Uncertainty Quantification, Volume 3: Proceedings of the 33rd IMAC, A Conference and Exposition on Structural Dynamics, 2015*; Springer International Publishing: Cham, Switzerland, 2015; pp. 171–178. [CrossRef]
9. Yang, B.; Li, D.; Yang, H.; Hu, Y.; Yang, P. Vibrational fatigue and reliability of package-on-package stacked chip assembly. *Microelectron. J.* **2019**, *92*, 104609. [CrossRef]
10. Xiao, H.; Zhang, D.W.; Shi, Y.G.; Sun, Z.H. Research on the Effect of Pre-Stress on Bending Stiffness of the Ball Screw Based on ANSYS. *Adv. Mater. Res.* **2013**, *690–693*, 1784–1787. [CrossRef]
11. Feron, J.; Latteur, P. Implementation and propagation of prestress forces in pin-jointed and tensegrity structures. *Eng. Struct.* **2023**, *289*, 116152. [CrossRef]
12. Yang, H.; Xu, J.; Wang, G.; Yang, Z.; Li, Q. Prestress Modal Analysis and Optimization of Cantilever Supported Rotor under the Unbalanced Axis Force and Moving Mass. *Appl. Sci.* **2022**, *12*, 4940. [CrossRef]
13. Orlowska, A.; Graczykowski, C.; Galezia, A. The effect of prestress force magnitude on the natural bending frequencies of the eccentrically prestressed glass fibre reinforced polymer composite beams. *J. Compos. Mater.* **2018**, *52*, 2115–2128. [CrossRef]
14. Hu, Y.; Xie, Y.; Peng, R.; Zhang, S. Modal Analysis of Rail Transit Gearbox. In *IOP Conference Series: Earth and Environmental Science*; IOP Publishing: Bristol, UK, 2019; Volume 267, p. 042055. [CrossRef]
15. Wang, Y. Lightweight steering equipment based on prestressed modal analysis. *J. Vibroeng.* **2024**, *26*, 370–382. [CrossRef]
16. Li, T.; Li, M.; Zhou, L.; Wang, H.; Hu, W.; Zhang, X. *Mechanical Property and Sealing Performance Analysis of the Diaphragm Compressor Cylinder Head for the Hydrogen Refueling Station Under the Mutation Impact Load*; Tech Science Press: Henderson, NV, USA, 2023. [CrossRef]
17. Liu, F.; Gong, R.; Duan, Z.; Wang, Z.; Zhou, J. Research on vibration reliability of solder joint based on modal experiment of PCBA. *J. Mater. Sci. Mater. Electron.* **2025**, *36*, 43. [CrossRef]
18. Huang, Z.; Fang, J.; Wang, K. Dynamics of flexible rotor-shaft assembly with consideration of contact behaviors. *Proc. Inst. Mech. Eng. Part C J. Mech. Eng. Sci.* **2015**, *229*, 859–868. [CrossRef]
19. Liu, G.; Hong, J.; Wu, W.; Sun, Y. Investigation on the influence of interference fit on the static and dynamic characteristics of spindle system. *Int. J. Adv. Manuf. Technol.* **2018**, *99*, 1953–1966. [CrossRef]
20. Fan, Y.; Ding, H.; Li, M.; Li, J. Modal Analysis of a Thick-Disk Rotor with Interference Fit Using Finite Element Method. *Math. Probl. Eng.* **2018**, *2018*, 5021245. [CrossRef]
21. Ce, R.; He, Z.; Li, D.; Yang, Z.; Xue, G. Dynamic Modeling and Analysis of Stack Giant Magnetostrictive Actuator. *Sens. Actuators A Phys.* **2018**, *276*, 205–218. [CrossRef]
22. Peng, J.; Fan, Y.; Li, S. Modeling of armature-flapper assembly in a hydraulic servo valve by finite element analysis and acoustic excitation measurements. In Proceedings of the 2015 International Conference on Fluid Power and Mechatronics (FPM), Harbin, China, 5–7 August 2015; pp. 301–306. [CrossRef]
23. Peng, J.; Zhang, Y.; Li, S.; Bao, W.; Tanaka, Y. Identification Algorithm and Improvement of Modal Damping Ratios for Armature Assembly in a Hydraulic Servo-Valve with Magnetic Fluid. *Energies* **2023**, *16*, 3419. [CrossRef]

actuators

MDPI

Article

# The Design of a Turning Tool Based on a Self-Sensing Giant Magnetostrictive Actuator

Dongjian Xie [1,2], Qibo Wu [1], Yahui Zhang [1], Yikun Yang [1], Bintang Yang [1,*] and Cheng Zhang [3,*]

1 State Key Laboratory of Mechanical System and Vibration, School of Mechanical Engineering, Shanghai Jiao Tong University, Shanghai 200240, China; xie_dongjian@163.com (D.X.); bogerger@sjtu.edu.cn (Q.W.); zhangyh92@sjtu.edu.cn (Y.Z.); yyk199164@sjtu.edu.cn (Y.Y.)
2 Aerospace System Engineering Shanghai, Shanghai 201109, China
3 Facility Design and Instrumentation Institute, China Aerodynamics Research and Development Center, Mianyang 621000, China
* Correspondence: btyang@sjtu.edu.cn (B.Y.); carloz@foxmail.com (C.Z.)

**Abstract:** Smart tools are limited by actuation–sensing integration and structural redundancy, making it difficult to achieve compactness, ultra-precision feed, and immediate feedback. This paper proposes a self-sensing giant magnetostrictive actuator-based turning tool (SSGMT), which enables simultaneous actuation and output sensing without external sensors. A multi-objective optimization model is first established to determine the key design parameters of the SSGMT to improve magnetic transfer efficiency, system compactness, and sensing signal quality. Then, a dynamic hysteresis model with a Hammerstein structure is developed to capture its nonlinear characteristics. To ensure accurate positioning and a robust response, a hybrid control strategy combining feedforward compensation and adaptive feedback is implemented. The SSGMT is experimentally validated through a series of tests including self-sensing displacement accuracy and trajectory tracking under various frequencies and temperatures. The prototype achieves nanometer-level resolution, stable output, and precise tracking across different operating conditions. These results confirm the feasibility and effectiveness of integrating actuation and sensing in one structure, providing a promising solution for the application of smart turning tools.

**Keywords:** giant magnetostrictive material; self-sensing actuator; magnetic–mechanical–thermal multi-physical field coupling; precision actuation; sensing technology

## 1. Introduction

Machining is widely used in the final molding of metal or non-metal parts in industry. With the continuous improvement in the complexity and geometric accuracy of design products, the application of new materials, and the shortening of product life cycle, the level of intelligence and automation of manufacturing equipment is facing major challenges [1]. With the development of technology, new mechatronics integrated in intelligent manufacturing equipment have gradually developed to meet the needs of the times. Compared with ordinary manufacturing equipment, intelligent manufacturing equipment is different in that it integrates different types of actuators and sensors to improve the equipment's sensitivity to and control of the machining process, as well as ensuring machining accuracy and machining system robustness [1–3].

As the appendage of manufacturing equipment, tools are also being developed to achieve precision and intelligence in cutting processing. Based on traditional tools, intelligent tools are integrated with precision drive and sensing technologies [4,5]. To ensure cutting functions, intelligent tools can achieve real-time adjustment of machining parameters [6,7], online monitoring of their own working conditions [8], and communication and interaction with numerical control systems [5].

Limited by the narrow installation space and specific design performance requirements, intelligent tools have specific requirements for the appearance of the driver geometry, output scale, operating frequency, and output power. Electromagnetic actuators are widely used in the field of intelligent tools, such as the voice coil motor driven by Lorentz force [9] and the micro solenoid actuator driven by Maxwell force [10]. In the process of miniaturization, the physical properties of the electromagnetic drive and the key structural guiding elements are prone to adverse scaling effects, and as such, the output force will be proportional to the fourth square of the structural size [11]. Drivers based on smart materials are widely used in the field of smart tools because of their superior performance in response speed, output energy, robustness, and easy control. Currently, piezoelectric materials (PZTs) [11], magnetorheological liquids (MRFs) [12], shape memory alloys (SMAs) [13], and giant magnetostrictive materials (GMMs) [14] are widely used.

The typical application of the integration of drive and sensing functions in smart tools is to use the sensor with the driver to assist the cutting tool for machining, or to install the cutting tool directly on the output end of the driver for machining. A representative of this class of applications is the fast tool servo (FST) system. FST systems play an important role in precision turning of micro-features or free-form surfaces of diamond tools, with common structures including piezoelectric [15] or voice coil drivers [16], flexure hinges, and position feedback sensors. In addition, in order to adapt to more complex design requirements and machining requirements of different drive frequencies, normal stress electromagnetic drives are also gradually applied to FST systems. Huang et al. [17] proposed a normal stress electromagnetic-driven FST system, in which flexible hinges were used to amplify the output displacement of the electromagnetic driver, and capacitive displacement sensors were installed inside the FST system to detect the output displacement. By modifying the conventional manufacturing device, the intelligent tool can further expand the process performance of the device on the basis of retaining the original processing capacity. Yoshioka et al. [18] integrated a giant magnetostrictive actuator (GMA) inside a milling machine spindle to process complex surfaces and fine patterns alongside conventional milling. However, in current smart tools, the drive and external sensors function as separate components, resulting in a bulky structure and complex installation, along with calibration errors and inconsistent responses due to diverse sensing mechanisms and vibration coupling. This study aims to fundamentally address issues of structural redundancy, matching difficulties, and vibration interference by integrating actuation and self-sensing capabilities into a single device, thereby enhancing the precision machining performance of smart tools.

To address the above challenges, this paper presents a self-sensing giant magnetostrictive actuator-based turning tool (SSGMT), in which both actuation and displacement sensing are realized within a unified structure. A coupled multi-physical model is developed to characterize the magneto–mechanical–thermal behavior of the actuator, and a Hammerstein-like dynamic hysteresis model is introduced to capture its nonlinear input–output characteristics. Based on this model, a constrained multi-objective optimization framework is established, balancing energy efficiency, magnetic field strength, mechanical stiffness, and system mass. The key structural parameters are optimized accordingly, and a prototype of the SSGMT is fabricated for subsequent validation. In addition, a hybrid control scheme combining inverse feedforward compensation and adaptive feedback is designed to suppress nonlinear hysteresis and enhance trajectory tracking accuracy.

The remainder of this paper is organized as follows. Section 2 introduces the working principle and mechanical structure of the SSGMT. Section 3 presents the multi-objective optimization model for key structural parameters and its solution. Section 4 describes the design of the nonlinear dynamic model and the composite control strategy. Section 5 reports the experimental validation results including displacement resolution, self-sensing accuracy, and tracking performance under various thermal and dynamic conditions. Finally, Section 6 summarizes the main conclusions and discusses future prospects.

## 2. Working Principle and Structure Scheme

The self-sensing giant magnetostrictive actuator (SSGMA) proposed in this paper has the ability to achieve linear feed with micro- and nano-level precision. Due to the magnetic–mechanic–thermal coupling characteristics of giant magnetostrictive materials, the SSGMA can detect its own output displacement and force during execution and sense the change in working environment temperature [19,20]. Based on detailed research on the characteristics of the SSGMA, the multi-physical self-sensing actuator is presented. In order to apply the SSGMA mechanism to practical engineering, this section proposes and designs a self-sensing giant magnetostrictive actuator-based turning tool (SSGMT) based on the SSGMA. Figure 1 shows the mechanical structure diagram of the designed SSGMT. The external structure of the SSGMT is shown in Figure 1a, which includes three parts: fixed parts, executive parts and machining tools. Among them, the fixed parts are responsible for the connection with the lathe chuck, and the executive parts are connected through the pin; the processing tool is fixed on the executive parts by a special screw, and the specific model of the processing tool needs to be selected according to the actual processing needs.

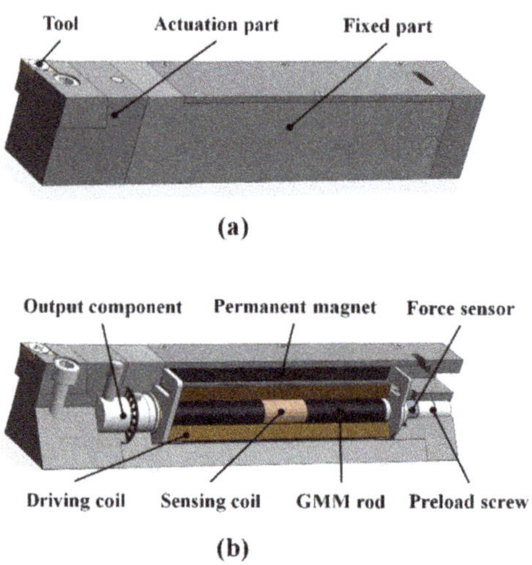


**Figure 1.** Mechanical structure of SSGMT: (**a**) external components; (**b**) internal structure.

The internal structure of the fixed component is shown in Figure 1b. The internal structure of the giant magnetostrictive self-sensing driver is the same as that of the giant magnetostrictive rod. The sensor coil is installed in the middle of the giant magnetostrictive rod, and the two are completely wrapped inside the drive coil. A magnetic yoke is placed around the driving coil to form a complete closed magnetic circuit with a permanent magnet that provides a biased magnetic field, so that the magnetic field evenly passes through the giant magnetostrictive rod. At the end of the fixed part, the preload nut fixes the force sensor to the rear end of the giant magnetostrictive rod. By adjusting the preload nut, the prestress applied to the giant magnetostrictive rod can be adjusted. The force sensor is used to detect the magnitude of the prestress and at the same time to monitor the load borne during SSGMT operation in order to calibrate the self-sensing signal. The front end of the giant magnetostrictive rod is equipped with an output component, which is connected with the actuator through a pin, and a guide ring is placed around the output component to reduce non-axial movement during execution. In the actual working process, the external magnetic field of the giant magnetostrictive rod is provided by the permanent magnet and the drive coil, and the sensing signal is extracted from the sensor coil in real time.

## 3. Optimization of Structural Parameters

### *3.1. Parameter Analysis*

For the actual engineering application scenarios of lathe cutting, the optimization design principles for the SSGMT should include the following: (1) reducing energy consumption, mainly the loss of magnetic energy, to ensure that the output of the GMM rod provides a uniform and stable driving magnetic field; (2) improving the quality factor of the self-sensing signal to ensure that it can effectively perceive its own state while stabilizing the output; (3) on the premise of ensuring the strength and stiffness of the mechanical system, reducing the size and quality of the mechanical structure, on the one hand, to adapt to the use of the actual machine tool space and, on the other hand, to avoid bringing too much additional redundancy to the machine tool and affecting its effective function. This section will optimize the design of key components under the premise of ensuring SSGMT output capability.

For the SSGMT, the analysis of the whole drive system from the perspective of energy flow includes three key links, namely, the electromagnetic conversion link, magnetic energy transfer link, and electromechanical output link, as shown in Figure 2. In the electromagnetic conversion process, the electric energy input to the system is converted into magnetic energy through the drive coil, so the parameters of the drive coil directly affect the efficiency of electromagnetic conversion. The energy lost in this process is mainly due to coil magnetic leakage loss and heating loss. In the magnetic energy transfer link, the magnetic energy is transferred in the magnetic circuit to magnetize the GMM rod, so the eddy current loss caused by the magnetic yoke in this link is related. Finally, the magnetic energy is converted into mechanical energy, and the displacement and force output by the GMM rod are transferred to the outside of the system through the mechanical structure to carry out work. Therefore, the energy lost in this process is related to the magneto conversion efficiency of the GMM rod, and the size of the GMM rod and the equivalent stiffness of the associated components will also affect the output performance of the SSGMT. In addition, some energy is converted into inductive electrical signals through the sensing coil, which directly affects the signal quality of the self-sensing signal; so, it is also necessary to optimize the sensor coil design.

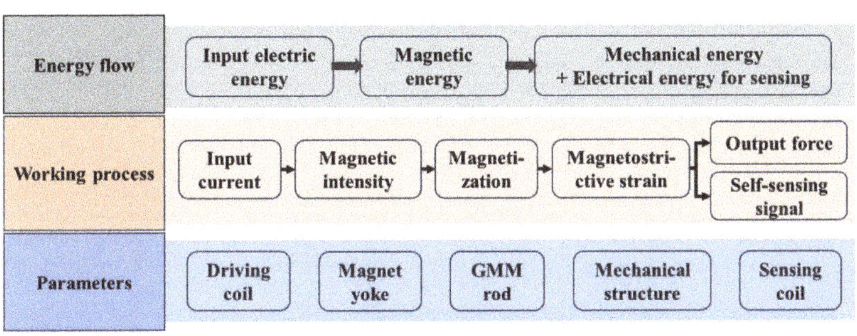


**Figure 2.** A schematic diagram of the energy flow during the operation of the SSGMT.

In order to build an optimal design model, theoretical analysis and parameter modeling are carried out for the key links involved in the above SSGMT working process:

I. Drive coil energy loss analysis

In the electromagnetic conversion process of the SSGMT, the drive coil mainly converts external input electric energy into magnetic energy. The system's magnetic circuit which forms part of the drive process is shown in Figure 3a. The total magnetic flux in the system $\varnothing_{all}$ is generated by the joint action of the drive coil and permanent magnet and transmitted through the magnetic circuit formed by the magnetic yoke; part of the magnetic flux $\varnothing_{rod}$ passes through the GMM rod. The GMM rod is magnetized and deformed, and part of $\varnothing_{leak}$ is lost as the magnetic leakage flux of the driving coil. In addition, there is a certain degree of magnetic leakage in the air, which is ignored in the analysis due to its small magnitude. According to the above process, the equivalent magnetic circuit model is established, as shown in Figure 3b. The number of turns and the input current of the drive coil are $N_{ca}$ and $I_c$, the coercive magnetic field and length of the permanent magnet are $H_p$ and $l_p$, and the magnetic leakage impedances of the drive coil, GMM rod, and magnetic circuit are $R_{leak}$, $R_{rod}$, and $R_{path}$, respectively. In order to increase the magnetic flux through the GMM rod, the value of $R_{leak}$ should be increased as much as possible, which also means that the magnetic leakage inductance of the drive coil should be reduced as much as possible.

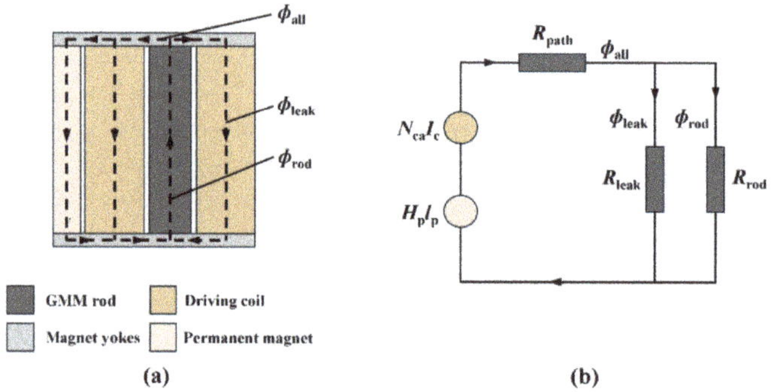


**Figure 3.** (**a**) Magnetic circuit diagram and (**b**) equivalent magnetic circuit model of driving process for SSGMT.

The driving coil is in the form of a conventional solenoid coil. The inner diameter of the coil is $r_{ca1}$, the outer diameter is $r_{ca2}$, the length is $l_{ca}$, and the radius of the enamelled wire composed of the driving coil is $d_{ca}$. The GMM rod is placed inside the drive coil with a radius of $r_m$ and a length of $l_m$. The number of turns $N_{ca}$ of the drive coil can be calculated directly from the size parameter of the coil, which is specifically expressed as follows:

$$N_{ca} = \frac{l_{ca}(r_{ca2} - r_{ca1})}{\eta_{c1}\eta_{c2}d_{ca}^2} \tag{1}$$

$\eta_{c1}$ is the axial winding coefficient of the solenoid coil, which is usually 1.05. $\eta_{c2}$ is the radial winding coefficient of the solenoid coil, usually 1.15. When the input current is $I_c$, the input magnetic field $H_c$ provided by the $N_{ca}$ turn drive coil can be expressed as follows [21]:

$$H_c = G_{ca}N_{ca}I_c\sqrt{\frac{\pi(\alpha_{ca} + 1)}{l_{ca}r_{ca1}(\alpha_{ca} - 1)}} \tag{2}$$

where $G_{ca}$ is the geometric shape parameter of the drive coil, which can be expressed as follows:

$$G_{ca} = \frac{1}{5}\sqrt{\frac{2\pi\beta_{ca}}{\alpha_{ca}^2 - 1}}\ln\left(\frac{\alpha_{ca} + \sqrt{\alpha_{ca}^2 + \beta_{ca}^2}}{1 + \sqrt{1 + \beta_{ca}^2}}\right) \tag{3}$$

In the above formula, $\alpha_{ca} = r_{ca2}/r_{ca1}$ and $\beta_{ca} = l_{ca}/2r_{ca1}$; according to the Biot–Savart law, the magnetic leakage inductance $L_{ca-leak}$ of the drive coil with the above parameters can be calculated as follows:

$$L_{ca-leak} = \mu_0\pi^2 G_{ca}^2 N_{ca}^2\left[\frac{r_m(\gamma_{ca}^2 - 1)(\alpha_{ca} + 1)}{\gamma_{ca}(\alpha_{ca} - 1)} + \frac{r_{ca1}(\alpha_{ca} + 1)(\alpha_{ca} + 3)}{6}\right] \tag{4}$$

where $\gamma_{ca} = r_{ca1}/r_m$. For the design in this section, the inner diameter of the drive coil $r_{ca1}$ is approximately $r_m$. Therefore, through optimal design of the structure parameters of the drive coil, the output magnetic field strength is stronger and the magnetic leakage loss is smaller under the same input current. According to the magnetic energy theory [22], the loss of the drive coil due to magnetic leakage per unit time can be calculated as follows:

$$W_{cl} = \frac{1}{2}I_c^2 L_{ca-leak} \tag{5}$$

In addition to magnetic leakage loss, coil loss due to heat also has a huge impact. In actual work, the current input to the SSGMT drive coil is composed of two parts: one is the low-frequency and high-amplitude current used to drive the GMM rod, and the other is the high-frequency and low-amplitude current used for sensing excitation. Therefore, considering the heat loss of the coil, the coil impedance should be considered. Among them, the static resistance $R_{c-dc}$ caused by the DC current and the AC resistance $R_{c-ac}$ caused by the AC current due to the "skin effect" are included. According to the literature [23], by ignoring the inductive reactance of the coil with less influence, the heating loss of the drive coil per unit time can be expressed as follows:

$$W_{ch} = I_c^2\sqrt{R_{c-dc}^2 + R_{c-ac}^2} \tag{6}$$

The static resistance $R_{c-dc}$ is related to the structural parameters of the drive coil and can be expressed as follows:

$$R_{c-dc} = \frac{N_{ca}^2\rho_{cw}\pi(\alpha_{ca} + 1)}{\lambda_c l_{ca}(\alpha_{ca} - 1)} \tag{7}$$

$\rho_{cw}$ is the winding resistivity of the copper coil, generally $1.72 \times 10^{-8}$ Ωm; $R_{c-dc}$ is the form factor of the copper coil, and $\pi/4$ is taken when the copper section is circular. The AC resistance $R_{c-ac}$ is related to the current frequency fc of the input drive coil and can be calculated as follows:

$$R_{c-ac} = \frac{l_{caw}\sqrt{\pi\mu_0\rho_{cw}f_c}}{\pi d_{ca}} \tag{8}$$

where $l_{caw}$ is the total length of the winding of the copper coil, which can be approximately expressed as follows:

$$l_{caw} = N_{ca}\pi(r_{ca1} + r_{ca2}) \tag{9}$$

Based on the above analysis, the energy loss caused by the drive coil in the electromagnetic conversion process can be expressed as follows:

$$W_{ca} = W_{cl} + W_{ch} \tag{10}$$

Therefore, on the premise of ensuring the output performance of the SSGMT, the coil parameters are optimized to reduce the value of Formula (10) and reduce the energy loss caused by driving the coil.

II. Drive magnetic field strength analysis

As shown in Figure 3b, after optimizing the magnetic flux leakage impedance caused by the coil, the functions of the magnetic yoke and GMM rod are mainly considered in the magnetic circuit, and the magnetic flux leakage caused by air is ignored. Therefore, the magnetic field strength acting on the GMM rod can be calculated as follows [23]:

$$H_{rod} = \frac{N_{ca}I_c + H_p l_p}{\mu_m A_m \left( \frac{l_m}{\mu_m A_m} + \frac{2}{\mu_{iron} t_y} \right)} \tag{11}$$

where $\mu_m$ represents the permeability of the GMM rod, $\mu_{iron}$ represents the permeability of the permeable yoke, and Am and ty represent the cross-sectional area of the GMM rod and the thickness of the permeable yoke, respectively. When the input current is constant, in order to ensure the output performance of the GMM rod, the lower limit of the magnetic field strength should be limited in the optimization design.

III. Energy loss analysis of the magnetic yoke

In order to ensure that the magnetic field generated by the drive coil and permanent magnet can effectively pass through the GMM rod, a permeable yoke is added to the design to form a complete loop. According to the literature [24], the loss caused by the yoke under the action of the sinusoidal excitation current mainly includes hysteresis loss and eddy current loss. The yoke loss is related to the properties of the material itself, and the determining factor is the overall quality. The eddy current loss caused by the greater mass will increase accordingly; so, the quality of the yoke components should be reduced in the design. The eddy current loss is related to the size of the permeability yoke. In order to ensure that the magnetic field passing through the GMM rod is uniform, an end yoke is arranged at both ends of the GMM rod. Therefore, the eddy current loss $W_{yl}$ caused by the yoke can be approximated as follows [25]:

$$W_{yl} = \frac{\pi^2 t_y^2 B_m^2 f_c}{6\rho_{iron}} \cdot \left( \frac{1}{2}\pi r_{ca2}^2 + 2r_{ca2}^2 + w_p h_p \right) t_y \tag{12}$$

where $\rho_{iron}$ is the resistivity of the material composed of the magnetic yoke, the material is permalloy, and the resistivity $\rho_{iron} = 0.1\ \mu\Omega\cdot\text{m}$; $f_c$ represents the current frequency, $t_y$ represents the thickness of the permeable yoke, and $w_p$ and $h_p$ represent the width and thickness of the permanent magnet, respectively. $B_m$ represents the peak value of magnetic induction intensity inside the permeability yoke, which mainly considers the influence of dynamic current. Specifically, it can be calculated according to the following formula [23]:

$$B_m = \frac{N_{ca}I_c}{\left( \frac{1}{2}\pi r_{ca2}^2 + 2r_{ca2}^2 + w_p h_p \right)\left( \frac{l_m}{\mu_m A_m} + \frac{2}{\mu_{iron} t_y} \right)} \tag{13}$$

Based on the above analysis, the optimal design of the structural size of the yoke ensures that the eddy current loss generated by the magnetic conductivity loop is reduced when the input current is constant, and the effective magnetic energy transferred to the GMM rod is increased.

IV. Analysis of sensor coil design parameters

The sensor coil is directly wound around the GMM rod as a single-layer spiral coil, and its structural parameters include the inner diameter of the coil $r_{cs1}$, the outer diameter

$r_{cs2}$, the length $l_{cs}$, and the radius $d_{cs}$ of the enameled wire forming the coil. As for the analysis of the action of the sensor coil, the induced voltage extracted by the sensor coil as a self-sensing signal is mainly affected by two key factors. One is the cross-sectional area $A_{cs}$ of the changing magnetic flux, which is related to the radius of the GMM rod:

$$A_{cs} = \pi r_m^2 \tag{14}$$

The other is the number of turns of the sensing coil $N_{cs}$; the more turns, the stronger the induction signal, which can be expressed as follows:

$$N_{cs} = \frac{l_{cs}(r_{cs2} - r_{cs1})}{\eta_{c1}\eta_{c2}d_{cs}^2} \tag{15}$$

In order to avoid affecting the layout inside the driver, the sensing coil is provided with only one layer, whose inner diameter $r_{cs1}$ is approximately $r_m$. Based on the above analysis, in order to ensure the strength of the self-sensing signal, the design parameters should be optimized to ensure that the number of turns should be reduced as far as possible on the basis of the design index to avoid the influence of the sensing coil on other parts.

V. Natural frequency analysis of the system

For the SSGMT execution system, its core component, the GMM rod, generally represents the output displacement under the coupling action of the driving magnetic field and the prestressed force exerted by the preloaded disc spring. Therefore, the natural frequency of the execution system can be approximately calculated as follows:

$$f_m = \frac{1}{2\pi}\sqrt{\frac{K_m}{m}} = \frac{1}{2\pi}\sqrt{\frac{k_m + k_{disc}}{m}} \tag{16}$$

where $k_m$ and $k_{disc}$ are the stiffness of the GMM rod and the prepressed disc spring, respectively, and $m$ is the overall mass of the execution system. The natural frequency of the core components of the execution system directly affects the upper limit of the drive frequency, so this value needs to be as large as possible to meet the design requirements. Since the stiffness of the GMM rod is related to material properties and external input, the key is to design the spring stiffness accordingly, and it is necessary to select a suitable spring model to meet the design size requirements.

VI. Quality analysis of key components

In order to improve the overall performance of the SSGMT, the overall quality of the system should be reduced as far as possible to ensure the use of its functions. According to the previous analysis, this is conducive to reducing the magnetic loss, but also a smaller volume can boost the overall efficiency of the operation. For the core output components of the tool, the mass of the SSGMT can be approximately considered to be equal to the total mass of the GMM rod, drive coil, sensing coil, and permeability yoke; so, it can be calculated as follows:

$$m = m_m + m_{ca} + m_{cs} + 2m_y \tag{17}$$

Among them, $m_m$ is the mass of the GMM rod, $m_{ca}$ is the mass of the drive coil, $m_{cs}$ is the quality of the sensing coil, and $m_y$ represents the quality of the magnetic yoke, which can be calculated as follows:

$$m_m = \rho_m \pi r_m^2 l_m \tag{18}$$

$$m_{ca} = \rho_c N_{ca} \frac{\pi^2 d_{ca}^2 (r_{ca1} + r_{ca2})}{2} \tag{19}$$

$$m_{cs} = \rho_c N_{cs} \frac{\pi^2 d_{cs}^2 (r_{cs1} + r_{cs2})}{2} \quad (20)$$

$$m_y = \rho_y \left[ \pi r_{ca2}^2 + 4r_{ca2}(r_{ca2} + t_p) \right] t_y \quad (21)$$

where $\rho_m$, $\rho_c$, and $\rho_y$ represent the densities of the GMM rod, copper coil, and magnetic yoke, respectively. One of the goals of the optimal design should be to reduce the total mass, i.e., the value of Equation (17).

### 3.2. Multi-Objective Parameter Optimization Model and Results

The SSGMT is designed to achieve precise feed self-sensing drive control, while taking into account the effects of vibration in the actual machining environment of the lathe, to minimize energy consumption, and to improve the quality of the system. Based on the analysis and modeling of the energy conversion and transfer links of the SSGMT, the following constrained multi-objective optimization design model is established:

$$\begin{cases} \max(H_{rod}, N_{cs}, f_m) \\ \min\left(W_{ca}, W_{yl}, m\right) \end{cases} \quad (22)$$
$$s.t.\ \textbf{\textit{Constaints I.}} \sim \textbf{\textit{VI.}}$$

There are 11 key parameters involved in the model, including the following: the radius $r_m$ of the GMM rod and its length $l_m$; the outer diameter of the drive coil $r_{ca2}$; the length $l_{ca}$ and the radius of the enamelled wire $d_{ca}$; the length of the permanent magnet $l_p$, as well as its width $w_p$ and thickness $h_p$; the length of the sensor coil $l_{cs}$; the radius of the enamelled wire $d_{cs}$; the thickness of the magnetic yoke $t_y$; and the stiffness of the disc spring $k_{disc}$.

Considering the actual use of the SSGMT and its size constraints, the selected boundary of the optimization design parameters can be preliminarily determined. In actual use, the SSGMT on one hand can inhibit the spindle vibration of the lathe, while on the one hand, it can carry out linear feed. Here, we set the output stroke of the SSGMT to greater than 100 μm in order to meet the above design requirements, and the GMM rod has a length $l_m$ of 100 mm. According to the literature [23], the length of the drive coil $l_{ca}$ affects the uniformity of the magnetic field passing through the GMM rod, so it should be greater than the length of the GMM rod. The outer diameter of the drive coil $r_{ca2}$ is limited by the installation size of the lathe cutter head. According to previous research in [20], in order to measure the self-sensing drive performance of the SSGMT, the magnetic field provided by the drive coil for a short time should reach 100 KA/m, and the bias magnetic field provided by the permanent magnet should exceed 20 KA/m to ensure the working performance of the SSGMT. Taking into account the assembly relationship, after the length of the drive coil lca and the outer diameter $r_{ca2}$ are determined, the size of the permanent magnet is also determined. The radius $d_{ca}$ of the drive coil enamelled wire is limited by the maximum drive current, and the radius $d_{cs}$ of the sensing coil enamelled wire is actually limited by the inner diameter of the drive coil. The stiffness disc of the disc spring is related to the assembly size, which can be determined by combining the radius $r_m$ of the GMM rod and the outer diameter $r_{ca2}$ of the coil. Therefore, the actual parameters that need to be optimized include $r_m$, $r_{ca2}$, $l_{ca}$, $d_{ca}$, $l_{cs}$, and $t_y$.

The solution of (22) is actually to solve a constrained multi-objective optimization problem. Multi-objective optimization refers to the realization of multiple objectives in a specific design scenario, but in the actual application scenario, there are mutual constraints between each objective, so the optimization of all objectives cannot be achieved at the same time. When one of the goals is optimized, it will affect the achievement of the other goals, so it is difficult to find a single optimal solution. Therefore, the purpose of optimal design is to find a compromise among all goals, so that the overall solution achieves the best possible outcome for all goals. At present, there are many related optimization algorithms, including the genetic algorithm, the particle swarm optimization algorithm, the simulated annealing algorithm, and so on [26]. In this section, the particle swarm optimization (PSO) algorithm is used to solve the multi-objective optimization problem. The PSO algorithm is an iterative optimization algorithm inspired by bird predation behavior. The PSO algorithm first generates several random particles and then iteratively searches for the optimal solution. Each particle updates itself by tracking individual extreme values and global extreme values, and finally achieves global target optimization [26]. Based on the PSO algorithm, Matlab optimization toolbox solution is adopted, and the final optimization results are shown in Table 1.

**Table 1.** Optimization results of SSGMT design parameters.

| Parameter | Value Range | Result | Selected Value |
|---|---|---|---|
| $r_m$ (mm) | [4, 8] | 4.6 | 5 |
| $r_{a2}$ (mm) | [8, 15] | 10.7 | 11 |
| $l_{ca}$ (mm) | [100, 110] | 104.6 | 105 |
| $d_{ca}$ (mm) | [0.25, 0.5] | 0.38 | 0.4 |
| $l_{cs}$ (mm) | [18, 25] | 20.3 | 20 |
| $t_y$ (mm) | [1.5, 3] | 2.2 | 2 |

## 4. Controller Design

To achieve ultra-precision actuation based on smart materials, it is essential to implement appropriate control algorithms. Currently, widely adopted control strategies include feedback control, feedforward control, and hybrid control approaches [27]. The Preisach–Ishlinskii (PI) model characterizes the input–output relationship of hysteresis through the use of play operators and density functions. Due to its capacity for fast and accurate modeling of asymmetric hysteresis behaviors, the PI model has been extensively utilized in the field [28]. In this section, an enhanced PI model based on polynomial operators is employed to describe the nonlinear hysteresis characteristics of GMM. A unilateral play operator is utilized to represent the hysteresis relationship between the input $u$ and output $v$, and its discrete formulation can be expressed as follows:

$$\begin{gathered} H_{r_H}[u](k) = \begin{cases} h_{r_H}(u(0), r_H, 0) \; k = 0 \\ h_{r_H}(u(k), r_H, H_{r_H}[u](k)) \; k \geq 1 \end{cases} \\ h_{r_H}(u, r_H, s) = \max(P(u) - r_H, \min(P(u), s)) \end{gathered} \tag{23}$$

where $r_H = [r_{H0}, r_{H1}, \ldots, r_{Hm}]^T$ denotes the threshold vector of the unilateral play operators, and $k$ represents a discrete time variable. The hysteresis phenomenon can be characterized using a polynomial operator $P(u)$, which is formulated as follows:

$$P(u) = w_p^T \cdot p(u) \tag{24}$$

where $p(u) = \left[u^l, u^{l-1}, \ldots, u^0\right]^T$ *and* $w_p = \left[w_{p0}, w_{p1}, \ldots, w_{pl}\right]^T$ are the polynomial basis vector and the corresponding weight vector, respectively. Accordingly, the improved PI model incorporating the polynomial operator can be expressed as follows:

$$v(k) = w_H^T \cdot H_{r_H}[w_p^T \cdot p(u)](k) \tag{25}$$

where $w_H = [w_{H0}, w_{H1}, \ldots, w_{Hm}]^T$ denotes the weight vector associated with the unilateral play operators. To compensate for the hysteresis effect, it is necessary to construct an inverse model of hysteresis nonlinearity. The structure of this inverse model is similar to that described in Equations (23) and (24). The final output $\hat{u}$ of the inverse model based on the polynomial operator-enhanced PI model can be expressed as follows:

$$\hat{u}(k) = \hat{w}_H^T \cdot H_{\hat{r}_H}[\hat{w}_p^T \cdot p(v)](k) \tag{26}$$

where $H_{\hat{r}H}$ denotes the set of unilateral play operators used in the inverse model, and $\hat{w}_H$ and $\hat{w}_p$ represent the corresponding weight vectors for the play operators and the polynomial operators, respectively. It is worth noting that, to reduce computational complexity in practical applications, the parameters of the inverse model are identified directly from experimental data, rather than being obtained through analytical inversion.

To accurately characterize the frequency-dependent hysteresis nonlinearity of GMM, ref. [19] proposes a modeling approach based on the Hammerstein model structure. Building upon this, the present study further employs an online identification strategy to model the dynamic characteristics of the SSGMT with high precision. The structural diagram of the specific model identification algorithm is shown in Figure 4. In this framework, $y(k)$ and $\hat{y}(k)$ denote the actual output of the SSGMT and the output of the identified model, respectively, while $y_e(k)$ represents the modeling error between the actual system output and the model output.

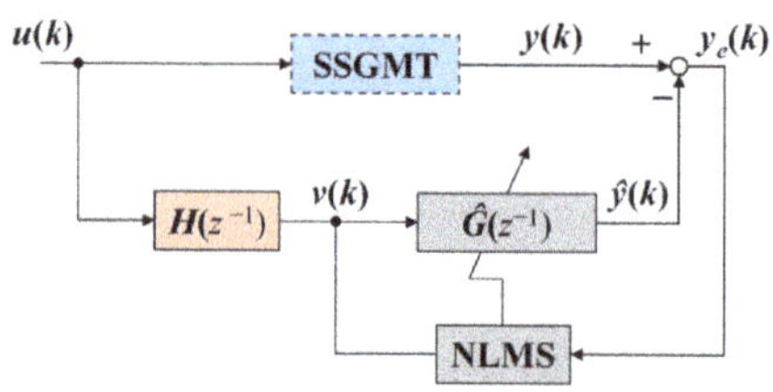


**Figure 4.** Block diagram of frequency-dependent Hammerstein model of SSGMT.

Building upon previous research, a composite control strategy is proposed, which integrates inverse feedforward compensation based on the polynomial-enhanced PI model with a feedforward controller constructed using an adaptive filter. The block diagram of the control algorithm is shown in Figure 5. In this diagram, $d(k),\ v(k),\ y(k),\ and\ e(k)$ represent the desired trajectory, the control current, the actual output of the SSGMT, and the tracking error, respectively. The self-sensing signal is first processed through a lock-in amplifier (denoted as $N_{fp}$) and subsequently fed into a generalized regression neural network (GRNN) model to yield the actual output $y(k)$ of the SSGMT.

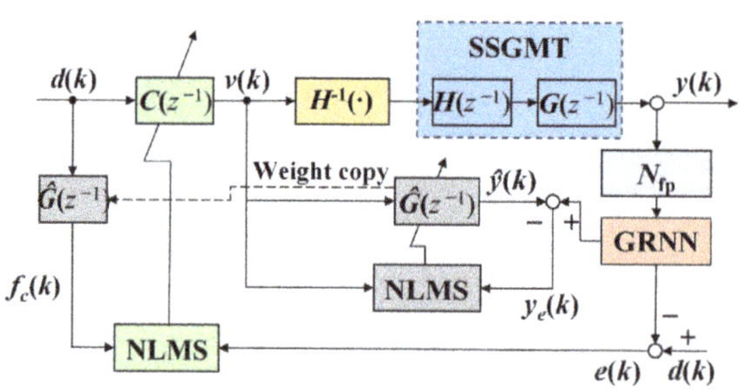


**Figure 5.** Block diagram of trajectory tracking controller for SSGMT.

In the control algorithm, the hysteresis compensator $H^{-1}(\cdot)$ is cascaded with the plant. As a result, after hysteresis compensation, the estimated model of the compensated system becomes $\hat{G}(z^{-1}) = H^{-1}(\cdot) \cdot H(z^{-1}) \cdot G(z^{-1}) = G(z^{-1})$. As illustrated in Figure 5, the tracking error $e(k)$ in the z-domain is expressed as follows:

$$e(k) = d(k) - y(k) = \left[1 - C(z^{-1})G(z^{-1})\right]d(k) \tag{27}$$

When $e(k) = 0$, the actual output of the SSGMT perfectly follows the desired trajectory, indicating zero tracking error. The hysteresis compensator $H^{-1}(\cdot)$ is implemented using the polynomial operator-based PI inverse model identified offline in this section, while the dynamic model $\hat{G}(z^{-1})$ of the SSGMT is constructed via the online identification method based on the normalized least mean squares (NLMS) algorithm. The specific update algorithm for $\hat{G}(z^{-1})$ is given as follows:

$$\boldsymbol{w}^g(n+1) = \boldsymbol{w}^g(n) + 2\mu_g y_e(n)\frac{\boldsymbol{v}(n)}{\|\boldsymbol{v}(n)\|^2 + \tau_g} \tag{28}$$

Here, $\boldsymbol{w}^g(n)$ denotes the weight vector of the estimated system model $\hat{G}(z^{-1})$ and is defined as $\boldsymbol{w}^g(n) = [w^g{}_0(n), w^g{}_1(n), \ldots, w^g{}_{K-1}(n)]^T$. The parameters $\mu_g$ and $\tau_g$ represent the step-size constant for adjusting the adaptive rate of $\hat{G}(z^{-1})$ and a small positive constant used to prevent division by zero, respectively. The input signal $\boldsymbol{v}(n)$ refers to the compensated version of $\boldsymbol{u}(n)$ after processing by $H(z^{-1})$ and is given by $\boldsymbol{v}(n) = [v(n), v(n-1), \ldots, v(n-K+1)]^T$.

The feedforward controller $C(z^{-1})$ adopts a finite impulse response (FIR) filter structure which can be expressed as follows:

$$C(z^{-1}) = \sum_{j=0}^{K-1} w_j^c(k)z^{-j} \tag{29}$$

where $w_j^c$ denotes the weight coefficients of the FIR filter that defines $C(z^{-1})$. These coefficients are updated using the NLMS algorithm. Accordingly, the iterative update formula for the weight vector of $C(z^{-1})$ is given by the following:

$$\boldsymbol{w}^c(n+1) = \boldsymbol{w}^c(n) + 2\mu_c e(n)\frac{\boldsymbol{f}_c(n)}{\|\boldsymbol{f}_c(n)\|^2 + \tau_c} \tag{30}$$

Here, $\boldsymbol{w}^c(n) = \left[w_0^c(n), w_1^c(n-1), \ldots, w_{K-1}^c(n)\right]^T$ represents the weight vector of the feedforward controller at time step $n$. The parameters $\mu_c$ and $\tau_c$ are the step-size constant for adjusting the adaptive rate of $C(z^{-1})$ and a small positive regularization constant to avoid division by zero, respectively. The input signal $\boldsymbol{f}_c(n)$ is generated by filtering the desired trajectory $d(n)$ through the identified model $\hat{G}(z^{-1})$, $\boldsymbol{f}_c(n) = [f_c(n), f_c(n-1), \ldots, f_c(n-K+1)]^T$ and is defined as follows:

$$f_c(k) = \sum_{j=0}^{K-1} d(k-j)w_j^g(k) \tag{31}$$

where $k = n, n-1, \ldots, n-K+1$, and the weight vector of $\hat{G}(z^{-1})$ at time $k$ is denoted as $\boldsymbol{w}^g(k) = \left[w_0^g(k), w_1^g(k), \ldots, w_{K-1}^g(k)\right]^T$.

## 5. Experimental Verification

### 5.1. Experimental Setup

According to the overall scheme proposed in Section 2 and the optimized system structure parameters in Section 3, a prototype of the SSGMT was made, and corresponding basic performance verification experiments were carried out. The device connection diagram of the experimental system is shown in Figure 6a. The SSGMT prototype shell is made of

304 stainless steel with good magnetic insulation and is fixed on a bench vice connected to the optical platform (AVIC Century ZPT-G-Y). Figure 6b shows the appearance of the prototype. The embedded real-time controller CompactRIO-9082 (National Instruments, Austin, USA), on one hand, extracts the SSGMT self-sensing signal collected and processed by the lock-in amplifier MFLI (Zurich Instruments AG, Zurich, Switzerland), and, on the other hand, controls the current input from the power amplifier NF BP4610 (NF Circuit Design Block, Yokohama, Japan) to the SSGMT. The output displacement of the SSGMT prototype is collected by a laser sensor KEYENCE LK-G10 (Keyence Corporation, Osaka, Japan). A type K thermocouple temperature sensor (Cole-Parmer, Vernon Hills, USA) and a force sensor LDC-08 (Shenzhen Ligent Sensor Technology Co., Ltd., Shenzhen, China) are installed inside the SSGMT prototype to detect the load and temperature of the GMM rod in real time.

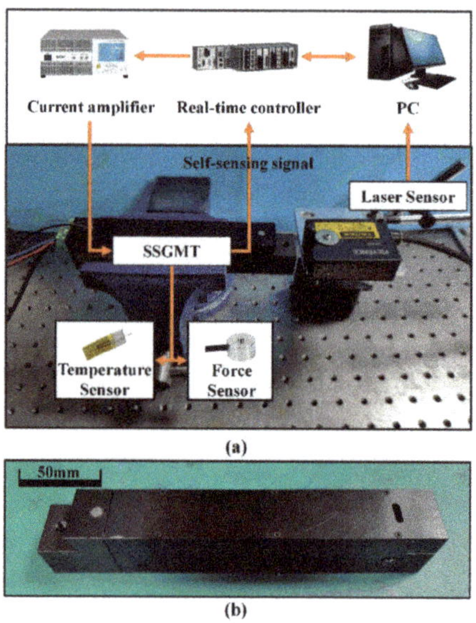


**Figure 6.** The basic performance verification experiment of the SSGMT: (**a**) device connection diagram; (**b**) the prototype of the SSGMT.

The core component of the SSGMT is the self-sensing actuator. Unlike conventional actuators, the SSGMT not only executes with precision but also continuously monitors its own output displacement, load force, and operating temperature. Next, the performance of the self-sensing actuator in the previously developed SSGMT will be tested.

### *5.2. Self-Sensing Output Displacement Performance Test*

At the conventional laboratory temperature (22 °C), the amplitude–frequency characteristic curves of the self-sensing signal were tested under a swept $H_s$ for different values of $H_a$, as shown in Figure 7. In the experiment, the amplitude of $H_s$ was set to 100 A/m, and its frequency ($f_s$) was swept from 750 Hz to 1250 Hz. As $H_a$ increased from 10 kA/m to 100 kA/m, due to the ΔE effect of the GMM, the system's resonance frequency ($f_{sr}$) first decreased and then increased, with the minimum $f_{sr}$ occurring at $H_a$ = 20 kA/m. Comparing the different $H_a$ conditions, the amplitude of the self-sensing signal reached its maximum at $f_s = f_{sr}$, and as $f_{sr}$ increased, the corresponding maximum amplitude of the self-sensing signal gradually decreased.

conditions and waveform types (measured after 3 s of continuous tracking) are summarized in Table 3.

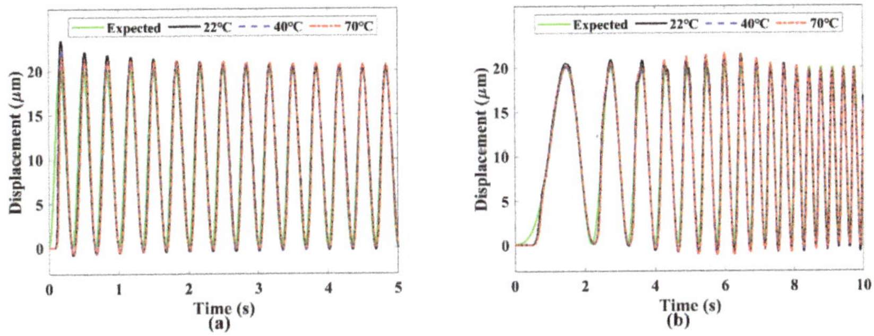


**Figure 14.** The trajectory tracking effect of the SSGMT on (**a**) a 2 Hz fixed-frequency waveform and (**b**) a 0.1–5 Hz sweep waveform with expected displacement at different temperatures.

**Table 3.** Trajectory tracking steady-state errors of SSGMT at different temperatures.

| Temperature (°C) | Sinusoidal Trajectory | | Sweep Trajectory | |
|---|---|---|---|---|
| | RMS (μm) | RRMS (%) | RMS (μm) | RRMS (%) |
| 22 | 0.976 | 2.37 | 1.056 | 2.60 |
| 40 | 1.038 | 2.53 | 1.129 | 2.79 |
| 70 | 1.113 | 2.70 | 1.259 | 3.11 |

The above experimental results demonstrate that the SSGMT is capable of achieving precise trajectory tracking under different temperatures and input frequencies. The research provides a theoretical and practical foundation for subsequent applications in lathe operations.

## 6. Conclusions

This paper proposes a self-sensing giant magnetostrictive actuator-based turning tool (SSGMT) to address the challenges of compactness, precision, and real-time feedback in intelligent machining. A comprehensive multi-physical model was established to capture the coupled magnetic, mechanical, and thermal behaviors of the actuator. Additionally, a dynamic hysteresis model based on a Hammerstein structure was developed to describe the rate-dependent nonlinear response of the system. To ensure optimal system performance under physical and structural constraints, a multi-objective optimization framework was constructed to guide the design of key parameters, balancing actuation force, energy efficiency, sensing signal quality, and structural compactness.

Based on the optimized parameters, a prototype of the SSGMT was fabricated and tested. Experimental evaluations demonstrated high-resolution self-sensing displacement capabilities with sub-100 nm precision and robust performance across varying thermal and dynamic conditions. A hybrid control scheme combining inverse hysteresis compensation and adaptive feedforward control was implemented to improve trajectory tracking accuracy. The system showed effective suppression of nonlinear distortion and maintained high positioning precision in both static and dynamic tracking tests. These results validate the feasibility and effectiveness of integrating actuation and sensing within a unified structure and provide a promising foundation for deploying self-sensing actuators in next-generation ultra-precision machining tools.

Future research will prioritize enhancing the self-sensing signal quality through advanced coil topologies and adaptive signal processing techniques to achieve sub-10 nm resolution under dynamic machining loads. Concurrently, efforts will focus on extending the operational bandwidth beyond 100 Hz via real-time hysteresis compensation algorithms for complex surface machining applications. Further validation in industrial environments—including robustness testing against coolant immersion, electromagnetic interference, and thermal transients (20–120 °C)—will be conducted to bridge laboratory prototypes and production-ready systems. The fundamental SSGMT architecture will also be adapted to milling tools and multi-axis smart tooling platforms.

**Author Contributions:** Conceptualization, methodology, D.X. and B.Y.; experiment, D.X. and Y.Z.; formal analysis, Y.Y.; writing—original draft preparation, Q.W.; writing—review and editing, D.X. and Q.W.; visualization, D.X.; supervision, B.Y.; project administration, B.Y. and C.Z. All authors have read and agreed to the published version of the manuscript.

**Funding:** This work is supported by the National Key R&D Program of China (2022YFB3902703), the National Natural Science Foundation of China (52173239), the Specific Project of SJTU for "Invigorating Inner Mongolia through Science and Technology" (2022XYJG0001-01-04), and the Startup Fund for Young Faculty at SJTU (24X010500146).

**Data Availability Statement:** The data are contained within the article.

**Acknowledgments:** The authors would like to express their sincere gratitude to the journal editors and reviewers for their meticulous work and valuable advice.

**Conflicts of Interest:** The authors declare no conflicts of interest.

## References

1. Bleicher, F.; Biermann, D.; Drossel, W.-G.; Moehring, H.-C.; Altintas, Y. Sensor and Actuator Integrated Tooling Systems. *CIRP Ann.* **2023**, *72*, 673–696. [CrossRef]
2. Zhou, C.; Guo, K.; Sun, J. An Integrated Wireless Vibration Sensing Tool Holder for Milling Tool Condition Monitoring with Singularity Analysis. *Measurement* **2021**, *174*, 109038. [CrossRef]
3. Bleicher, F.; Ramsauer, C.M.; Oswald, R.; Leder, N.; Schoerghofer, P. Method for Determining Edge Chipping in Milling Based on Tool Holder Vibration Measurements. *CIRP Ann.* **2020**, *69*, 101–104. [CrossRef]
4. Zhang, P.; Gao, D.; Lu, Y.; Ma, Z.; Wang, X.; Song, X. Cutting Tool Wear Monitoring Based on a Smart Toolholder with Embedded Force and Vibration Sensors and an Improved Residual Network. *Measurement* **2022**, *199*, 111520. [CrossRef]
5. Mohamed, A.; Hassan, M.; M'Saoubi, R.; Attia, H. Tool Condition Monitoring for High-Performance Machining Systems—A Review. *Sensors* **2022**, *22*, 2206. [CrossRef]
6. Mishra, D.; Pattipati, K.R.; Bollas, G.M. Intelligent Feature Engineering for Monitoring Tool Health in Machining. In *Industry 4.0 Driven Manufacturing Technologies*; Kumar, A., Kumar, P., Liu, Y., Eds.; Springer Nature: Cham, Switzerland, 2024; pp. 37–67, ISBN 978-3-031-68271-1.
7. Deshpande, S.; Araujo, A.C.; Lagarrigue, P. *Identification of Material Layer During Helical Milling Process on Aerospace Stacks Using Spindle Power Data*; Elsevier: Amsterdam, The Netherlands, 2025.
8. Hu, S.; Yang, Y.; Bao, W.; Nie, Y. Research on the Control System and Method of Cutting Arms for Digital Twin Driven Excavation Equipment. In Proceedings of the 2024 9th International Conference on Intelligent Computing and Signal Processing (ICSP), Xi'an, China, 19–21 April 2024; pp. 1475–1480.
9. Huber, J.E.; Fleck, N.A.; Ashby, M.F. The Selection of Mechanical Actuators Based on Performance Indices. *Proc. R. Soc. London. Ser. A Math. Phys. Eng. Sci.* **1997**, *453*, 2185–2205. [CrossRef]
10. Ceyssens, F.; Sadeghpour, S.; Fujita, H.; Puers, R. Actuators: Accomplishments, Opportunities and Challenges. *Sens. Actuators A Phys.* **2019**, *295*, 604–611. [CrossRef]
11. Wang, S.; Rong, W.; Wang, L.; Xie, H.; Sun, L.; Mills, J.K. A Survey of Piezoelectric Actuators with Long Working Stroke in Recent Years: Classifications, Principles, Connections and Distinctions. *Mech. Syst. Signal Process.* **2019**, *123*, 591–605. [CrossRef]
12. De Vicente, J.; Klingenberg, D.J.; Hidalgo-Alvarez, R. Magnetorheological Fluids: A Review. *Soft Matter* **2011**, *7*, 3701–3710. [CrossRef]

13. Maiwald, D.; Bucht, A.; Zorn, W.; Drossel, W.-G. Development Approach of Actuators for High Loads Based on Shape Memory Alloys. In Proceedings of the ASME 2020 Conference on Smart Materials, Adaptive Structures and Intelligent Systems, Virtual, Online, 15 September 2020.
14. Yoshioka, H.; Shinno, H.; Sawano, H. A Newly Developed Rotary-Linear Motion Platform with a Giant Magnetostrictive Actuator. *CIRP Ann.* **2013**, *62*, 371–374. [CrossRef]
15. Zhu, Z.; Zhou, X.; Liu, Z.; Wang, R.; Zhu, L. Development of a Piezoelectrically Actuated Two-Degree-of-Freedom Fast Tool Servo with Decoupled Motions for Micro-/Nanomachining. *Precis. Eng.* **2014**, *38*, 809–820. [CrossRef]
16. Li, H.; Ibrahim, R.; Cheng, K. Design and Principles of an Innovative Compliant Fast Tool Servo for Precision Engineering. *Mech. Sci.* **2011**, *2*, 139–146. [CrossRef]
17. Huang, W.-W.; Li, L.; Zhu, Z.; Zhu, L.-M. Modeling, Design and Control of Normal-Stressed Electromagnetic Actuated Fast Tool Servos. *Mech. Syst. Signal Process.* **2022**, *178*, 109304. [CrossRef]
18. Yoshioka, H.; Kojima, K.; Toyota, D. Micro Patterning on Curved Surface with a Fast Tool Servo System for Micro Milling Process. *CIRP Ann.* **2020**, *69*, 325–328. [CrossRef]
19. Xie, D.; Yang, Y.; Zhang, Y.; Yang, B. Precision Positioning Based on Temperature Dependence Self-Sensing Magnetostrictive Actuation Mechanism. *Int. J. Mech. Sci.* **2024**, *272*, 109174. [CrossRef]
20. Xie, D.; Yang, Y.; Yang, B. Self-Sensing Magnetostrictive Actuator Based on ΔE Effect: Design, Theoretical Modeling and Experiment. *Smart Mater. Struct.* **2022**, *31*, 055007. [CrossRef]
21. Kim, H.-S.; Lee, K.-I.; Lee, K.-M.; Bang, Y.-B. Fabrication of Free-Form Surfaces Using a Long-Stroke Fast Tool Servo and Corrective Figuring with on-Machine Measurement. *Int. J. Mach. Tools Manuf.* **2009**, *49*, 991–997. [CrossRef]
22. Zhong, E.; Wang, S.; Zhai, C.; Li, W. Improved Non-Singular Fast Terminal Sliding Mode Control with Hysteresis Compensation for Piezo-Driven Fast Steering Mirrors. *Actuators* **2025**, *14*, 170. [CrossRef]
23. Sun, X.; Liu, Y.; Hu, W.; Sun, T.; Hu, J. Design Optimization of a Giant Magnetostrictive Driving System for Large Stroke Application Considering Vibration Suppression in Working Process. *Mech. Syst. Signal Process.* **2020**, *138*, 106560. [CrossRef]
24. Sun, X.; Wei, H.; Liu, Y.; Wang, Z.; Jun, H. Integrated Design and Simulation of a Giant Magnetostrictive Driving Positioning and Vibration Suppression System. *J. Syst. Simul.* **2021**, *33*, 358–365. [CrossRef]
25. Barbisio, E.; Fiorillo, F.; Ragusa, C. Predicting Loss in Magnetic Steels under Arbitrary Induction Waveform and with Minor Hysteresis Loops. *IEEE Trans. Magn.* **2004**, *40*, 1810–1819. [CrossRef]
26. Wang, D.; Tan, D.; Liu, L. Particle Swarm Optimization Algorithm: An Overview. *Soft Comput.* **2018**, *22*, 387–408. [CrossRef]
27. Liu, C.; Jing, X.; Daley, S.; Li, F. Recent Advances in Micro-Vibration Isolation. *Mech. Syst. Signal Process.* **2015**, *56–57*, 55–80. [CrossRef]
28. Kuhnen, K. Modeling, Identification and Compensation of Complex Hysteretic Nonlinearities: A Modified Prandtl-Ishlinskii Approach. *Eur. J. Control.* **2003**, *9*, 407–418. [CrossRef]

actuators

MDPI

Article

# Gas–Solid Coupling Dynamic Modeling and Transverse Vibration Suppression for Ultra-High-Speed Elevator

**Jiacheng Jiang, Chengjin Qin *, Pengcheng Xia and Chengliang Liu**

School of Mechanical Engineering, Shanghai Jiao Tong University, Shanghai 201100, China; dan_xsy@sjtu.edu.cn (J.J.); xpc19960921@sjtu.edu.cn (P.X.); chlliu@sjtu.edu.cn (C.L.)
* Correspondence: qinchengjin@sjtu.edu.cn

**Abstract:** When in operation, ultra-high-speed elevators encounter transverse vibrations due to uneven guide rails and airflow disturbances, which can greatly undermine passenger comfort. To alleviate these adverse effects and boost passenger comfort, a gas–solid coupling dynamic model for ultra-high-speed elevator cars is constructed, and a vibration suppression approach is proposed. To start with, the flow field model of the elevator car-shaft under different motion states is simulated, and the calculation formula of air excitation is derived. Next, by incorporating the flow field excitation into the four degrees of freedom dynamic model of the separation between the car and the frame, a transverse vibration model of the elevator car based on gas–solid coupling is established. Finally, an LQR controller is used to suppress elevator transverse vibration, and a multi-objective optimization algorithm is applied to optimize the parameters of the weight matrix to obtain the optimal solution of the LQR controller. A set of controllers with moderate control cost and system performance meeting the requirements was selected, and the effectiveness of the controller was verified. Compared with other methods, the proposed LQR-based method has greater advantages in suppressing the transverse vibration of ultra-high-speed elevators. This work provides an effective solution for enhancing the ride comfort of ultra-high-speed elevators and holds potential for application in the vibration control of high-speed transportation systems.

**Keywords:** ultra-high-speed elevator; gas–solid coupling; multi-objective genetic algorithm; LQR controller; vibration suppression

## 1. Introduction

With growing building height, ordinary elevators are unable to meet the needs of people in super high-rise buildings anymore. Faster and safer elevators have become urgently needed for such buildings. The emergence of ultra-high-speed elevators has resolved the aforementioned needs. Nevertheless, the higher the elevator's speed during operation, the more intense the vibration, and passengers standing inside the car can sense a decline in comfort [1]. Meanwhile, it also diminishes the elevator's lifespan. In fact, the most notable factor causing this situation is the transverse vibration of the elevator car, which is chiefly caused by two factors: the guide system and airflow interference.

To alleviate adverse effects and upgrade passenger comfort, a multitude of methods has been advanced. Presently, vibration reduction for ultra-high-speed elevator cars chiefly depends on two means: passive and active vibration attenuation. Given that passive vibration attenuation is usually inflexible and insufficiently effective, active suppression has turned into the primary method for vibration control [2]. Zhang et al. [3] devised an active car shock absorber incorporating a linear motor, formulated a five degrees of freedom space vibration model, and implemented a back propagation neural network PID controller in conjunction with a linear prediction model. Tusset et al. [4] probed into the transverse response of vertical transportation with nonlinearities under guide rail deformation excitation and adopted an LQR control strategy using MR dampers to uplift passenger comfort. In Zhang et al.'s [5] investigation, an electro-hydraulic active guide shoe was conceived to tackle car transverse vibration triggered by guide unevenness excitation. The Takagi–Sugeno fuzzy reasoning method was utilized to approximate the nonlinear car system model, and an adaptive gain $H_\infty$ output feedback control strategy was proposed, hinging on the parallel distributed compensation rule, to curtail transverse vibration. Su et al. [6] posited a transient response feedback control strategy with specified performance, centering on problems like subpar transient convergence and the real-time performance of current active control algorithms for car system transverse vibration. Ge et al. [7] tendered a fuzzy sliding mode-based active disturbance rejection (ADR) control method, adept at governing the horizontal vibration of high-speed elevator car systems attributable to factors such as irregular guide rails and the nonlinear damping of springs in guide shoes.

In the study of high-speed elevator transverse vibration mentioned in the above references, active control methods are mainly used to reduce transverse vibration. Yet, ultra-high-speed elevators are influenced not just by rail excitation but also by airflow disturbances. Reference [8] points out that the greater the elevator speed, the more significant the airflow impact. Currently, an increasing number of scholars have investigated the airflow disturbance of high-speed elevators. Zhang et al. [9] looked into the airflow characteristics in various hoistways. Qiao et al. [10] performed the theoretical modeling of super-high-speed elevators based on Bernoulli's principle for unsteady flow and analyzed the parameter impact of the car-induced airflow. Zhang et al. [11] came up with a multi-region dynamic layering method to research the hoistway airflow and the rational ventilation hole opening. Cui et al. [12] set up and numerically simulated a full-scale 3D model of the shaft with different blocking ratios, aiming to probe into the aerodynamic features of the elevator car's high-speed movement and the airflow law in the shaft. Qiu et al. [13] presented an elevator car air pressure compensation method grounded in internal–external flow field (IE-FF) coupling analysis, which is beneficial for adaptively tracking the ideal car interior air pressure curve and controlling air pressure fluctuation.

Currently, some studies have taken the gas–solid coupling method into account. Qiu et al. [14] created the high-speed elevator car transverse vibration gas–solid coupling model, integrating guide excitations and air excitations, and set up the numerical car-hoistway interface region model. Shi et al. [8] developed a gas–solid coupling model for the whole high-speed elevator operating process using the finite volume method, Lagrange's theorem, and so on. The actual testing of a 7 m/s high-speed elevator confirmed the feasibility of the model and modeling method.

Recent advances in elevator vibration control predominantly adopt two isolated approaches: Mechanical-focused methods [3–7] actively suppress rail-induced vibrations but neglect aerodynamic coupling effects, which should not be ignored, especially at high speeds [8,12]. Aerodynamic-focused methods [9–13] optimize airflow fields yet lack integration with real-time vibration control systems. This disconnect creates a critical gap: unmodeled fluid–structure interactions may destabilize controllers at ultra-high speeds. Our work bridges this by unifying high-fidelity CFD with LQR control, providing a new framework for control under gas–solid coupling.

This paper uses a more intuitive and straightforward approach, namely, the numerical fitting of aerodynamic load, to avoid the complexity and inefficiency of calculations. Considering gas–solid coupling, the formula linking aerodynamic load to working conditions is derived via simulation. This aerodynamic load is then incorporated into the dynamic equation to finish gas–solid coupling modeling. Moreover, this paper innovatively applies a multi-objective genetic algorithm to optimize LQR control parameters, effectively suppressing ultra-high-speed elevator lateral vibration.

The key contributions of this study are outlined as follows:

1 A gas–solid coupling dynamic model for ultra-high-speed elevators is established, and a control method based on the LQR has been proposed.
2 Simulation analysis of the flow field characteristics inside high-speed elevator shafts is executed, leading to a calculation formula for air excitation corresponding to the elevator car motion state, with an R-square as high as 0.999.
3 The key weight matrix Q and R parameters in the LQR controller are optimized utilizing a multi-objective genetic algorithm, considering both control performance and cost of the elevator.
4 When the speed is 6 m/s, 8 m/s, and 10 m/s, compared with PID control, the proposed LQR-based method can significantly reduce the transverse acceleration by 7.55%, 8.47%, and 10.27%. The higher the speed, the more effective the proposed LQR-based method is.

The rest of this paper is structured as follows: Section 2 introduces external excitations, including rail excitation and air excitation. Section 3 is the dynamic modeling and state equation representation of ultra-high-speed elevators. Then, in Section 4, the method for suppressing transverse vibrations is mainly presented. Section 5 compares the obtained results with various situations to validate the effectiveness of the proposed method. Finally, the conclusion is presented in Section 6.

## 2. Modeling of External Excitation

Manufacturing and installation errors unavoidably lead to rail unevenness, which is a primary cause of vibration in ultra-high-speed elevators. Moreover, during the ascending or descending of ultra-high-speed elevator cars within the hoistway, they are subjected to varying aerodynamic pressures from the airflow inside the shaft, which intensifies the transverse vibrations. Therefore, this section focuses on analyzing two main elements influencing the transverse vibration of ultra-high-speed elevators: guide excitation and air excitation. The guide excitation is illustrated by white noise signals filtered by a low-pass filter, whereas the air excitation is acquired through Fluent simulation and formula fitting.

### 2.1. Air Excitation Modeling

#### 2.1.1. Turbulence Model

During operation, the ultra-high-speed elevator is restricted from performing lifting movements in the shaft, and the air around the ultra-high-speed elevator in the shaft is driven by the elevator's operation. The air part inside the shaft, excluding the elevator, is analyzed as a whole flow field, which exhibits a low Mach number (Ma < 0.3) and high Reynolds number (Re > 2300) state [15]. This flow field can be analyzed as incompressible turbulence. Therefore, the flow field inside the shaft can be described utilizing the Navier–Stokes equations, which include two parts: momentum conservation and mass conservation. The equations and coefficients of the turbulence model are provided in Appendix A.

#### 2.1.2. Numerical Simulation of the Airflow Within the Shaft

Figure 1 presents the model of the elevator shaft and car. In an ideal scenario, the elevator car should be positioned on the vertical centerline of the shaft. If the car and car frame have uniform shape and mass distribution, the flow field's pressure and velocity distributions around them would be symmetrical. This symmetry would cause the air-generated excitations on the car to cancel out in all directions, resulting in a zero net force on the car. However, achieving this ideal state is challenging. Installation errors and the car's operational vibrations often cause it to deviate from the centered, symmetrical position. Consequently, the car experiences additional air excitations.

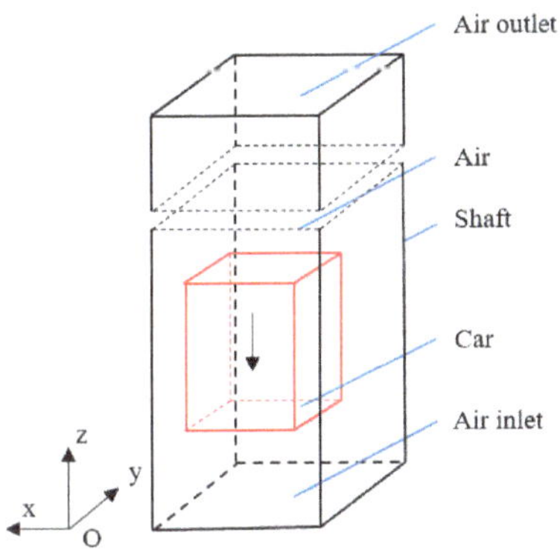


**Figure 1.** A 3D model of the elevator shaft and car.

SIMPLE is a classic semi-implicit method. This paper utilizes the second-order upwind scheme and SIMPLE algorithm for steady-state flow field simulation in Ansys Fluent.

In order to analyze the factors affecting air excitation under asymmetric car positions, the three-dimensional model needs to be established based on Figure 1 and simulated and solved utilizing Ansys Fluent 2022 R2. For the convenience of calculation, the subsequent simplifications are implemented:

(1) The elevator car and the elevator frame are considered to be symmetrically installed with a rectangular shape, and their centers of mass are located on the centerline of the hoistway.
(2) The guide shoes and shock absorbers are simplified to a spring–damper system.
(3) The influence of the hoisting ropes on the car's transverse vibrations is neglected.

The boundary conditions for solving the flow field are set as velocity inlet and boundary outlet, and no-slip conditions are set for all outer surfaces of the elevator car and the inner wall of the shaft.

Boundary conditions for airflow inlet:

$$u_y = v_s, u_x = 0, \frac{\partial p}{\partial x} = 0 \tag{1}$$

Boundary conditions for airflow outlet:

$$\frac{\partial v_y}{\partial y} = \frac{\partial u_x}{\partial x} = 0 \tag{2}$$

As the elevator operates within the hoistway, the airflow exerts aerodynamic drag on the car's outer surface. As the car moves within the shaft, the variation in the flow field region complicates the solution of the flow field [16]. Based on the principle of relative motion, in the simulation, air enters from one side of the shaft entrance at the same speed as the car, flows through the car, and exits from the shaft outlet [17]. The top and bottom of the car are positioned 5 m away from the upper and lower openings of the hoistway, respectively. The boundary conditions for the flow field solution are set as velocity inlet and boundary outlet, with all the elevator car's outer surfaces and the shaft's inner walls set to no-slip conditions.

In order to speed up the calculation, after dividing the entire flow field into grids according to the default method, additional grid refinement processing was performed on the car wall and shaft boundary [18,19] to ensure more accurate calculation results at the boundary.

Utilize the Fluent module for solving, adjust the convergence criterion to $1 \times 10^{-5}$, and the maximum number of iteration steps is 1000.

The equivalent transverse concentrated force, and concentrated moment $M_z$ have the most significant impact on the lateral vibration. By integrating the air pressure and viscous stress acting on the surface of the car body, the equivalent transverse concentrated force and concentrated moment $M_z$, acting on the center of the car, can be obtained as follows [20]:

$$F_y = \int_{\text{Scar}} ps_{car} + \int_{\text{Scar}} \tau d|S_{car}| \tag{3}$$

$$M_z = \int (R_a - R_o) dF_x \tag{4}$$

In the equation, $p$ is the airflow pressure, $\tau$ is the viscous stress, Scar is the area of the outer surface of the car, $R_a$ represents the position of the action surface unit, and $R_o$ represents the position of the car center of mass.

As noted in reference [21], the elevator car's aerodynamic load is closely related to its operating speed and position deviation. Therefore, the simulation (uniform speed) will focus on the transverse force and overturning moment experienced by the elevator car's center of mass under varying conditions of transverse displacement, deflection angle displacement, and rated speed. Ultimately, this study will explore the relationships between transverse force, overturning moment, transverse displacement, deflection angle displacement, and rated speed [22].

In this study, the transverse displacements ($y$) of the car body are taken as 2 mm, 4 mm, 6 mm, 8 mm, and 10 mm, combined with the running speeds ($v$) of 4 m/s, 6 m/s, 8 m/s, 10 m/s, and 12 m/s, respectively. The deflection angulars ($\theta$) are taken as 0.5°, 1°, 1.5°, 2°, and 2.5°, combined with the running speeds (v) of 4 m/s, 6 m/s, 8 m/s, 10 m/s, and 12 m/s, respectively. The results obtained are shown in Tables 1 and 2.

Utilizing the data in Tables 1 and 2, plot the variation in aerodynamic load with different transverse displacements and running speeds of the car, as shown in Figures 2 and 3.

**Table 1.** Equivalent aerodynamic load of the car under different speeds and transverse displacement.

| Velocity/m·s$^{-1}$ | Aerodynamic Load/N | Transverse Displacement/$10^{-3}$m | | | | |
|---|---|---|---|---|---|---|
| | | 2 | 4 | 6 | 8 | 10 |
| 4 | $F_y$/N | 1.443328 | 2.820529 | 4.433297 | 5.847365 | 7.3607995 |
| | $M_z$/N·m | 0.618003 | 1.0108336 | 1.9515373 | 2.0938242 | 3.1336382 |
| 6 | $F_y$/N | 3.295254 | 6.7790241 | 10.148966 | 13.594114 | 16.945068 |
| | $M_z$/N·m | 1.350664 | 2.5176488 | 4.2651951 | 5.1110839 | 6.820766 |
| 8 | $F_y$/N | 5.714808 | 11.594293 | 17.849078 | 27.131752 | 30.916064 |
| | $M_z$/N·m | 2.568629 | 4.3573907 | 7.6928151 | 10.613392 | 13.341611 |
| 10 | $F_y$/N | 9.473283 | 18.867489 | 27.128074 | 39.512653 | 48.962687 |
| | $M_z$/N·m | 3.877174 | 7.3896428 | 11.632615 | 15.696517 | 20.812077 |
| 12 | $F_y$/N | 13.207638 | 26.184425 | 39.162721 | 52.128905 | 65.367814 |
| | $M_z$/N·m | 5.472908 | 11.094941 | 16.469281 | 22.089928 | 27.486092 |

**Table 2.** Equivalent aerodynamic load of the car at different speeds and deflection angles.

| Velocity/m·s$^{-1}$ | Aerodynamic Load | Deviation Angle/° | | | | |
|---|---|---|---|---|---|---|
| | | 0.5 | 1 | 1.5 | 2 | 2.5 |
| 4 | $F_y$/N | 17.251313 | 34.505746 | 52.871793 | 71.051558 | 91.897572 |
| | $M_z$/N·m | 7.868474 | 14.97594 | 22.961873 | 30.512319 | 37.802049 |
| 6 | $F_y$/N | 38.561698 | 77.617425 | 118.40733 | 159.92379 | 205.83671 |
| | $M_z$/N·m | 17.462975 | 34.716924 | 51.357129 | 68.417756 | 84.439622 |
| 8 | $F_y$/N | 68.351175 | 137.96199 | 209.92972 | 284.15439 | 364.88847 |
| | $M_z$/N·m | 30.941648 | 59.233599 | 90.979506 | 121.32131 | 149.41584 |
| 10 | $F_y$/N | 106.55557 | 215.40127 | 327.46022 | 443.79444 | 569.06771 |
| | $M_z$/N·m | 48.171635 | 92.279483 | 141.81271 | 189.18662 | 232.75207 |
| 12 | $F_y$/N | 153.12809 | 309.96903 | 471.02111 | 638.66358 | 818.20037 |
| | $M_z$/N·m | 69.106236 | 132.5533 | 203.83158 | 271.97316 | 334.42207 |

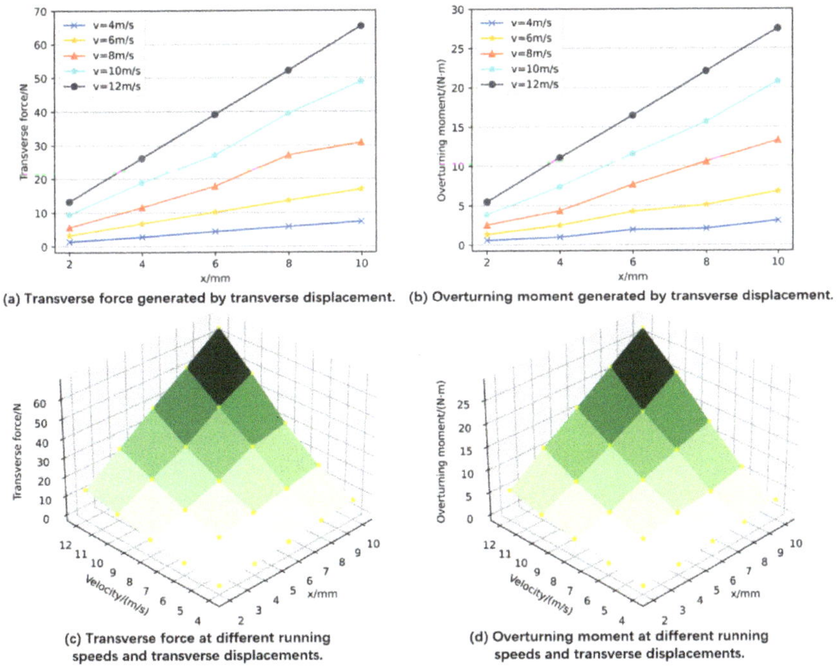


(a) Transverse force generated by transverse displacement. (b) Overturning moment generated by transverse displacement.

(c) Transverse force at different running speeds and transverse displacements.

(d) Overturning moment at different running speeds and transverse displacements.

**Figure 2.** Relationship between car transverse displacement and aerodynamic load.

Figure 2 demonstrates that, as the car's transverse displacement and speed increase, both the transverse force and overturning moment on the car's center of mass exhibit a consistent upward trend. The transverse force and overturning moment exhibit a linear relationship with the transverse displacement. Similarly, both transverse force and overturning moment are approximately proportional to the deflection angle.

By fitting the curve in Figure 4, it can be demonstrated that there is a high R-squared in the quadratic fitting. The ratios of transverse force and overturning moment to transverse displacement at different rated speeds are designated as the influence coefficients $C_{fx}$ and $C_{mx}$. Consider taking $C_{fx}$ and $C_{mx}$ as quadratic polynomials of $v$ and solving the coefficients of the polynomials utilizing the equation shown in Equation (11). Equation (12) is the result obtained by fitting.

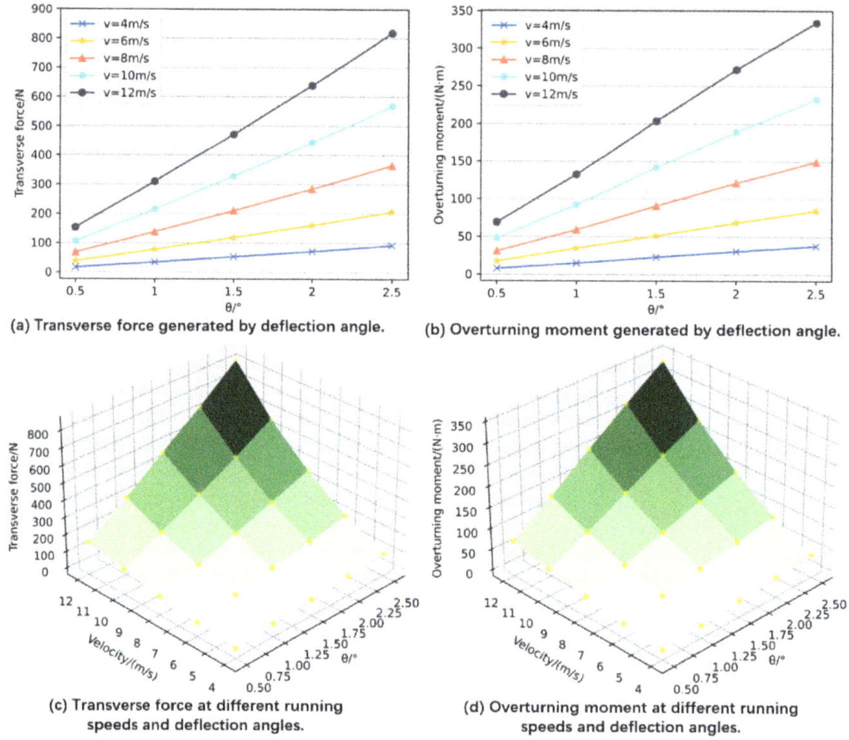


(a) Transverse force generated by deflection angle. (b) Overturning moment generated by deflection angle. (c) Transverse force at different running speeds and deflection angles. (d) Overturning moment at different running speeds and deflection angles.

**Figure 3.** Relationship between deflection angle and aerodynamic load.

$$
\begin{aligned}
C_{fx} &= a_1 v^2 + b_1 v + c_1 \\
C_{mx} &= a_2 v^2 + b_2 v + c_2
\end{aligned}
\tag{5}
$$

$$
\begin{aligned}
F_y &= \left(35.93v^2 + 160.94v - 521.67\right)y \\
M_y &= \left(16.84v^2 + 40.88v - 153.83\right)y \\
F_\theta &= \left(124.65v^2 + 11.48v - 17.85\right)\theta \\
M_\theta &= \left(52.90v^2 + 12.55v - 17.94\right)\theta
\end{aligned}
\tag{6}
$$

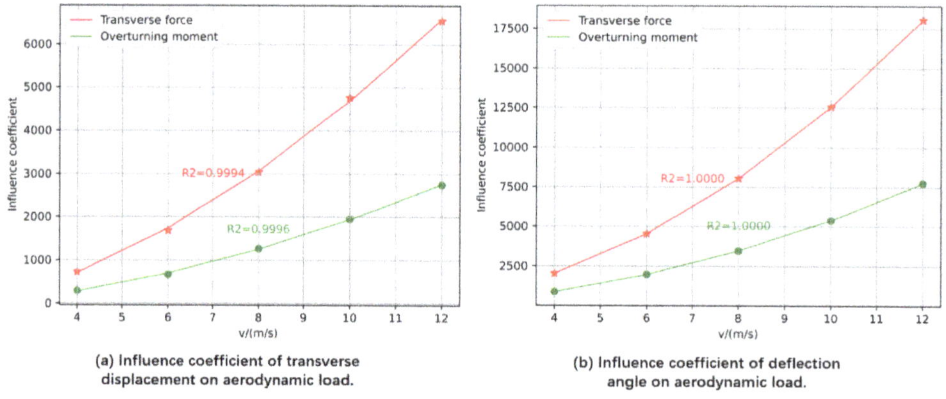


(a) Influence coefficient of transverse displacement on aerodynamic load. (b) Influence coefficient of deflection angle on aerodynamic load.

**Figure 4.** Quadratic polynomial fitting of the influence coefficients.

The influence of air disturbance in the elevator system can be simplified as transverse force and overturning moment acting on the elevator car, and the equivalent aerodynamic load $F_g$ and aerodynamic torque $M_g$ will be applied to the ultra-high-speed elevator car, thereby establishing a transverse vibration model for the ultra-high-speed elevator car.

### 2.2. Guide Excitation Modeling

In elevator installation, achieving symmetrical alignment on both sides is crucial. However, guide rail manufacturing and installation errors can impact their flatness, leading to displacement excitations from bending, tilting, and step variations [14]. These excitations can be theoretically combined to approximate real-world scenarios, though parameter selection is challenging. For easier calculations, following reference [23], real guide rail excitations mainly occur in the low-frequency range below 10 Hz, determined by the guide rail's properties and installation section length [24]. Thus, a low-pass filter processed white noise signal is chosen as the guide rail excitation, retaining only the low-frequency components [25], as shown in Equation (13).

$$w_d = \frac{2\pi \times 10}{s + 2\pi \times 10} \tag{7}$$

## 3. Gas–Solid Coupling Dynamic Modeling of Ultra-High-Speed Elevator

This section establishes a dynamic model of the car's transverse vibration caused by external excitations, including rail excitation and fluctuating aerodynamic pressure, and develops the corresponding state-space equations.

### 3.1. Dynamic Modeling of Elevator Transverse Vibration

The transverse vibration of elevators primarily involves the car system and the guidance system. The car system comprises the car and its frame, linked by supporting rubber, while the guidance system consists of guide rails and shoes. Guide rails stabilize the car's position, counteracting tilting and swaying from traction rope torsion and asymmetric loads. Active guide shoe controllers, along with spring–damping systems, dampen car vibrations, enhancing operational safety and comfort [26,27].

Guide shoes are pivotal in suppressing elevator transverse vibrations. Active guide shoes reduce vibrations by applying forces opposite to them, a common method in elevator vibration control. In dynamic modeling, guide shoes can be simplified as a spring–damping system, as shown in Figure 5. The ARG (active roller guide) in Figure 5 denotes the active control mechanism that applies active control force to the car.

To simplify subsequent calculations, similar variables are merged into a matrix:

$$Z_c = \begin{bmatrix} z_{cu} \\ z_{cd} \end{bmatrix}, X_c = \begin{bmatrix} x_c \\ \theta_c \end{bmatrix}, Z_f = \begin{bmatrix} z_{fu} \\ z_{fd} \end{bmatrix}, X_f = \begin{bmatrix} x_{fc} \\ \theta_{fc} \end{bmatrix} \tag{8}$$

$z_{cu}$ and $z_{cd}$, respectively, represent the transverse displacement of the upper and lower parts of the elevator car;

$x_c$ and $\theta_c$, respectively, represent the transverse displacement and deflection angle of the center of mass position of the elevator car;

$z_{fu}$ and $z_{fd}$, respectively, represent the transverse displacement of the upper and lower parts of the car frame;

$x_{fc}$ and $\theta_{fc}$, respectively, represent the transverse displacement and deflection angle of the center of mass position of the car frame.

$Z_c$ and $Z_f$, as well as $X_c$ and $X_f$, can be represented by the equations shown in Equation (9).

$$Z_c = H_c X_c,\ Z_f = H_f X_f \tag{9}$$

where

$$H_c = \begin{bmatrix} 1 & L_{cu} \\ 1 & -L_{cd} \end{bmatrix}, H_f = \begin{bmatrix} 1 & L_{fu} \\ 1 & -L_{fd} \end{bmatrix} \tag{10}$$

define

$$H_{fc} = \begin{bmatrix} 1 & L_{fcu} \\ 1 & -L_{fcd} \end{bmatrix}, H_t = \begin{bmatrix} 0 & 0 \\ 0 & 1 \end{bmatrix} \tag{11}$$

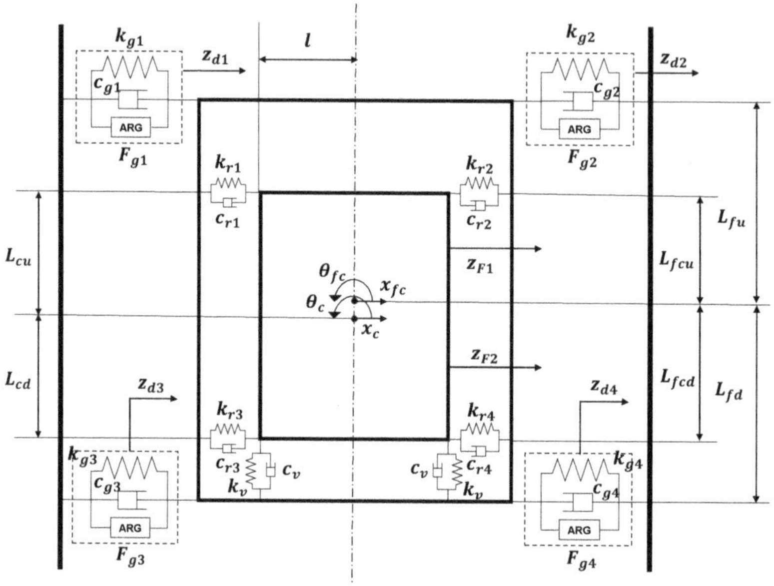


**Figure 5.** Dynamic model of transverse vibration of an ultra-high-speed elevator.

Considering airflow disturbance, the transverse dynamic equation of gas–solid coupling in the ultra-high-speed elevator car is as follows:

$$\begin{aligned} M_f \ddot{X}_f = & H_f{}^T \left( F_{gl} + K_{gl} Z_{dl} + K_{gr} Z_{dr} + C_{gl} \dot{Z}_{dl} + C_{gr} \dot{Z}_{dr} \right) \\ & - H_f{}^T \left( \left( K_{gl} + K_{gr} \right) Z_f + \left( C_{gl} + C_{gr} \right) \dot{Z}_f + F_{gr} \right) \\ & + H_{fc}{}^T \left( (K_{rl} + K_{rr}) Z_c + (C_{rl} + C_{rr}) \dot{Z}_c \right) \\ & - H_{fc}{}^T \left( (K_{rl} + K_{rr}) H_{fc} H_f{}^{-1} Z_f + (C_{rl} + C_{rr}) H_{fc} H_f{}^{-1} \dot{Z}_f \right) \\ & + 2l^2 H_t \left( k_v \left( H_c{}^{-1} Z_c - H_f{}^{-1} Z_f \right) + c_v \left( H_c{}^{-1} \dot{Z}_c - H_f{}^{-1} \dot{Z}_f \right) \right) \end{aligned} \tag{12}$$

$$\begin{aligned} M_c \ddot{X}_c = & H_c{}^T \left( Z_{dF} - (K_{rl} + K_{rr}) Z_c - (C_{rl} + C_{rr}) \dot{Z}_c \right) \\ & + H_c{}^T \left( (K_{rl} + K_{rr}) H_{fc} H_f{}^{-1} Z_f + (C_{rl} + C_{rr}) H_{fc} H_f{}^{-1} \dot{Z}_f \right) \\ & - 2l^2 H_t \left( k_v \left( H_c{}^{-1} Z_c - H_f{}^{-1} Z_f \right) + c_v \left( H_c{}^{-1} \dot{Z}_c - H_f{}^{-1} \dot{Z}_f \right) \right) \end{aligned} \tag{13}$$

The coefficient matrices are derived from the dynamic model parameters (see Appendix A for full matrix definitions).

*3.2. Establishment of System State Equation*

Utilizing x as the state variable equation of the system,

$$x = \begin{bmatrix} Z_f & Z_c & \dot{Z}_f - H_f M_f {H_f}^T \left( C_{gl} Z_{dl} + C_{gr} Z_{dr} \right) & \dot{Z}_c \end{bmatrix}^T \tag{14}$$

The calculation formulas for guide rail excitation, air excitation, and active control force matrix $d_g$, $d_F$, and $u_g$ are as follows:

$$\begin{aligned} d_g &= \begin{bmatrix} Z_{dl} \\ Z_{dr} \end{bmatrix} \\ d_F &= Z_{dF} \\ u_g &= \begin{bmatrix} F_{gl} \\ F_{gr} \end{bmatrix} \end{aligned} \tag{15}$$

Then, the transverse vibration dynamic Equation (13) of the elevator car system can be written in the form of the system state Equation (16):

$$\dot{x} = A(p)x + B_{1g}(p)d_g + B_{1F}(p)d_F + B_{2g}(p)u_g \tag{16}$$

The complete coefficient matrix is detailed in the Appendix A.

The output z can be represented by Equation (17).

$$\begin{cases} z = \begin{bmatrix} \ddot{Z}_c & Z_{sd} & Z_{cd} \end{bmatrix}^T \\ Z_{sd} = Z_f - \frac{Z_{dl} + Z_{dr}}{2} \\ Z_{cd} = Z_c - Z_f \end{cases} \tag{17}$$

where $\ddot{Z}_c$ represents the transverse vibration acceleration experienced by the top and bottom sections of the elevator car, $Z_{sd}$ represents the displacement of the frame's top and bottom from the central axis, and $Z_{cd}$ represents the relative lateral displacement between the elevator car and the elevator frame.

This will result in the following output Equation (18):

$$z = C_z(p)x + D_{zg1}(p)d_g + D_{zF1}(p)d_F + D_{zg2}(p)u_g \tag{18}$$

The complete coefficient matrix is also detailed in Appendix A.

Next, it is essential to include the excitation from the changing flow field during motion into the calculations, employing the concept of time discretization [28], and to separately solve for the flow field excitation and the elevator model at each time step [29], as is shown in Figure 6.

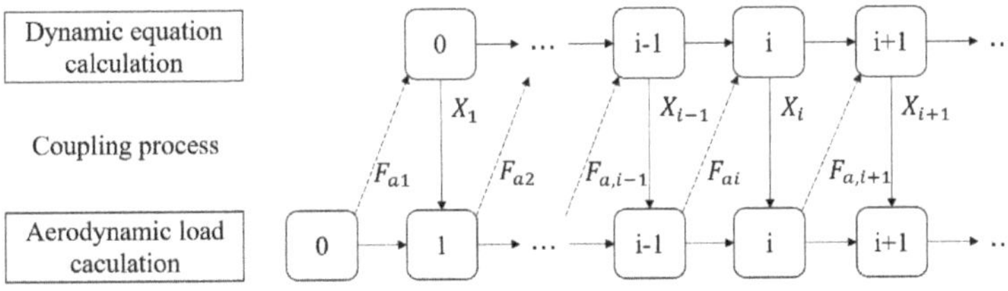


**Figure 6.** Gas-solid coupling calculation step diagram.

## 4. Proposed LQR-Based Vibration Suppression Method

Since LQR can account for multiple control objectives at once and attain a multi-objective balance by optimizing the cost function, thereby adapting to the dynamics of complex systems, this paper uses the LQR-based method to suppress transverse vibration. Meanwhile, for better results, a multi-objective optimization algorithm was used to optimize the weight matrix parameters, obtaining the optimal LQR controller solution. The control process diagram of the gas–solid coupling model is shown in Figure 7.

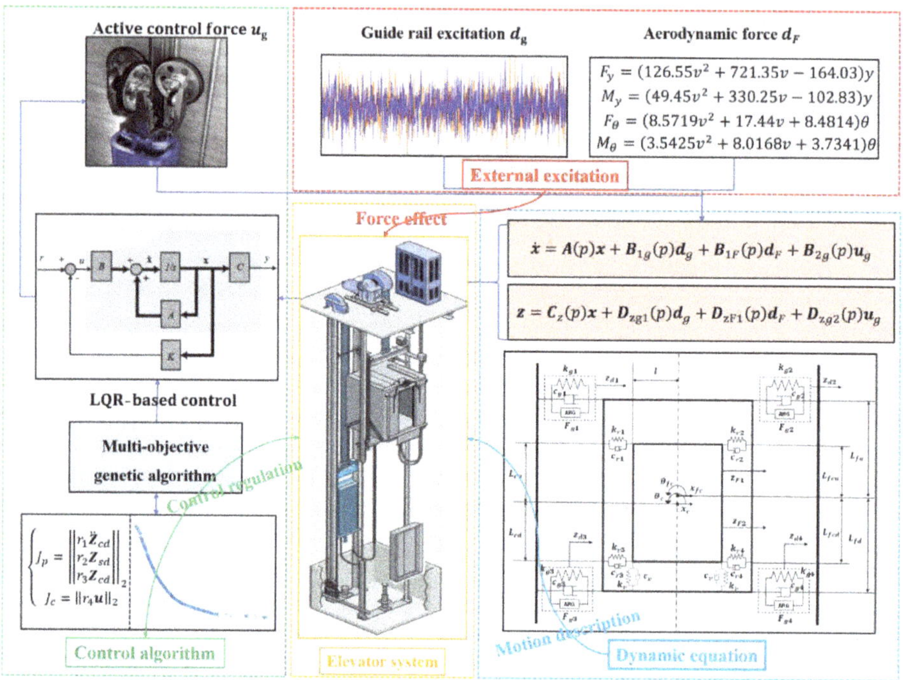


**Figure 7.** Process of the gas–solid coupling.

### 4.1. Linear Quadratic Regulator

LQR (Linear Quadratic Regulator) belongs to the use of linear quadratic optimal control. Before applying LQR design, it is necessary to define a time-domain continuous performance functional. The following Equation (19) is commonly used for infinite time-domain performance functions:

$$J = \frac{1}{2}\int_0^\infty \left[x(t)^T Q x(t) + u(t)^T R u(t)\right] dt \tag{19}$$

Optimal control aims to identify a controller that minimizes the performance function. The state cost weight matrix Q and control cost weight matrix R, typically diagonal, are to be designed. And these two terms each have their own weight coefficients. Under the same error, the larger the weight, the greater the proportion of this term in $J$ [30,31].

The optimal control law $u = -Kx$ is obtained by solving the algebraic Riccati equation:

$$\begin{gathered} A^T P + PA + Q - PBR^{-1}B^T P = 0 \\ K = R^{-1}B^T P \end{gathered} \tag{20}$$

The derivation process is detailed in Appendix A.

As per LQR theory, the closed-loop system is asymptotically stable if (A, B) is controllable and Q and R are positive definite [32,33].

### 4.2. Multi-Objective Genetic Algorithm

The multi-objective genetic algorithm (MOGA) involves several key definitions [34]:

Definition 1: Pareto Solution

Consider a multi-objective optimization problem with objective function $F(X) = \{f_1(x), f_2(x), f_3(x), \ldots f_k(x)\}$ where $k \geq 2$. For two solution vectors $x_0$ and $x_1$ in the solution space, $x_0$ dominates$x_1$ if there exists an $i \in [1,k]$ such that $f_i(x_0) > f_i(x_1)$, and for all $i \in [1,k]$, $f_i(x_0) > f_i(x_1)$. A solution is a Pareto optimal solution if no other solution in the space is better. The set of all such solutions is the Pareto optimal set, expressed as follows:

$$PS = \{u \in \Omega | \nexists v \in \Omega, \mathrm{v} \succ \mathrm{u}\} \tag{21}$$

Here, u is a solution vector, and $\Omega$ is the solution space.

Definition 2: Non-inferior solution sorting.

This sorting is crucial in MOGA as it guides the optimization search direction. The process is as follows:

For solution space S, pairwise comparison yields the non-inferior set $S_1$.

Update $S = S - S_1$, and repeat to find $S_2$.

Continue until S is empty.

Solutions in $S_1$ are first-level non-inferior; those in $S_2$ are second-level, etc. $S_1$ contains the optimal solutions.

In each generation, selection, crossover, and mutation operations were applied to the population to generate a new one. Then, fitness calculation and non-dominated solution ranking were performed, repeating until completion.

This approach has been successfully applied in complex dynamic systems such as milling stability prediction [35], where Pareto-optimal solutions balance conflicting objectives like machining efficiency and vibration suppression.

### 4.3. Design and Parameter Optimization of LQR Controller

Based on the gas–solid coupling dynamic model of the ultra-high-speed elevator established in Section 3.2, the cost function is established as follows:

$$J = \int_0^\infty x^T Q x + u^T R u \, dt \tag{22}$$

Q and R are both weight matrices, where $Q = diag(q_1, q_2, q_3, q_4, q_5, q_6, q_7, q_8)$, $R = diag(r_1, r_2, r_3, r_4)$. The size of the diagonal elements in matrices $Q$ and $R$ determines the weight of each corresponding element in $x(t)$ and $u(t)$ during the optimization process, respectively.

The main evaluation criteria for vibration suppression in ultra-high-speed elevators are the car's transverse acceleration and displacement and the displacement relative to the frame. To meet these criteria, the weight matrix Q's corresponding elements can be set to higher values for better control effects. Some scholars have successfully optimized LQR controllers using optimization algorithms [36,37].

Adjusting Q and R matrices can tailor the system's control performance. However, manual tuning becomes impractical when there are too many parameters, making the design process complex. Therefore, leveraging the multi-objective genetic algorithm from Section 4.2, we can design a fitting fitness function to optimize the two weighting matrices, achieving an LQR controller with desired performance.

Firstly, design the fitness function as follows:

$$\begin{cases} J_p = \left\| \begin{matrix} r_1\ddot{\mathbf{Z}}_{cd} \\ r_2\mathbf{Z}_{sd} \\ r_3\mathbf{Z}_{cd} \end{matrix} \right\|_2 \\ J_c = \|r_4\mathbf{u}\|_2 \end{cases} \quad (23)$$

where $r_1$, $r_2$, $r_3$ and $r_4$ are regularization coefficients, and the performance cost $J_p$ is calculated from the H2 norm of the $\ddot{\mathbf{Z}}_{cd}$, $\mathbf{Z}_{sd}$ and $\mathbf{Z}_{cd}$. The control cost $J_c$ is calculated from the H2 norm of the control force.

A simulation control system for ultra-high-speed elevators is built using the Simulink toolbox. In MATLAB, the LQR controller is solved using the lqr() function and the system state equation:

$$K = lqr(A, B, Q, R) \quad (24)$$

The multi-objective genetic algorithm was used to optimize the weight matrix Q and R parameters. Since the optimization's main goal is to suppress the car's transverse acceleration, the weight matrix parameters' upper and lower limits were set as follows:

$l_b = [10^{-6}, 10^{-6}, 10^{-6}, 10^{-6}, 10^{-6}, 10^{-6}, 10^{-6}, 10^{-6}, 10^{-8}, 10^{-8}, 10^{-8}, 10^{-8}]$,

$u_b = [10^2, 10^2, 10^6, 10^6, 10^2, 10^2, 10^6, 10^6, 10^{-5}, 10^{-5}, 10^{-5}, 10^{-5}]$.

The maximum allowable control cost $J_{c\ max}$ was set to 600. The population size was 400, and the maximum number of generations was 500. The regularization coefficients are $r_1 = 10^3, r_2 = 10, r_3 = 2 \times 10^{-2}, r_4 = 1 \times 10^{-3}$. Optimize and iterate the parameters in the weight matrix to obtain the Pareto optimal solution curve of the parameters in the weight matrix, as shown in Figure 8.

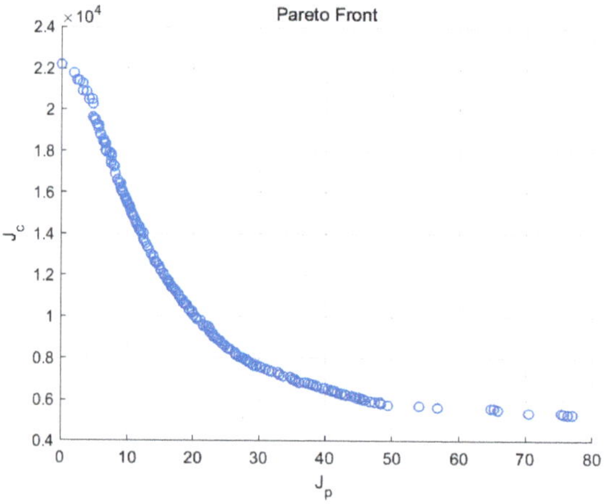


**Figure 8.** Multi-objective optimization results of LQR weight matrices Q and R.

In Figure 8, the horizontal axis denotes the control system's performance indicators, and the vertical axis indicates the control cost. The curve is a Pareto curve made up of optimized controller parameters. Curves closer to the vertical axis mean smaller system output errors, while those nearer to the horizontal axis mean lower control costs. The elevator structural parameters are shown in Table 3.

**Table 3.** Elevator structure parameter values.

| Parameters | Unit | Value |
|---|---|---|
| $m_c$ | kg | 1100 |
| $I_c$ | kg·m$^2$ | 1600 |
| $m_{fc}$ | kg | 2400 |
| $I_{fc}$ | kg·m$^2$ | 8600 |
| $k_v$ | N/m | 900,000 |
| $c_v$ | N·s/m | 3,600,000 |
| $k_{ri}(i=1,2)$ | N/m | 200,000 |
| $k_{ri}(i=3,4)$ | N/m | 170,000 |
| $c_{ri}(i=1,2)$ | N·s/m | 2300 |
| $c_{ri}(i=3,4)$ | N·s/m | 1800 |
| $k_{gi}(i=1\ldots4)$ | N/m | 120,000 |
| $c_{gi}(i=1\ldots4)$ | N·s/m | 2000 |
| $L_{cu}$ | m | 1.6 |
| $L_{cd}$ | m | 1.4 |
| $L_{fu}$ | m | 3.3 |
| $L_{fd}$ | m | 4.5 |
| $L_{fcu}$ | m | 1.25 |
| $L_{fcd}$ | m | 1.75 |
| $l$ | m | 0.5 |

## 5. Results and Discussion

### 5.1. Comparison Results of Control Methods

In order to compare the performance of the LQR controller before and after optimization, empirical weight matrices are used as a benchmark. Additionally, traditional PID control and no control are selected for comparison to confirm the method's efficiency.

From the optimization results using the MOGA, a weight matrix with a control cost of approximately 45 is chosen as the optimized LQR parameter. The corresponding Q and R matrices are given by the following:

$Q = \text{diag}\,[57.327, 40.264, 335{,}248.304, 370{,}400.807, 54.426, 67.531, 566{,}579.656, 822593.37]$

$R = \text{diag}\,[4.782\times10^{-6}, 3.076\times10^{-7}, 3.773\times10^{-6}, 3.4050\times10^{-7}]$

Based on experience and control objectives, this paper manually sets another set of weight matrices to solve the LQR controller:

$Q_1 = \text{diag}\,[0, 1, 100, 100, 1, 1, 100, 100]$

$R_1 = \text{diag}\,[1\times10^{-6}, 1\times10^{-6}, 1\times10^{-6}, 1\times10^{-6}]$

Multivariable PID is used to compare with LQR-based methods. Since the input state variable is an $8\times1$ vector and the control output is a $4\times1$ vector, the proportion P, integral I, and derivative D in multivariate PID are a $4\times8$ matrix. Here, the multivariable PID controller employs gain matrices. After multiple iterations of simulation, the relationship between the gain matrix $M$ and $P$, $I$, and $D$ is set as follows:

$$K_P = 2M,\ K_I = 3M,\ K_D = 0.01M \tag{25}$$

where the base matrix $M = \begin{bmatrix} 10^5 & 10^5 & -10^5 & -10^5 & 10^4 & 10^3 & 10^3 & 10^3 \\ 10^6 & 10^6 & -10^6 & -10^6 & 10^4 & 10^4 & 10^5 & 10^5 \\ 10^5 & -10^5 & 10^5 & 10^5 & -10^4 & -10^4 & -10^4 & -10^4 \\ 10^6 & -10^5 & 10^6 & 10^6 & -10^4 & -10^4 & -10^5 & -10^5 \end{bmatrix}$.

The relevant coefficients are explained as follows:

Force-dominant rows (1–2): High-magnitude terms ($10^5$–$10^6$) in columns 1–4 resist excitation. Displacement-focused rows (3–4): Negative coupling terms compensate for car–frame interactions. Velocity compensation: Lower-gain terms ($10^3$–$10^4$) in columns 5–8 match vibration energy at different positions.

After selecting the elevator speed, Equation (6) is used to obtain air excitation, which is then added to the control model. According to the weight matrices Q, R, $Q_1$, and $R_1$, the feedback gain matrices K and $K_1$ are calculated using the lqr() function in MATLAB 2023b. These matrices are then substituted into the same control model to obtain the numerical simulation results of the LQR controller before and after optimization.

The results at speed $v = 10$ m/s are as follows: Figures 9–11 show the simulation results of the output $z$ (the transverse vibration acceleration of the bottom of the car, the displacement of the bottom of the frame off the centerline, the relative transverse displacement between the bottom car and the bottom frame) at elevator speed $v = 10$ m/s.

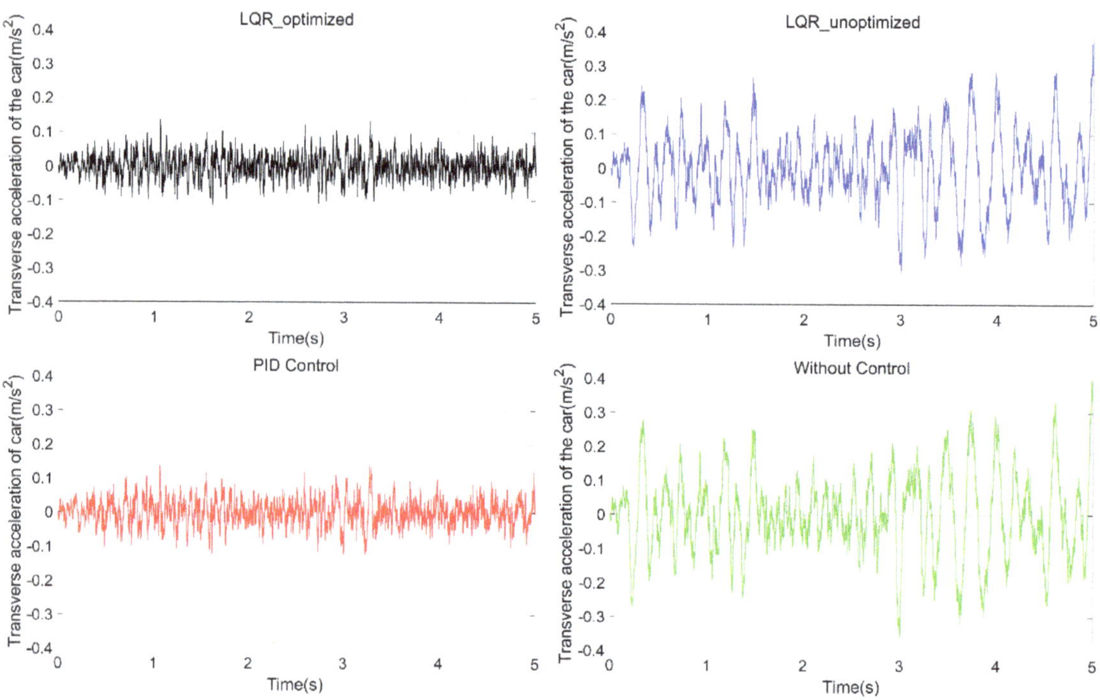


**Figure 9.** Comparison of the transverse acceleration of the car.

Figure 9 compares the lateral acceleration of the elevator under four control methods. It shows that the optimized LQR controller effectively reduces the lateral acceleration. Figure 10 compares the displacement of the car frame off the centerline. The three active control methods have almost no effect on controlling this displacement because this part's proportion in the Q weight setting is not large, and the car frame offset is almost not suppressed. Figure 11 shows the comparison of the displacement of the car relative to the frame. The peak value obtained by the optimized LQR method is the smallest among the four control cases.

The root mean square values of system performance indicators, control costs, lateral acceleration, etc., under four control methods are shown in Tables 4–6 (at elevator speeds of 10 m/s, 8 m/s, and 6 m/s, respectively).

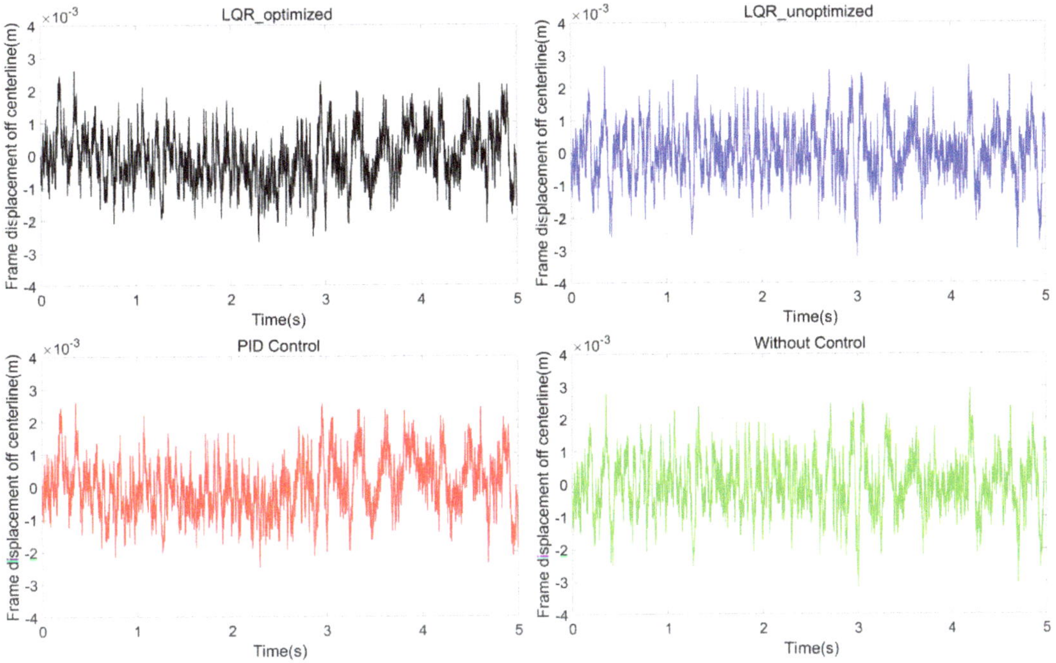


**Figure 10.** Comparison of the displacement of the car frame off the centerline.

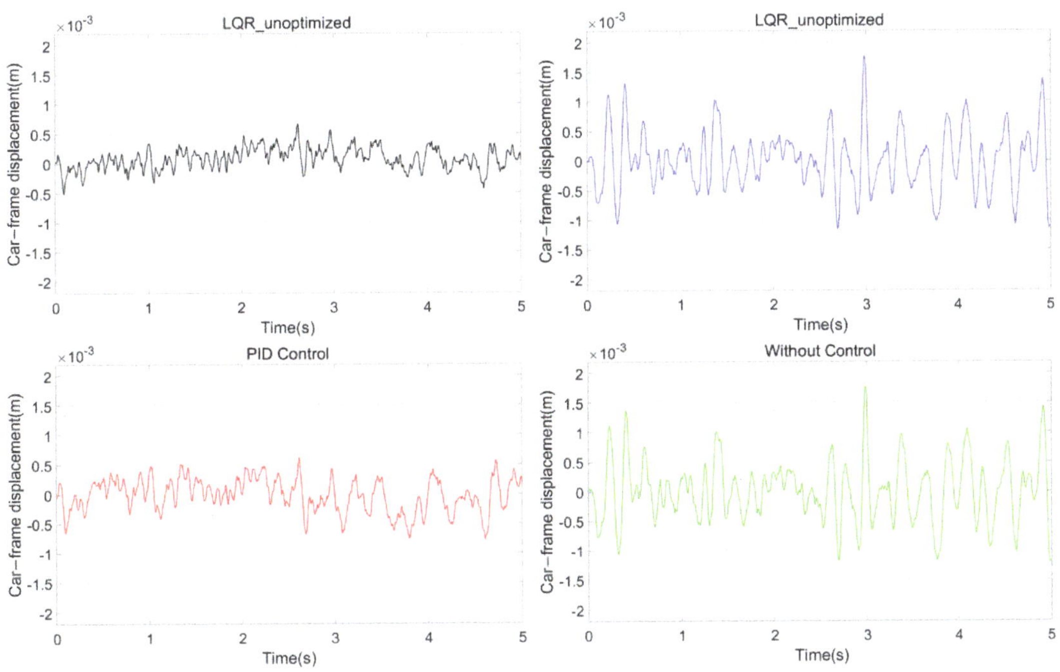


**Figure 11.** Comparison of the displacement of the car relative to the frame.

**Table 4.** Comparison of system output performance before and after optimization (v = 10 m/s).

| | Optimized LQR Control | Unoptimized LQR Control | PID Control | Without Control |
|---|---|---|---|---|
| Performance index ($J_p$) | $5.192 \times 10^3$ | $1.621 \times 10^4$ | $5.787 \times 10^3$ | $2.141 \times 10^4$ |
| Control cost ($J_c$) | 41.943 | 4.726 | 24.832 | 0 |
| Displacement of the car frame off the centerline ($Z_{sd}$) | $8.653 \times 10^{-4}$ | $8.705 \times 10^{-4}$ | $8.931 \times 10^{-4}$ | $8.894 \times 10^{-4}$ |
| Displacement of the car relative to the frame ($Z_{cd}$) | $2.006 \times 10^{-4}$ | $5.538 \times 10^{-4}$ | $3.278 \times 10^{-4}$ | $5.702 \times 10^{-4}$ |
| Transverse acceleration at the bottom of the car ($\ddot{Z}_c$) | 0.037 | 0.114 | 0.041 | 0.130 |
| Control force ($F_{g1}$) | 90.558 | 17.502 | 54.633 | 0 |
| Control force ($F_{g2}$) | 114.787 | 17.502 | 91.056 | 0 |
| Control force ($F_{g3}$) | 215.523 | 16.261 | 168.792 | 0 |
| Control force ($F_{g4}$) | 194.744 | 16.261 | 168.792 | 0 |

**Table 5.** Comparison of system output performance before and after optimization (v = 8 m/s).

| | Optimized LQR Control | Unoptimized LQR Control | PID Control | Without Control |
|---|---|---|---|---|
| Performance index ($J_p$) | $4.583 \times 10^3$ | $1.838 \times 10^4$ | $5.007 \times 10^3$ | $2.141 \times 10^4$ |
| Control cost ($J_c$) | 40.302 | 5.342 | 24.887 | 0 |
| Displacement of the car frame off the centerline ($Z_{sd}$) | $8.706 \times 10^{-4}$ | $8.724 \times 10^{-4}$ | $9.082 \times 10^{-4}$ | $9.361 \times 10^{-4}$ |
| Displacement of the car relative to the frame ($Z_{cd}$) | $2.131 \times 10^{-4}$ | $5.470 \times 10^{-4}$ | $3.349 \times 10^{-4}$ | $5.961 \times 10^{-4}$ |
| Transverse acceleration at the bottom of the car ($\ddot{Z}_c$) | 0.032 | 0.130 | 0.035 | 0.151 |

**Table 6.** Comparison of system output performance before and after optimization (v = 6 m/s).

| | Optimized LQR Control | Unoptimized LQR Control | PID Control | Without Control |
|---|---|---|---|---|
| Performance index ($J_p$) | $4.161 \times 10^3$ | $1.638 \times 10^4$ | $4.495 \times 10^3$ | $2.176 \times 10^4$ |
| Control cost ($J_c$) | 38.871 | 5.641 | 24.172 | 0 |
| Displacement of the car frame off the centerline ($Z_{sd}$) | $8.453 \times 10^{-4}$ | $8.670 \times 10^{-4}$ | $8.721 \times 10^{-4}$ | $9.957 \times 10^{-4}$ |
| Displacement of the car relative to the frame ($Z_{cd}$) | $2.336 \times 10^{-4}$ | $5.636 \times 10^{-4}$ | $3.535 \times 10^{-4}$ | $6.155 \times 10^{-4}$ |
| Transverse acceleration at the bottom of the car ($\ddot{Z}_c$) | 0.029 | 0.116 | 0.032 | 0.154 |

### *5.2. Multidimensional Performance Evaluation*

In order to intuitively compare the pros and cons of each method, radar charts are built using performance index, control cost, and three transverse vibration-related outputs as evaluation indicators. The radar charts for three different speeds are presented in Figures 12–14. Since smaller values of the performance index, transverse acceleration, frame displacement off the centerline, and car–frame displacement indicate better methods, their reciprocals are used and combined with cost control as radar chart indicators. As shown in Figures 12–14, the optimized LQR control method has the largest range, with all indicators surpassing those of other methods. PID control is the second-best, underperforming compared to the optimized LQR in some aspects. The unoptimized LQR control method ranks third in terms of performance. All three methods outperform the no-control approach.

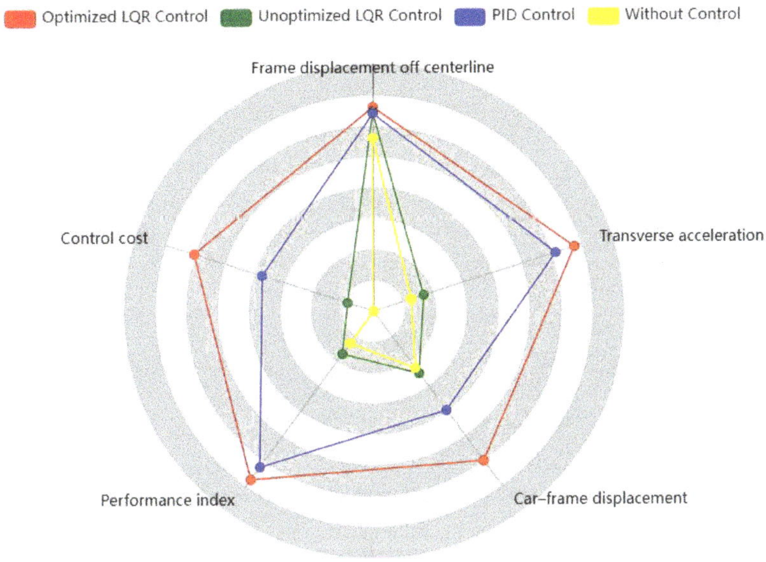


**Figure 12.** Radar chart of performance indicators at v = 6 m/s.

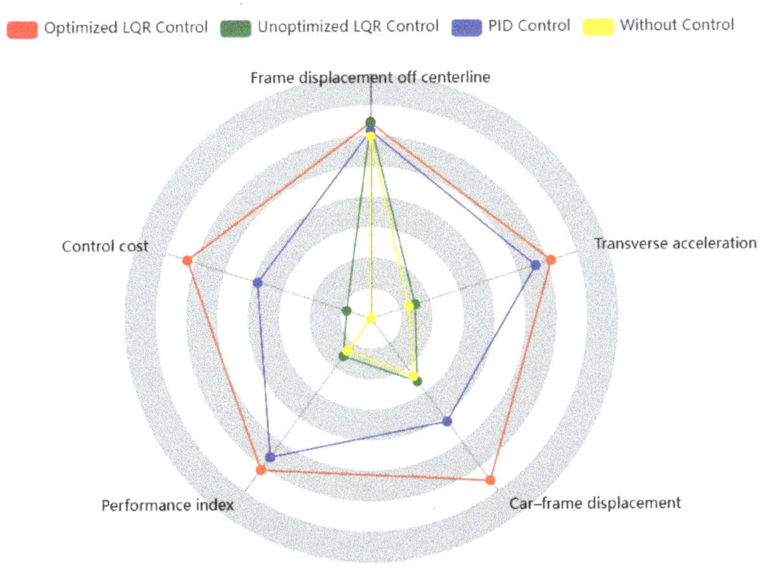


**Figure 13.** Radar chart of performance indicators at v = 8 m/s.

Taking the 10 m/s scenario as an example, comparison between the optimized and unoptimized LQR controller reveals the following: The optimized LQR controller achieved a 25.18% reduction in transverse acceleration, a 0.60% decrease in car frame displacement off the centerline, and a 63.78% decrease in relative car–frame displacement. Additionally, the root mean square values of the active control forces were all under 250 N, with the maximum control force not exceeding 850 N, as shown in Figure 15. These data indicate that the optimized LQR controller can more effectively reduce the elevator's transverse vibration acceleration and the displacement of the car relative to the frame compared to its unoptimized counterpart. The optimized LQR controller has proven to be highly effective in controlling the elevator's transverse acceleration, thus affirming the efficiency of the optimization approach in designing LQR controllers. By establishing appropriate objective functions and weight matrices, the optimal LQR controller can be derived.

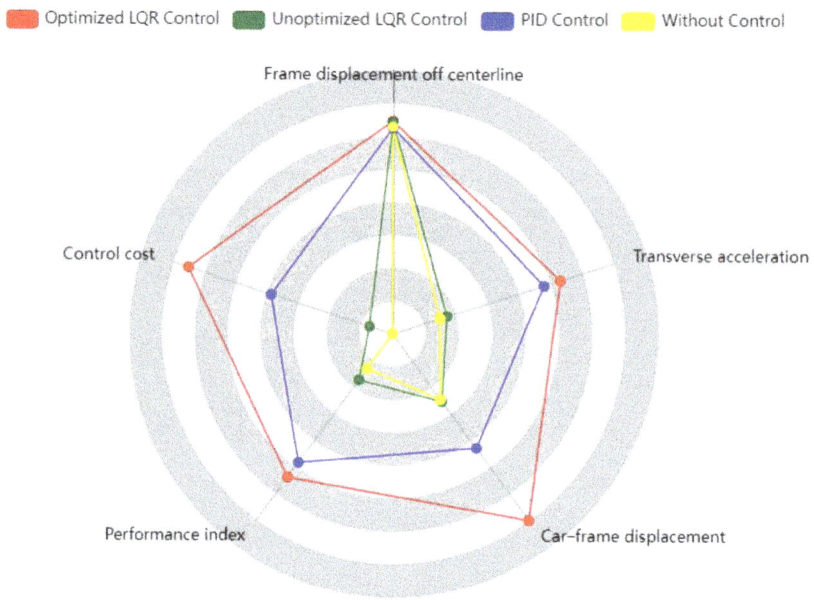


**Figure 14.** Radar chart of performance indicators at v = 10 m/s.

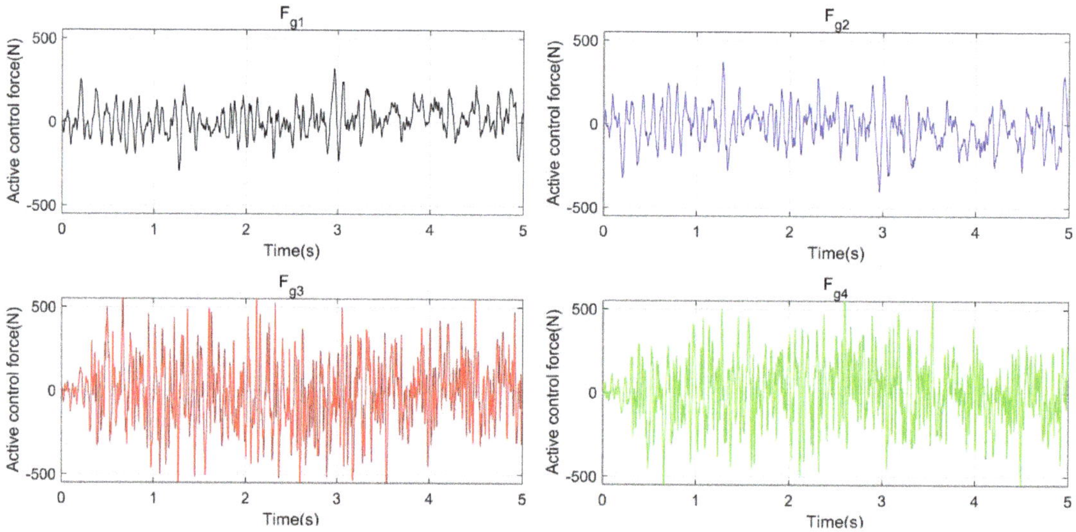


**Figure 15.** Active control force variation diagram (first 5 s).

When compared to PID control, the LQR controller demonstrated greater effectiveness in reducing the following: (1) transverse acceleration by 10.27%, (2) car–frame displacement by 3.12%, and (3) relative car–frame displacement by 38.8%. These results indicate that the proposed LQR-based method is superior to traditional PID control. In the context of ultra-high-speed elevators, which are highly nonlinear, strongly coupled, and characterized by variable parameters, PID controllers may struggle to adapt to rapidly changing control requirements due to their lack of predictive capabilities. Consequently, they may not provide sufficient control accuracy and stability for systems that demand rapid response and high precision. In contrast, the LQR is better equipped to handle multivariable control problems in such complex systems, offering faster response times and higher control accuracy.

### 5.3. Speed Sensitivity Research

To analyze the impact of air excitation on transverse vibration across different speeds, we have drawn transverse vibration diagrams for various speeds, presented in Figures 16–19. To clearly illustrate how transverse vibration varies with speed, the left figures display normalized values at three different velocities using the maximum–minimum normalization method, while the right figures show the true values. Given the significant differences in numerical ranges, logarithmic scales are utilized. In these figures, PI denotes performance index, CC denotes control cost, FDOC denotes frame displacement off centerline, CFD denotes car–frame displacement, and TA denotes transverse acceleration. Since the without control method has a control cost of 0, there is no CC in Figure 19.

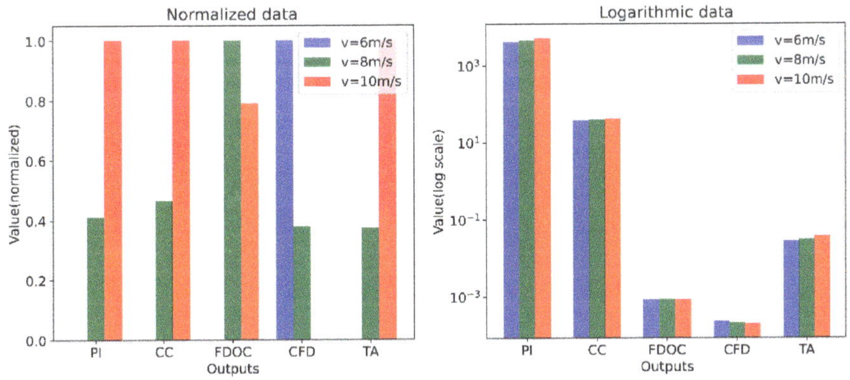


**Figure 16.** Comparison of optimized LQR control for transverse vibration at different speeds.

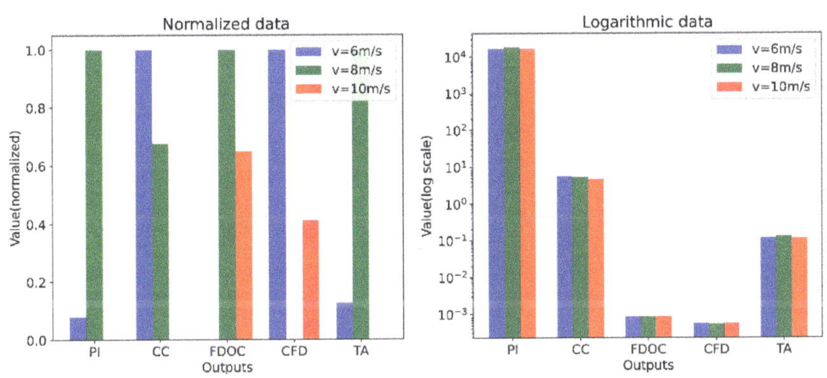


**Figure 17.** Comparison of unoptimized LQR control for transverse vibration at different speeds.

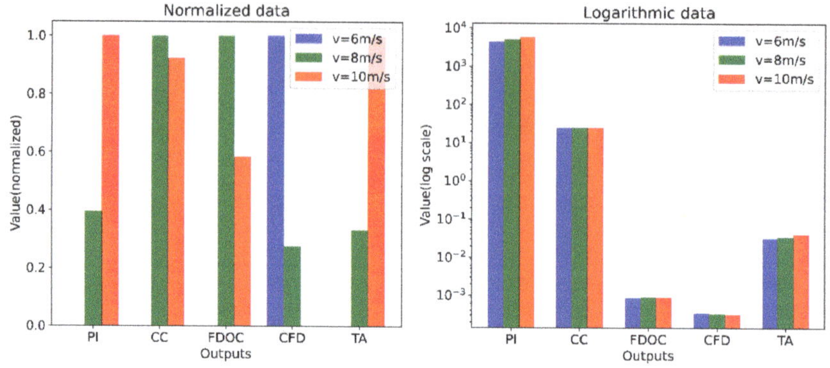


**Figure 18.** Comparison of LQR control for transverse vibration at different speeds.

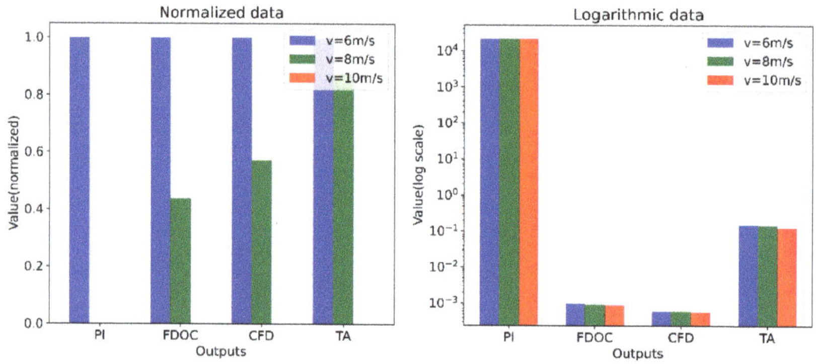


**Figure 19.** Comparison without control for transverse vibration at different speeds.

From Figures 16–19, it is evident that, generally, higher speeds result in greater control costs and performance indices. When comparing LQR and PID control methods at different speeds, at v = 10 m/s, 8 m/s, and 6 m/s, the LQR outperforms PID in reducing transverse vibration acceleration by 10.27%, 8.47%, and 7.55%, respectively. Similarly, the LQR surpasses PID in suppressing car–frame displacement by 38.8%, 36.3%, and 33.9% at these speeds.

Overall, as elevator speed increases, airflow disturbance becomes more intense, exerting a greater impact on the elevator. Correspondingly, LQR control demonstrates better effectiveness in suppressing transverse vibration. However, from Figures 17–19, it can be observed that, when the control method is suboptimal, transverse acceleration may slightly decrease with increasing speed. This can be explained as follows:

- Although speed increase typically heightens air resistance, in some cases, it can stabilize the airflow between the elevator car and hoistway, thereby reducing aerodynamic-induced vibration.
- At high speeds, the interaction between airflow and the guide shoe–guide rail contact may become closer, potentially mitigating vibrations caused by guide rail unevenness.

### 5.4. Limitations and Robustness Analysis

Though the MOGA-LQR performs well at different speeds, its engineering application boundaries still need to be explored. This section quantitatively analyzes the engineering applicability boundary of the proposed method, focusing on three dimensions: optimization robustness, computational cost, and nonlinear interference impacts.

We conducted tests on four sets of MOGA optimization algorithms for different population sizes and the number of generations and found that the final results showed almost consistent trends, as shown in Figure 20 (with an initial parameter x of 0). Meanwhile, there is not much difference in the LQR coefficients of the same points on different graphs. The results confirm MOGA's strong convergence characteristics for our optimization problem.

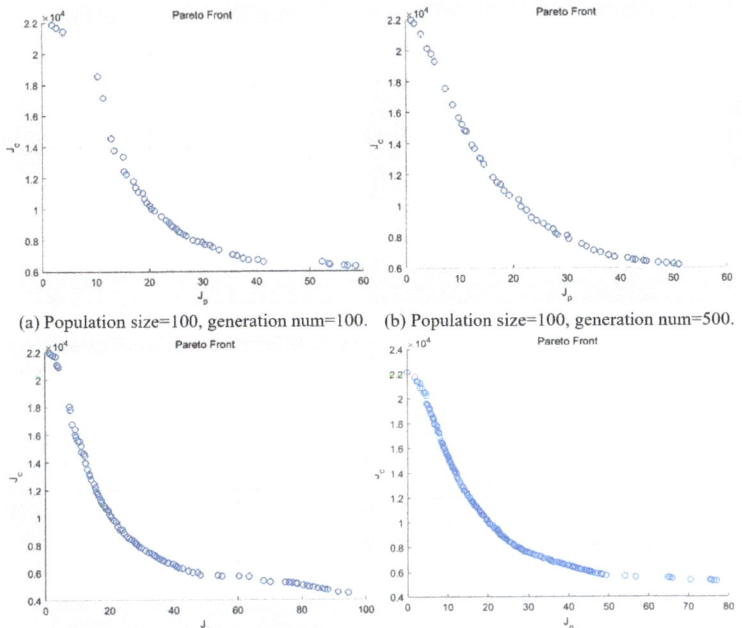


(a) Population size=100, generation num=100. (b) Population size=100, generation num=500.

(c) Population size=200, generation num=100. (d) Population size=400, generation num=500.

**Figure 20.** The results of the MOGA in different population sizes and the number of generations.

The computational cost of the multi-objective genetic algorithm (MOGA) optimization is primarily determined by three key factors: population size ($N$), number of generations ($G$), and fitness evaluation time per individual ($t_f$). The time consumption of genetic operations (crossover/selection/mutation, etc.) is much smaller than simulation and can be ignored. The total optimization time can be expressed as follows:

$$T_{total} \approx N \times G \times t_f + T_A \quad (26)$$

where $N$ is the population size, $G$ is the number of generations (maximum 500 generations), $t_f$ is the time per fitness evaluation, and $T_A$ is the algorithm operation time.

The fitness evaluation time constitutes the dominant computational cost, as each evaluation requires running the full Simulink model simulation. Our analysis reveals scales with both elevator speed and model complexity:

1 Speed dependency: We tested 10 times each at 6 m/s, 8 m/s, and 10 m/s, and the time results for each run are as follows: At 6 m/s: $t_f = [0.871] \pm [0.017]$ s. At 8 m/s: $t_f = [0.874] \pm [0.027]$ s;. At 10 m/s: $t_f = [0.883] \pm [0.016]$ s. It can be observed that higher operating speeds do not significantly increase aerodynamic complexity and do not prolong the simulation time for each evaluation.

2 Model complexity impact: Obviously, more complex models will lead to an increase in simulation time. We tested 10 times using the LQR method, PID method, and

uncontrolled conditions ($v = 10$ m/s), and the time results for each run are as follows: LQR: $t_f = [0.883] \pm [0.016]$ s; PID: $t_f = [0.891] \pm [0.018]$ s; without control: $t_f = [0.788] \pm [0.011]$ s. In 10 tests, the PID method consumed slightly longer time than the LQR method, which is related to the differentiation and integration in the calculation process, but this part has little impact on the overall time. In short, the more complex the control method (the more complex the model), the longer the time required for a single Simulink calculation, and the longer the time consumed by the entire MOGA.

Current modeling uses Gaussian white noise filtered through a low-pass filter to represent rail excitation. However, long-term wear alters three critical statistical properties of rail irregularities:

1 Mean shift increases baseline vibration.
2 Variance expansion ($\sigma^2$ grows after numerous cycles) amplifies stochastic disturbances.
3 Peak deformation intensification causes intermittent high-amplitude impulses (>3$\sigma$ events will increase).

These changes significantly alter the guiding incentives, and one approach is to monitor the data of the guide rail in real time to obtain relevant attributes. However, real-time monitoring remains challenging because laser displacement sensors cannot be permanently installed in hoistways due to contamination risks, and strain gauges measure and achieve wireless transmission in the presence of vibration under the shaft.

Future implementations could embed relevant sensors in guide shoes to track wear progression and dynamically adjust the excitation model's parameters ($\mu$, $\sigma$, kurtosis).

The quadratic approximation in Equation (6) effectively captures steady-state effects but neglects irregular flow phenomena in turbulence, such as turbulence intermittency and transient vortex shedding. The relevant content deserves further research, such as establishing more reasonable models or developing adaptive turbulence correction methods to achieve better results.

There are currently nonlinear problems, and deep neural networks are good at handling these problems. The fusion of physics-based models and data-driven neural networks can enhance the riding experience of elevators. In addition, it can also be used to predict the lifespan of elevators, assess the risks of key components, and more.

The Nonlinear Energy Sink (NES) is a promising vibration control device [38]. Combining the NES with the proposed LQR-based active control method could create a hybrid control strategy that leverages the strengths of both approaches. While the LQR controller actively adjusts the guide shoes to counteract disturbances, the NES passively absorbs and dissipates residual vibrational energy. This combination could enhance the overall effectiveness of the vibration suppression system, and this is also one of the directions that we will try in the future.

In the future, elevators will become more intelligent. Typically equipped with accelerometers and gyroscopes to measure vibration acceleration and angular velocity, elevators can enhance the accuracy and reliability of vibration monitoring and suppression through data fusion algorithms [39] that integrate data from these sensors.

## 6. Conclusions

This paper incorporates active control force into the elevator system via active guide shoes, constructing a four-degree-of-freedom ultra-high-speed elevator transverse vibration dynamic model. The airflow during operation is analyzed as incompressible viscous turbulence within the shaft. Fluent is adopted to analyze the flow field, iteratively solving until convergence to identify the air excitation. Quantitative analysis determines the relationship between air excitation and operating speed, car transverse displacement, and car deflection angle. Then, shaft airflow disturbances are transformed into transverse deflection forces and torques on the car, establishing a gas–solid coupling dynamic model for ultra-high-speed elevators. Using this model, the state equation is derived to solve the LQR controller with MATLAB tools. Simultaneously, a multi-objective genetic algorithm optimizes the key weight matrices Q and R of the LQR controller, using system performance and control cost as objective functions. The Pareto solution curves for parameters in these matrices are obtained, and suitable parameters are selected to solve the optimal LQR controller. Numerical simulations via Simulink show that, compared to PID control, the LQR-based method reduces transverse acceleration by 7.55%, 8.47%, and 10.27% at 6 m/s, 8 m/s, and 10 m/s, respectively, verifying the effectiveness of the controller, and the effectiveness of the proposed method increases with the increase in speed. This work provides an effective solution for enhancing the ride comfort of ultra-high-speed elevators and holds potential for application in the vibration control of high-speed transportation systems. While this study establishes a numerical framework for gas–solid coupling vibration control, further investigations are warranted: (1) experimental validation of the proposed model on full-scale ultra-high-speed elevator systems, particularly under real-world disturbances (e.g., shaft turbulence), to verify simulation accuracy; (2) integration of passive control mechanisms, such as Nonlinear Energy Sinks (NESs), to develop hybrid passive–active suppression systems.

**Author Contributions:** Conceptualization, C.L.; Methodology, J.J.; Software, J.J. and P.X.; Validation, J.J. and C.Q.; Formal analysis, P.X.; Investigation, J.J.; Resources, C.L.; Data curation, C.Q.; Writing—original draft preparation, J.J.; Writing—review and editing, J.J., P.X. and C.Q.; Visualization, J.J. and P.X.; Supervision, C.L.; Project administration, C.L.; Funding acquisition, C.L. All authors have read and agreed to the published version of the manuscript.

**Funding:** This research received no external funding.

**Data Availability Statement:** Dataset available on request from the authors.

**Acknowledgments:** This study was supported by the Natural Science Foundation of China (Grant No. 51935007).

**Conflicts of Interest:** The authors declare that they have no conflicts of interest.

## Nomenclature

| | |
|---|---|
| $m_c$ | Car quality |
| $I_c$ | Car's moment of inertia |
| $m_{fc}$ | Frame quality |
| $I_{fc}$ | Frame moment of inertia |
| $k_v$ | Stiffness coefficient of the supporting components at the bottom of the car and the frame |
| $c_v$ | Damping coefficient of the supporting components at the bottom of the car and the frame |
| $k_{ri}(i = 1,2)$ | Stiffness coefficient of the upper support components of the car and the frame |
| $k_{ri}(i = 3,4)$ | Stiffness coefficient of the lower supporting components of the car and the frame |
| $c_{ri}(i = 1,2)$ | Damping coefficient of the upper support components of the car and the frame |

| | |
|---|---|
| $c_{ri}(i=3,4)$ | Damping coefficient of the lower supporting components of the car and the frame |
| $k_{gi}(i=1\ldots4)$ | Stiffness coefficient of the rolling guide shoe spring |
| $c_{gi}(i=1\ldots4)$ | Damping coefficient of the rolling guide shoe spring |
| $L_{cu}$ | The vertical distance from the center of mass of the car to the top of the car |
| $L_{cd}$ | The vertical distance from the center of mass of the car to the bottom of the car |
| $L_{fu}$ | The vertical distance from the center of mass of the car frame to the top of the frame |
| $L_{fd}$ | The vertical distance from the center of mass of the car frame to the bottom of the frame |
| $L_{fcu}$ | The vertical distance from the center of mass of the car frame to the top of the car |
| $L_{fcd}$ | The vertical distance from the center of mass of the car frame to the bottom of the car |
| $l$ | The horizontal distance from the center of mass of the car to the side wall of the car |
| $F_{gi}(i=1\ldots4)$ | Active control force |
| $z_{di}(i=1\ldots4)$ | Guide rail displacement excitation |
| $z_{Fi}(i=1\ldots2)$ | Air excitation of the shaft flow field effect |
| $U$ | Air velocity vector |
| $\rho$ | Air density |
| $u/v/w$ | Velocity components of U in the $x$, $y$, and $z$ coordinate directions |
| $p$ | Pressure |
| $\mu$ | Dynamic viscosity of air |
| $\mu_t$ | Turbulent viscosity coefficient |
| $C_\mu$, $C_\mu$, $C_\mu$, $\sigma_k$, $\sigma_\varepsilon$ | Empirical constants |
| $\tau$ | Viscous stress |
| $R_a$ | The position of the action surface unit |
| $R_o$ | The position of the car's center of mass |
| $C_{fx}$ | Transverse force influence coefficient |
| $C_{mx}$ | Overturning moment influence coefficient |
| $\theta$ | Deflection angle |
| $y$ | Transverse displacement |
| $F_g$ | Equivalent aerodynamic load |
| $M_g$ | Equivalent aerodynamic torque |
| $Q$ | The state cost weight matrix |
| $R$ | Control cost weight matrix |

## Appendix A

Turbulence model:

For viscous incompressible flow, the governing equation is as follows:

The equation of mass equation:

$$\mathrm{div}(U) = 0 \tag{A1}$$

The equation of momentum:

$$\frac{\partial U}{\partial t} + div(uU) = div(\nu\ grad\ u) - \frac{1}{\rho}\frac{\partial p}{\partial x} \tag{A2}$$

$$\frac{\partial U}{\partial t} + div(vU) = div(v\ grad\ v) - \frac{1}{\rho}\frac{\partial p}{\partial y} \tag{A3}$$

$$\frac{\partial U}{\partial t} + div(wU) = div(vgradw) - \frac{1}{\rho}\frac{\partial p}{\partial z} \tag{A4}$$

In Equations (A2)–(A4), U is the air velocity vector, t is time, $\rho$ is air density, $u$, $v$, and $w$ are the velocity components of U in the $x$, $y$, and $z$ coordinate directions, $p$ is pressure, *div* is divergence, and *grad* is gradient.

The numerical simulation of turbulent flow fields needs a turbulence model. Right now, Reynolds-Averaged Navier–Stokes (RANS) is the engineering field's most extensively used turbulence numerical simulation approach. Among its turbulence models, the standard two-equation model sees the most use. Consequently, this paper employs RANS and the standard two-equation model to simulate the flow field within the crane runway [40].

Turbulent kinetic energy k equation:

$$\frac{\partial k}{\partial t} + \mathrm{div}(kU) = \frac{1}{\rho}\mathrm{div}\left[\left(\mu + \frac{\mu_t}{\sigma_k}\right)\mathrm{grad}k\right] - \varepsilon + \frac{\mu_t P_G}{\rho} \tag{A5}$$

Turbulent dissipation rate ε equation:

$$\frac{\partial \varepsilon}{\partial t} + \mathrm{div}(\varepsilon U) = \frac{1}{\rho}\mathrm{div}\left[\left(\mu + \frac{\mu_t}{\sigma_\varepsilon}\right)\mathrm{grad}\varepsilon\right] - C_2\frac{\varepsilon^2}{k} + \frac{C_1\mu_t\varepsilon P_G}{\rho k} \tag{A6}$$

where $\mu$ is the dynamic viscosity of air, $\mu_t$ is the turbulent viscosity coefficient, and $C_\mu$, $C_1$, $C_2$, $\sigma_k$, and $\sigma_\varepsilon$ are empirical constants; $C_\mu = 0.09$, $C_1 = 1.44$, $C_2 = 1.92$, $\sigma_k = 1.0, \sigma_\varepsilon = 1.3$.

The state-space equation:

The equation derived in Section 3.1 is as follows:

$$\begin{aligned} M_f\ddot{X}_f =& H_f{}^T\left(F_{gl} + K_{gl}Z_{dl} + K_{gr}Z_{dr} + C_{gl}\dot{Z}_{dl} + C_{gr}\dot{Z}_{dr}\right) \\ &- H_f{}^T\left(\left(K_{gl} + K_{gr}\right)Z_f + \left(C_{gl} + C_{gr}\right)\dot{Z}_f + F_{gr}\right) \\ &+ H_{fc}{}^T\left((K_{rl} + K_{rr})Z_c + (C_{rl} + C_{rr})\dot{Z}_c\right) \\ &- H_{fc}{}^T\left((K_{rl} + K_{rr})H_{fc}H_f{}^{-1}Z_f + (C_{rl} + C_{rr})H_{fc}H_f{}^{-1}\dot{Z}_f\right) \\ &+ 2l^2H_t\left(k_v\left(H_c{}^{-1}Z_c - H_f{}^{-1}Z_f\right) + c_v\left(H_c{}^{-1}\dot{Z}_c - H_f{}^{-1}\dot{Z}_f\right)\right) \end{aligned} \tag{A7}$$

$$\begin{aligned} M_c\ddot{X}_c =& H_c{}^T\left(Z_{dF} - (K_{rl} + K_{rr})Z_c - (C_{rl} + C_{rr})\dot{Z}_c\right) \\ &+ H_c{}^T\left((K_{rl} + K_{rr})H_{fc}H_f{}^{-1}Z_f + (C_{rl} + C_{rr})H_{fc}H_f{}^{-1}\dot{Z}_f\right) \\ &- 2l^2H_t\left(k_v\left(H_c{}^{-1}Z_c - H_f{}^{-1}Z_f\right) + c_v\left(H_c{}^{-1}\dot{Z}_c - H_f{}^{-1}\dot{Z}_f\right)\right) \end{aligned} \tag{A8}$$

where

$$M_c = diag(m_c,\ I_c), M_f = diag\left(m_f,\ I_f\right),$$

$$K_{gl} = diag(k_{g1},\ k_{g3}), K_{gr} = diag(k_{g2},\ k_{g4}),$$

$$C_{gl} = diag(c_{g1},\ c_{g3}), C_{gr} = diag(c_{g2},\ c_{g4}),$$

$$K_{rl} = diag(k_{r1},\ k_{r3}), K_{rr} = diag(k_{r2},\ k_{r4}),$$

$$C_{rl} = diag(c_{r1},\ c_{r3}), C_{rr} = diag(c_{r2},\ c_{r4}),$$

$$Z_{dl} = \begin{pmatrix} z_{d1} \\ z_{d3} \end{pmatrix},\ Z_{dr} = \begin{pmatrix} z_{d2} \\ z_{d4} \end{pmatrix}, Z_{dF} = \begin{pmatrix} z_{F1} \\ z_{F2} \end{pmatrix},$$

$$F_{gl} = \begin{pmatrix} F_{g1} \\ F_{g3} \end{pmatrix}, F_{gr} = \begin{pmatrix} F_{g2} \\ F_{g4} \end{pmatrix}.$$

The equation derived in Section 3.2 is as follows:

$$\dot{x} = A(p)x + B_{1g}(p)d_g + B_{1F}(p)d_F + B_{2g}(p)u_g \tag{A9}$$

The calculation formula for the coefficient matrix in the Equation (A9) is as follows:

$$\begin{cases} A(p) = \begin{bmatrix} 0 & 0 & I & 0 \\ 0 & 0 & 0 & I \\ a_{31} & a_{32} & a_{33} & a_{34} \\ a_{41} & a_{42} & a_{43} & a_{44} \end{bmatrix} \\ a_{31} = -H_f M_f^{-1}\left(H_f^T\left(K_{gl} + K_{gr}\right) + H_{fc}^T(K_{rl} + K_{rr})H_{fc}H_f^{-1} + 2l^2 H_t k_v H_f^{-1}\right) \\ a_{32} = H_f M_f^{-1}\left(H_{fc}^T(K_{rl} + K_{rr}) + 2l^2 H_t k_v H_c^{-1}\right) \\ a_{33} = -H_f M_f^{-1}\left(H_f^T\left(C_{gl} + C_{gr}\right) + H_{fc}^T(C_{rl} + C_{rr})H_{fc}H_f^{-1} + 2l^2 H_t c_v H_f^{-1}\right) \\ a_{34} = H_f M_f^{-1}\left(H_{fc}^T(C_{rl} + C_{rr}) + 2l^2 H_t c_v H_c^{-1}\right) \\ a_{41} = H_c M_c^{-1}\left(H_c^T(K_{rl} + K_{rr})H_{fc}H_f^{-1} + 2l^2 H_t k_v H_f^{-1}\right) \\ a_{42} = -H_c M_c^{-1}\left(H_c^T(K_{rl} + K_{rr}) + 2l^2 H_t k_v H_c^{-1}\right) \\ a_{43} = H_c M_c^{-1}\left(H_c^T(C_{rl} + C_{rr})H_{fc}H_f^{-1} + 2l^2 H_t c_v H_f^{-1}\right) \\ a_{44} = -H_c M_c^{-1}\left(H_c^T(C_{rl} + C_{rr}) + 2l^2 H_t k_v H_c^{-1}\right) \end{cases}$$

$$\begin{cases} B_{1g}(p) = \begin{bmatrix} H_f M_f^{-1} H_f^T C_{gl} & H_f M_f^{-1} H_f^T C_{gr} \\ 0 & 0 \\ b_{31} & b_{32} \\ b_{41} & b_{42} \end{bmatrix} \\ b_{31} = H_f M_f^{-1}\left(H_f^T\left(K_{gl} - \left(C_{gl} + C_{gr}\right)H_f M_f^{-1} H_f^T C_{gl}\right)C_{gl}\right) \\ -H_{fc}^T(C_{rl} + C_{rr})H_{fc}M_f^{-1}H_f^T C_{gl} - 2l^2 H_t c_v M_f^{-1} H_f^T \\ b_{32} = H_c M_c^{-1}\left(H_c^T(C_{rl} + C_{rr})H_{fc}M_f^{-1}H_f^T C_{gl} - 2l^2 H_t c_v M_f^{-1} H_f^T C_{gl}\right) \\ b_{41} = H_f M_f^{-1}\left(H_f^T\left(K_{gr} - \left(C_{gl} + C_{gr}\right)H_f M_f^{-1} H_f^T C_{gr}\right)\right) \\ -H_{fc}^T(C_{rl} + C_{rr})H_{fc}M_f^{-1}H_f^T C_{gr} - 2l^2 H_t c_v M_f^{-1} H_f^T C_{gr} \\ b_{42} = H_c M_c^{-1}\left(H_c^T(C_{rl} + C_{rr})H_{fc}M_f^{-1}H_f^T C_{gr} + 2l^2 H_t c_v M_f^{-1} H_f^T C_{gr}\right) \end{cases}$$

where

$$B_{1F}(p) = \begin{bmatrix} 0 \\ 0 \\ 0 \\ H_c M_c^{-1} H_c^T \end{bmatrix}, B_{2g}(p) = \begin{bmatrix} 0 & 0 \\ 0 & 0 \\ H_f M_f^{-1} H_f^T & -H_f M_f^{-1} H_f^T \\ 0 & 0 \end{bmatrix}$$

The output z can be represented by Equation (A10):

$$z = C_z(p)x + D_{zg1}(p)d_g + D_{zF1}(p)d_F + D_{zg2}(p)u_g \tag{A10}$$

Here

$$\begin{cases} C_z(p) = \begin{bmatrix} c_{11} & c_{12} & c_{13} & c_{14} \\ I & 0 & 0 & 0 \\ -I & I & 0 & 0 \end{bmatrix} \\ c_{11} = H_c M_c^{-1}\left(H_c^T(K_{rl}+K_{rr})H_{fc}H_f^{-1} + 2l^2 H_t k_v H_f^{-1}\right) \\ c_{12} = -H_c M_c^{-1}\left(H_c^T(K_{rl}+K_{rr}) + 2l^2 H_t k_v H_c^{-1}\right) \\ c_{13} = H_c M_c^{-1}\left(H_c^T(C_{rl}+C_{rr})H_{fc}H_f^{-1} + 2l^2 H_t c_v H_f^{-1}\right) \\ c_{14} = -H_c M_c^{-1}\left(H_c^T(C_{rl}+C_{rr}) + 2l^2 H_t k_v H_c^{-1}\right) \end{cases}$$

$$\begin{cases} D_{zg1}(p) = \begin{bmatrix} d_{11} & d_{12} \\ -\frac{I}{2} & -\frac{I}{2} \\ 0 & 0 \end{bmatrix} \\ d_{11} = H_c M_c^{-1}\left(H_c^T(C_{rl}+C_{rr})H_{fc}M_f^{-1}H_f^T C_{gl} - 2l^2 H_t c_v M_f^{-1} H_f^T C_{gl}\right) \\ d_{12} = H_c M_c^{-1}\left(H_c^T(C_{rl}+C_{rr})H_{fc}M_f^{-1}H_f^T C_{gr} - 2l^2 H_t c_v M_f^{-1} H_f^T C_{gr}\right) \end{cases}$$

$$D_{zF1}(p) = \begin{bmatrix} H_c M_c^{-1} H_c^T \\ 0 \\ 0 \end{bmatrix} D_{zg2}(p) = \begin{bmatrix} 0 & 0 \\ 0 & 0 \\ 0 & 0 \end{bmatrix}$$

LQR algorithm derivation:

$$J = \frac{1}{2}\int_0^\infty \left[x(t)^T Q x(t) + u(t)^T R u(t)\right] dt \quad \text{(A11)}$$

To minimize the performance function, linear state feedback control is adopted, with a tracking reference signal of zero, and $K$ represents the controller to be solved. Substituting it into the cost function yields the following equation:

$$J = \frac{1}{2}\int_0^\infty x^T\left(Q + K^T R K\right) x dt \quad \text{(A12)}$$

Assuming there exists a constant matrix $P$ that satisfies

$$\frac{d}{dt}\left(x^T P x\right) = -x^T\left(Q + K^T R K\right) x \quad \text{(A13)}$$

By substituting Equation (A13) into Equation (A12), it has been derived that, as time $t$ approaches infinity, the system's state vector $x(t)$ approaches 0, resulting in the following equation:

$$J = -\frac{1}{2}\int_0^\infty \frac{d}{dt}\left(x^T P x\right) dt = \frac{1}{2} x(0)^T P x(0) \quad \text{(A14)}$$

Next, Equation (A15) is differentially expanded:

$$\dot{x}^T P x + x^T P \dot{x} + x^T Q x + x^T K^T R K x = 0 \quad \text{(A15)}$$

Substitute into Equation (A16), as follows:

$$x^T\left[(A - BK)^T P + P(A - BK) + Q + K^T R K\right] x = 0 \quad \text{(A16)}$$

The necessary condition for Equation (A17) to have a solution is that the part in parentheses must be equal to 0, which means the following:

Let $K = R^{-1} B^T P$, and obtain the Riccati equation:

$$A^T P + PA + Q - PBR^{-1} B^T P = 0 \quad \text{(A17)}$$

In modern control theory, Equation (A18) is a key Riccati equation. Once the parameters (A, B, Q, R) are known, P can be solved from this equation. Then, the control variable can be expressed as

$$\begin{aligned} u &= -Kx \\ K &= R^{-1}B^{T}P \end{aligned} \quad \text{(A18)}$$

## References


1. Jiang, X.; Rui, Y. Research on vibration Control of traction elevator. In Proceedings of the 2015 International Industrial Informatics and Computer Engineering Conference, Xi'an, China, 10–11 January 2015; Atlantis Press: Paris, France, 2015; pp. 2144–2147.
2. Jia, Y.; Zhang, Z.; Du, S.; Zhong, W.; Xu, Y.; Pu, C.; Páez, L.M.R.; Qian, P. Linear Active Disturbance Rejection Motion Control of a Novel Pneumatic Actuator with Linear-Rotary Compound Motion. *Int. J. Hydromechatron.* **2024**, *7*, 382–399. [CrossRef]
3. Zhang, Q.; Yang, Z.; Wang, C.; Yang, Y.; Zhang, R. Intelligent control of active shock absorber for high-speed elevator car. *Proc. Inst. Mech. Eng. Part C J. Mech. Eng. Sci.* **2019**, *233*, 3804–3815. [CrossRef]
4. Tusset, A.M.; Santo, D.R.; Balthazar, J.M.; Piccirillo, V.; Santos, L.C.D.; Brasil, R.M. Active vibration control of an elevator system using magnetorheological damper actuator. *Int. J. Nonlinear Dyn. Control* **2017**, *1*, 114–131. [CrossRef]
5. Zhang, R.; Tian, Q.; Chen, C. Adaptive Gain H∞ Output Feedback Control Strategy for Horizontal Vibration of High-Speed Elevator Car System Based on T-S Fuzzy Model. *J. Mech. Sci. Technol.* **2023**, *37*, 919–929. [CrossRef]
6. Su, X.; Zhang, R.; He, Q.; Qiu, T.; Liu, L. Transient Response Feedback Control for Horizontal Vibration of High-Speed Elevator Car with Prescribed Performance. *Proc. Inst. Mech. Eng. Part C J. Mech. Eng. Sci.* **2024**, *238*, 4917–4931. [CrossRef]
7. Ge, Q.; Zhang, L.; Li, H.; He, Q.; Zhang, R. Study on Fuzzy Sliding Mode Active Disturbance Rejection Control Method for Horizontal Vibration of High-Speed Elevator Car System. *J. Control Eng. Appl. Inform.* **2023**, *25*, 13–24. [CrossRef]
8. Shi, R.; Zhang, Q.; Zhang, R.; He, Q.; Liu, L. Analysis of the Horizontal Vibration Response under Gas–Solid Coupling throughout the Entire High-Speed Elevator Operating Process. *J. Vib. Control* **2023**, *29*, 4455–4465. [CrossRef]
9. Zhang, R.; Qiao, S.; Zhang, Q.; Liu, L. Universal Modeling of Unsteady Airflow in Different Hoistway Structures and Analysis of the Effect of Ventilation Hole Parameters. *J. Wind. Eng. Ind. Aerodyn.* **2019**, *194*, 103987. [CrossRef]
10. Qiao, S.; Zhang, R.; He, Q.; Zhang, L. Theoretical Modeling and Sensitivity Analysis of the Car-Induced Unsteady Airflow in Super High-Speed Elevator. *J. Wind. Eng. Ind. Aerodyn.* **2019**, *188*, 280–293. [CrossRef]
11. Zhang, Q.; Jing, H.; Qiao, S.; Zhang, R.; Liu, L. Three-Dimensional Numerical Simulation of Unsteady Airflow in High-Speed/Ultra-High-Speed Elevator Based on Multi-Region Dynamic Layering Method. *Mech. Based Des. Struct. Mach.* **2023**, *51*, 4406–4431. [CrossRef]
12. Cui, H.; Zhang, Q.; Zhang, R.; Zhang, L. Research on Aerodynamic and Flow Field Characteristics of High Speed Elevator with Different Shaft Blockage Ratio. *J. Phys. Conf. Ser.* **2020**, *1653*, 12024. [CrossRef]
13. Qiu, L.; Zhou, H.; Wang, Z.; Zhang, S.; Zhang, L.; Lou, W. High-Speed Elevator Car Air Pressure Compensation Method Based on Coupling Analysis of Internal and External Flow Fields. *Appl. Sci.* **2021**, *11*, 1700. [CrossRef]
14. Qiu, L.; Su, G.; Wang, Z.; Zhang, S.; Zhang, L.; Li, H. High-Speed Elevator Car Horizontal Vibration Fluid–Solid Interaction Modeling Method. *J. Vib. Control* **2022**, *28*, 2984–3000. [CrossRef]
15. Li, X.-D.; Wang, K. Optimization of Aerodynamic Characteristics of High-Speed Elevator. *Ha'Erbin Gong Ye Da Xue Xue Bao* **2009**, *41*, 82–86.
16. Bai, H.-L.; Shen, G.; So, A. Experimental-Based Study of the Aerodynamics of Super-High-Speed Elevators. *Build. Serv. Eng. Res. Technol.* **2005**, *26*, 129–143. [CrossRef]
17. Chen, Y.; Yang, L.; Fu, Z.; Chen, L.; Chen, J. Gas flow behavior and flow transition in elevator shafts considering elevator motion during a building fire. *Build. Simul.* **2018**, *11*, 765–771. [CrossRef]
18. Shi, L.; Liu, Y.; Jin, S.; Cao, Z. Numerical Simulation of Unsteady Turbulent Flow Induced by Two-Dimensional Elevator Car and Counter Weight System. *J. Hydrodyn.* **2007**, *19*, 720–725. [CrossRef]
19. Pierucci, M.; Frederick, M. Ride quality and noise in high speed elevators. *J. Acoust. Soc. Am.* **2008**, *123*, 3247. [CrossRef]
20. Zhang, R.; Liu, J.; Liu, M.; Zhang, Q.; He, Q. Gas–Solid Coupling Lateral Vibration Characteristics of High-Speed Elevator Based on Blockage Ratio. *J. Braz. Soc. Mech. Sci. Eng.* **2022**, *44*, 184. [CrossRef]
21. Wang, X.; Lin, Z.; Tang, P.; Ling, Z. Research of the blockage ratio on the aerodynamic performances of high speed elevator. In Proceedings of the 4th International Conference on Mechatronics, Materials, Chemistry and Computer Engineering, Xi'an, China, 12–13 December 2015.
22. Mangani, L.; Buchmayr, M.; Darwish, M.; Moukalled, F. A fully coupled OpenFOAM® solver for transient incompressible turbulent flows in ALE formulation. *Numer. Heat Transf. Part B Fundam.* **2017**, *71*, 313–326. [CrossRef]
23. Fu, W.J.; Zhu, C.M.; Ye, Q.T. Multi-objective Integrated Optimization Based on Evolutionary Strategy with a Dynamic Weighting Schedule. *J. Southeast Univ.* **2006**, *22*, 204–207.

24. Chen, B.S.; Cheng, Y.M. A structure-specified H∞ optimal control design for practical applications: A genetic approach. *IEEE Trans. Control Syst. Technol.* **1998**, *6*, 707–718. [CrossRef]
25. Wee, H.; Kim, Y.Y.; Jung, H.; Lee, G.N. Nonlinear rate-dependent stick-slip phenomena: Modeling and parameter estimation. *Int. J. Solids Struct.* **2001**, *38*, 1415–1431. [CrossRef]
26. Zhao, M.; Qin, C.; Tang, R.; Tao, J.; Xu, S.; Liu, C. An Acceleration Feedback-Based Active Control Method for High-Speed Elevator Horizontal Vibration. *J. Vib. Eng. Technol.* **2024**, *12*, 1943–1956. [CrossRef]
27. Tang, R.; Qin, C.; Zhao, M.; Xu, S.; Tao, J.; Liu, C. An Optimized Fractional-Order PID Horizontal Vibration Control Approach for a High-Speed Elevator. *Appl. Sci.* **2023**, *13*, 7314. [CrossRef]
28. Wang, Y.; Cai, J.; Li, Q.; Yin, H.; Yang, X. Diffuse Interface Simulation of Bubble Rising Process: A Comparison of Adaptive Mesh Refinement and Arbitrary Lagrange-Euler Methods. *Heat Mass Transf.* **2018**, *54*, 1767–1778. [CrossRef]
29. Walter, B.; Dumbser, M. High Order Accurate Direct Arbitrary-Lagrangian-Eulerian ADER-WENO Finite Volume Schemes on Moving Curvilinear Unstructured Meshes. *Comput. Fluids* **2016**, *136*, 48–66.
30. Pourazarm, P.; Modarres-Sadeghi, Y.; Lackner, M. A parametric study of coupled-mode flutter for MW-size wind turbine blades. *Wind. Energy* **2016**, *19*, 497–514. [CrossRef]
31. Bazilevs, Y.; Hsu, M.C.; Kiendl, J.; Akkerman, I.; Wright, S.; Takizawa, K.; Henicke, B.; Spielman, T.; Tezduyar, T.E. 3D simulation of wind turbine rotors at full scale. Part I: Geometry modeling and aerodynamics. *Int. J. Numer. Methods Fluids* **2011**, *65*, 207–235. [CrossRef]
32. Huang, S.; Li, R.; Li, Q. Numerical simulation on fluid-structure interaction of wind around super-tall building at high Reynolds number conditions. *Struct. Eng. Mech.* **2013**, *46*, 197–212. [CrossRef]
33. Zhang, Y.; Habashi, W.G.; Khurram, R.A. Predicting wind-induced vibrations of high-rise buildings using unsteady CFD and modal analysis. *J. Wind. Eng. Ind. Aerodyn.* **2015**, *136*, 165–179. [CrossRef]
34. Cunegatto, E.H.T.; Zinani, F.S.F.; Rigo, S.J. Multi-objective optimisation of micromixer design using genetic algorithms and multi-criteria decision-making algorithms. *Int. J. Hydromechatron.* **2024**, *7*, 224–249. [CrossRef]
35. Qin, C.; Tao, J.; Liu, C. Stability analysis for milling operations using an Adams-Simpson-based method. *Int. J. Adv. Manuf. Technol.* **2017**, *92*, 969–979. [CrossRef]
36. Qin, C.; Tao, J.; Liu, C. A novel stability prediction method for milling operations using the holistic-interpolation scheme. *Proc. Inst. Mech. Eng. Part C J. Mech. Eng. Sci.* **2019**, *233*, 4463–4475. [CrossRef]
37. Yuan, C.Y.; Li, K.; Zang, G.; Wang, X.; Shi, D.; Wu, G. Optimization of semi-active suspension LQR parameters based on local optimization with a skipping out particle swarm algorithm. In Proceedings of the International Conference on Mechanical Design and Simulation (MDS 2022), Wuhan, China, 18–20 March 2022; Volume 12261, pp. 1111–1116.
38. Aryan, S.; Moore, K.J. Identification of Multiple Local Nonlinear Attachments Using a Single Measurement Case. *J. Sound Vib.* **2021**, *513*, 116410.
39. Huang, G.; Qin, C.; Wang, H.; Liu, C. TBM rock fragmentation classification using an adaptive spot denoising and contour-texture decomposition attention-based method. *Tunn. Undergr. Space Technol.* **2025**, *161*, 106498. [CrossRef]
40. Zhang, S.; Zhang, R.; He, Q.; Cong, D. The Analysis of the Structural Parameters on Dynamic Characteristics of the Guide Rail–Guide Shoe–Car Coupling System. *Arch. Appl. Mech.* **2018**, *88*, 2071–2080. [CrossRef]

*actuators*

MDPI

*Article*

# NPS6D100—A 6D Nanopositioning System with Sub-10 nm Performance in a ⌀100 mm × 10 mm Workspace

**Steffen Hesse *,†, Alex Huaman †, Michael Katzschmann † and Ludwig Herzog**

IMMS Institut für Mikroelektronik- und Mechatronik-Systeme Gemeinnützige GmbH (IMMS GmbH), Ehrenbergstraße 27, 98693 Ilmenau, Germany; alex.huaman@imms.de (A.H.); michael.katzschmann@imms.de (M.K.); ludwig.herzog@imms.de (L.H.)

* Correspondence: steffen.hesse@imms.de

† These authors contributed equally to this work.

**Abstract:** This paper presents the development of a compact nanopositioning stage with long-range capabilities and six-degree-of-freedom (DOF) closed-loop control. The system, referred to as NPS6D100, provides ⌀100 mm planar and 10 mm vertical travel range while maintaining direct force transfer to the moving platform (or slider) in all DOFs. Based on an integrated planar direct drive concept, the system is enhanced by precise vertical actuation and full 6D output feedback control. The mechanical structure, drive architecture, guiding, and measurement subsystems are described in detail, along with experimental results that confirm sub-10 nm servo errors under constant setpoint operation and in synchronized multi-axis motion scenarios. With its scalable and low-disturbance design, the NPS6D100 is well suited as a nanopositioning platform for sub-10 nm applications in nanoscience and precision metrology.

**Keywords:** nanopositioning stage; sub-10 nm; multi-coordinate direct drive; 6D control

## 1. Introduction

The progression of nanotechnology continues to drive demand for increasingly precise and scalable positioning systems. In fields such as semiconductor manufacturing, nano-optics, and high-precision metrology, there exists an increasing demand for positioning systems capable to reach the single-digit nanometer precision [1,2]. Within this context, recent advances in nanoscience have enabled the nanotechnology field to operate at the sub-10 nm scale [3,4]. These advances in nanoscience require robust nanopositioning systems that combine ultra-high precision with fully actuated degrees of freedom (DOFs), in addition to offering reconfigurable and scalable operation depending on the task demands.

A significant contribution to nanotechnology is the development of integrated nanopositioning platforms that enable motion in all six DOFs—three translations ($x$, $y$, $z$) and three rotations about the corresponding axes ($\varphi_x$, $\varphi_y$, $\varphi_z$)—within a unified system architecture. The motion platform originally introduced as the NPPS100-6D [5,6], now referred to as the NPS6D100, represents a compact and scalable solution for 6D multicoordinate nanopositioning. The NPS6D100 constitutes a well-suited nanopositioning platform for sub-10 nm applications in nanoscience and sub-10 nm precision metrology. The system is highly relevant for medical applications that require precision nanopositioning during micro-surgical procedures [7,8]. The nanopositioner architecture integrates a planar direct drive for horizontal motion in $x$, $y$, and $\varphi_z$, upgraded with three lifting and actuating units (LAUs) for vertical displacement in $z$ and angular rotations about $\varphi_x$ and $\varphi_y$. Each LAU has a parallel arrangement combining a low-dynamic pneumatic actuator for weight force compensation and a high-dynamic voice-coil drive for precision motion; see [9,10]. In addition, the integration of a suitable guiding system provides virtually frictionless planar and vertical motion.

The pursuit of sub-10 nm precision has led to the development of integrated nanopositioning solutions. These approaches permit nanometer positioning across long ranges while maintaining robustness and high-precision performance. While many works target small travel ranges [11–13], this paper addresses long-range stages ($\geq$50 mm $\times$ 50 mm in the $xy$ plane) that maintain nanometer-level servo errors through careful stage architecture and precision engineering principles [14–16]. Various concepts exist in the literature, often using parallel kinematics with Lorentz actuators and roller or air bearings, as seen in the NanoCMM [17], and the high-precision 3D CMM [18]. Integrated direct-drive architectures [6,19], combined with air bearing guidance, show particular promise. At TU Ilmenau, the long-range NPMM-200 [20]—evolved from earlier stacked designs [21,22]—has demonstrated excellent 3D measurement capabilities [23,24]. The NFM-100 system [25,26] offers planar positioning over a ⌀100 mm. A similar platform, the NPP-1 [27], is commercially available from SIOS GmbH. Furthermore, initial investigation on the NPPS100-6D [5] extends the planar direct-drive concepts to include vertical actuation with long travel ranges of ⌀100 mm $\times$ 10 mm. The most significant exemplar is the NPS6D200 [28,29] that combines a 3D planar direct drive with vertical actuation featuring a long working range of ⌀200 mm $\times$ 25 mm with high-precision 6D closed-loop control and single-digit nanometer precision [30]. A comparative table to supplement the distinction between similar nanopositioners in the literature is shown in Table 1. In summary, we recognize the potential of the direct drive principle on the one hand, and the need to integrate vertical adjustment on the other hand. A combination of both without the necessity for coil lift integration closes the gap between the existing solutions and allows the investigation of the characteristics, the benefits, and the limitations of such motion stage architectures.

**Table 1.** Comparison of similar nanopositioning systems.

| | **NPS6D100** | **NPS6D200** | **NFM-100<br>NPP-1** |
|---|---|---|---|
| Planar travel range | ⌀100 mm | ⌀200 mm | ⌀100 mm |
| Vertical travel range | 10 mm | 25 mm | — |
| Closed loop control<br>Active DOF | 6D<br>$x, y, z, \varphi_x, \varphi_y, \varphi_z$ | 6D<br>$x, y, z, \varphi_x, \varphi_y, \varphi_z$ | 3D<br>$x, y, \varphi_z$ |
| Lifting modules | LAU10 [10] | LAU25 [28] | — |
| Planar coil fixation | Fixed | Coil lift functionality | Fixed |
| Measurement configuration | Cartesian plane mirrors<br>Single and double beam LIF (SP2000) | 120° triangular plane mirrors<br>Differential LIF (SP5000 DI) | 120° triangular plane mirrors<br>Differential LIF (SP5000 DI) |

The main difference from its larger counterpart—the NPS6D200—is that the NPS6D100 does not include a coil lift system. This results in a simpler and lighter construction with enhanced stiffness in the actuation chain and lower disturbance generation, while still maintaining full 6D actuation. The NPS6D100 features a long-range working volume—⌀100 mm $\times$ 10 mm—with high-precision 6D closed-loop control and sub-10 nm precision. The absence of the coil lift system, combined with the compact dimensions of the nanopositioner, enables improved positioning fidelity in nanofabrication and nanomeasurement applications. The NPS6D100 presents a metrology setup different from the NPS6D200 [28,29], which consists of three orthogonally arranged laser interferometers (LIFs). This Cartesian metrology configuration is particularly advantageous for tasks requiring direct alignment with the moving axes, ensuring direct positioning measurement, whereas the alignment follows the Abbe comparator principle [16]. In the broader context of nanopositioning systems, the NPS6D100 represents a versatile platform for applications that demand deterministic motion control and multi-axis stability over medium to long travel ranges. Its open architecture allows for easy integration with probing systems, optical sensors, or surface interaction tools, extending its applicability to nano-inspection, nanofabrication, and metrology instrumentation. This paper provides an overview of the NPS6D100, presenting its mechanical architecture, actuation principles, control system, and performance outcomes. The focus is placed on six-dimensional precise positioning within a long-range working volume, suitable for high-precision research and industrial applications.

This paper is organized in five sections. Following the introduction, Section 2 introduces the NPS6D100 and its overall design, covering the planar drive system, vertical actuation, and guiding mechanism. Section 3 presents the motion controller design. Section 4 evaluates the positioning performance achieved across various motion tasks. Section 5 closes this paper with conclusions and the outlook for future work.

## 2. Description of the Nanopositioning System

The NPS6D100 comprises a 6-DOF nanopositioning system designed to achieve sub-10 nm performance within a cylindrical workspace of $\varnothing$100 mm $\times$ 10 mm. The NPS6D100 builds on its predecessor—the NPPS100 platform [6,19]—by incorporating vertical lifting and tilting motion, extending its functionality toward full 6D actuation. The system architecture combines a planar electromagnetic direct drive with three lifting modules (or LAUs), enabling precise motion in $x$, $y$, $z$, $\varphi_x$, $\varphi_y$, and $\varphi_z$. The following subsections describe the mechanical structure, planar drive system, and vertical actuation.

### 2.1. System Description

The NPS6D100 consists of two main elements, i.e., the slider, which includes all the moving components, and the stator, which comprises all the frame-fixed parts [5]. The slider comprises a triangular monolithic Zerodur platform, providing extremely low thermal expansion and high stiffness in the actuation chain. Two mirrors are orthogonally aligned and bonded on the top of the slider, while a third mirror is placed on its bottom surface. These mirrors interact with LIFs, from SIOS GmbH, arranged in a Cartesian layout to directly measure displacement along the $x$, $y$, and $z$ axes. Angular displacements $\varphi_x$ and $\varphi_y$ are measured using a 2D autocollimator directed at the bottom surface of the slider, while $\varphi_z$ is determined using a dual-beam LIF configuration. The slider is supported on three LAUs, each placed at a corner of the slider. This three-point support forms a parallel kinematic structure that ensures a well-defined mechanical constraint while enabling actuation in all six DOF. Given the implementation of these lifting modules, the vertical working range covers up to 10 mm. To achieve control accuracy at the nanometer level, it is vital to reduce the impact of external disturbances. Therefore, the whole system is placed in an 18 t basement equipped with passive pneumatic vibration isolation. The system is also located inside an acoustic enclosure to shield it from air currents and acoustic noise in the laboratory. This enclosure also provides thermal decoupling from the laboratory air temperature, thereby stabilizing the thermal environment of the machine. An overview of the nanopositioning system and its components is shown in Figure 1 and the main mechanical parameters of the NPS6D100 are listed in Table 2.

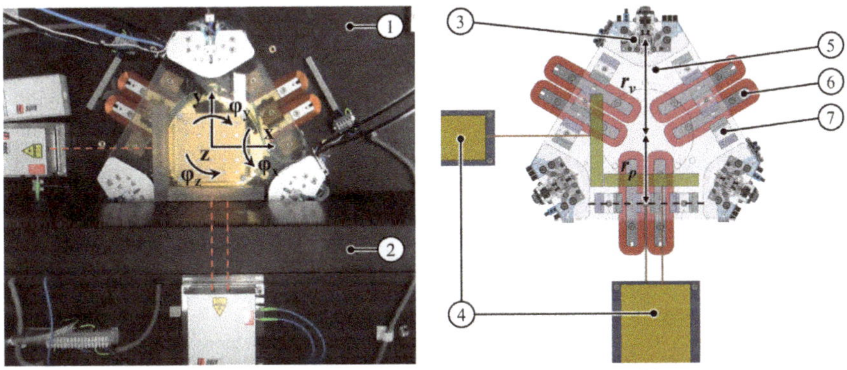


(**a**) (**b**)

**Figure 1.** NPS6D100 nanopositioning system (top view). Components: 1. granite stator, 2. granite portal with probe mount interface, 3. lifting module LAU10, 4. laser interferometer, 5. Zerodur slider with mirrors, 6. coils of the planar direct drive (fixed to the stator), and 7. magnets of the planar direct drive (moving with the slider). (**a**) Photograph of the NPS6D100. (**b**) Schematic of the NPS6D100.

**Table 2.** Key parameters of the NPS6D100 nanopositioning system.

| **Parameter** | **Value** |
|---|---|
| Planar travel range $(x, y)$ | $\varnothing$100 mm |
| Vertical travel range $(z)$ | 10 mm |
| Maximum planar velocity [1] | 50 mm/s |
| Maximum vertical velocity | 2 mm/s |
| Maximum acceleration | 250 mm/s$^2$ |
| Moving mass (slider) | 10 kg |
| Maximum payload | 2 kg |
| Overall system dimensions | 800 mm × 700 mm × 600 mm |

[1] This is the current maximum velocity setting. It may be increased if required.

The Cartesian configuration of the measurement system implies that each motion axis $(x, y, z)$ is measured independently along orthogonal directions. This setup simplifies the transformation of raw sensor data into stage coordinates and provides minimal coupling between degrees of freedom. For $\varphi_z$ specifically, the yaw angle is computed directly from the optical path difference between the two parallel beams of the $y$-axis LIF, which are laterally offset by 25.4 mm. Note that the motion in $y$-direction is measured using only a single beam from the dual-beam $y$-axis LIF. The beam paths of the LIFs—considering only a single beam of the $y$-axis LIF—intersect at a virtual common point, located near the center of the optical cube formed by the three mirror surfaces, see Figure 1. This intersection ensures compliance with Abbe's comparator principle [16]. The monolithic construction of the slider guarantees that the geometric relationships remain fixed over time. Since all DOFs are directly measured, no further linear transformation is required to compute the spatial position coordinates and orientation. Table 3 summarizes the resolution of the displacement and angular measurement systems.

A centralized dSPACE control hardware is used for rapid control prototyping and manages signal acquisition and interfacing with the LIFs, 2D autocollimator, linear current amplifiers, pneumatic supply units, and redundant local encoders, among other components, for initialization, shutting down, and fail-safe procedures. All electrical currents and pneumatic pressures are routed from the slider to the stator via structured cables and tube paths to minimize parasitic force disturbances.

Table 3. Measurement resolution of the feedback system.

| Parameter | Value |
|---|---|
| Resolution in $x, y, z$ | 20 pm |
| Resolution in $\varphi_x, \varphi_y$ | 40 nrad |
| Resolution in $\varphi_z$ | 0.8 nrad |

*2.2. Planar Drive System with Aerostatic Guiding*

The planar drive system consists of three Lorentz-force actuators arranged in a 120° triangular configuration. Each actuator is composed of a pair of planar coils mounted on the stator and a magnet array fixed to the underside of the slider. The permanent magnet arrays are aligned along the edges of the triangular slider, while the corresponding coil pairs are mounted on the granite stator surface. This is illustrated in Figure 1.

The forces generated by the three electromagnetic actuators are combined to produce arbitrary motion in the $x$ and $y$ directions, as well as a rotation $\varphi_z$ around the $z$-axis. The coils are energized through a motor commutation, producing horizontal forces. Each actuator contributes a force $F_i$, with $i \in \{\mathrm{A}, \mathrm{B}, \mathrm{C}\}$, which is resolved into Cartesian components and combined to produce planar forces in the $x$ and $y$ directions, as well as a torque around the $z$-axis. The resulting forces are calculated using the linear transformation

$$\begin{pmatrix} F_x \\ F_y \\ M_{\varphi_z} \end{pmatrix} = \begin{pmatrix} 1 & -\frac{1}{2} & -\frac{1}{2} \\ 0 & \frac{\sqrt{3}}{2} & -\frac{\sqrt{3}}{2} \\ r_p & r_p & r_p \end{pmatrix} \begin{pmatrix} F_A \\ F_B \\ F_C \end{pmatrix}, \qquad (1)$$

where $r_p$ is the distance from the geometric center of the slider to each Lorentz actuator.

As a result of vertical displacement, the air gap between the magnet array and the planar coils increases as the slider is lifted. However, this effect does not compromise the stability of the motion platform. Experimental data and finite element method (FEM) simulations for similar drives have confirmed that for vertical displacements of up to 10 mm, the drive provides adequate propulsion without the implementation of a coil lift system [5,28]. This effect of the force constant reduction was evaluated in an experimental identification of the planar drive's force constant, which revealed that it decreases from 1.6 N/A at a working height of 1 mm to 0.66 N/A at the maximum working height. In the actual system configuration, the maximum peak coil current is 3 A, which introduces a working height limit of 7 mm, up to which the full acceleration of 250 mm/s$^2$ is obtainable. Beyond 7 mm, the permissible acceleration steadily decreases to 200 mm/s$^2$ at the maximum working height.

To guarantee frictionless motion, the slider is supported on three aerostatic bearing pads, one located beneath each LAU (see Section 2.3). These pads provide a pressurized air gap of approximately 5 µm. The air flows through the surface, generating a stable pressure field. The use of air bearings eliminates mechanical contact, allowing smooth motion and stable dynamic behavior. Therefore, the 3D planar Lorentz-drive together with the aerostatic guidings provide a robust platform for high-precision motion in the $xy$-plane, with almost no friction and hysteresis nonlinearities. To complement the discussion, the reader may consult [5,6,19] and the references therein.

*2.3. Vertical Drive System with Aerostatic Guiding and Weight Force Compensation*

The vertical displacement (in the $z$-axis) and angular rotation ($\varphi_x$, $\varphi_x$ around the $x$, $y$-axis) of the NPS6D100 are realized through the propulsion of three identical LAUs [9,10]. Each lifting module integrates two parallel actuators—a voice coil and a pneumatic piston—with a high-stiffness aerostatic guiding and an auxiliary encoder for local displacement measurement. Each lifting module operates within a travel range of 10 mm and generates a force $F_j$, with $j \in \{1,2,3\}$, which is combined to produce a vertical force in the $z$-axis and torques around the $x$ and $y$ axes. These forces are computed through the linear transformation

$$\begin{pmatrix} F_z \\ M_{\varphi_x} \\ M_{\varphi_y} \end{pmatrix} = \begin{pmatrix} 1 & 1 & 1 \\ r_v & -\frac{1}{2}r_v & -\frac{1}{2}r_v \\ 0 & -\frac{\sqrt{3}}{2}r_v & \frac{\sqrt{3}}{2}r_v \end{pmatrix} \begin{pmatrix} F_1 \\ F_2 \\ F_3 \end{pmatrix}, \quad (2)$$

where $r_v$ is the distance from the geometric center of the slider to each lifting module.

The precise positioning force is generated by a Lorentz-force drive. The voice coil is embedded concentrically within the LAU and generates force via Lorentz interaction between the coil and the radial magnetic field generated by the permanent magnet. The electromagnetic drive is designed to generate dynamic forces and operates at zero average current during stationary conditions. This almost zero current is produced due to the application of a suitable control allocation strategy [31,32] and results in minimal heat emission in the measuring space. To reduce nonlinearities, the ferromagnetic material in the lifting module is minimized, resulting in a linear and thermally stable behavior.

The low-dynamic pneumatic actuator counteracts the static vertical load, which includes the weight of the slider and any mounted payload. This integrated subsystem consists of a pneumatic piston with an air pressure chamber featuring contactless sealing to minimize friction and wear. An equilibrium pressure of approximately 1 bar compensates one-third of the total slider weight per LAU, ensuring that the electromagnetic actuator does not bear big offset forces. To extend the functionality of the NPS6D100, the pneumatic actuator has a working pressure range of up to 2 bar.

Each LAU features an aerostatic bushing which provides almost frictionless vertical motion and an aerostatic bearing pad mounted on the bottom of the LAU for almost frictionless planar motion. The lifting module additionally includes a dedicated linear optical encoder. This sensor is used for internal measurement during initialization and fail-safe scenarios, where the LIF's feedback may be temporarily unavailable. The encoder provides vertical measurement with micrometer resolution. The control commands are distributed among the integrated electromagnetic and pneumatic actuators using a control allocation approach that separates the frequency ranges [32–34]. In this context, the pneumatic piston manages the low-frequency components, while the Lorentz-force actuator handles the high-frequency dynamics. A cross-sectional view of the LAU is shown in Figure 2.

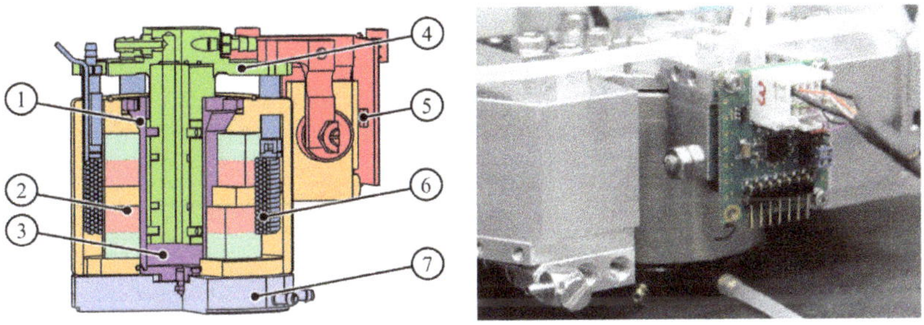


(a) (b)

**Figure 2.** Lifting and actuating module LAU10. Components: 1. air bushing as vertical guiding, 2. voice coil actuator, magnet (fixed), 3. pressure chamber of pneumatic weight force compensation, 4. moving part (mounted to the slider), 5. encoder measurement system, 6. voice coil actuator, coil (moving), and 7. planar air bearing. (**a**) LAU10 sectional view [10]. (**b**) Photograph of the LAU10.

## 3. Control System

The control hardware used in the NPS6D100 is based on a dSPACE platform, which supports real-time control development and data acquisition. As shown in Figure 3, the closed-loop system runs using a 10 kHz sampling rate. During each cycle, 6-DOF feedback data ($q_p$, $q_v$) are acquired via the analog and digital inputs. These signals are subsequently processed to compute the setpoints ($i_p^*$, $p_v^*$, $i_v^*$) for the actuators. The target values for the planar actuators ($i_p^*$) and the vertical actuators ($p_v^*$, $i_v^*$) are transferred to hardware interfaces consisting of linear current amplifiers and high-precision pressure regulators. These components convert the control signals ($i_p^*$, $p_v^*$, $i_v^*$) into real quantities ($i_p$, $p_v$, $i_v$).

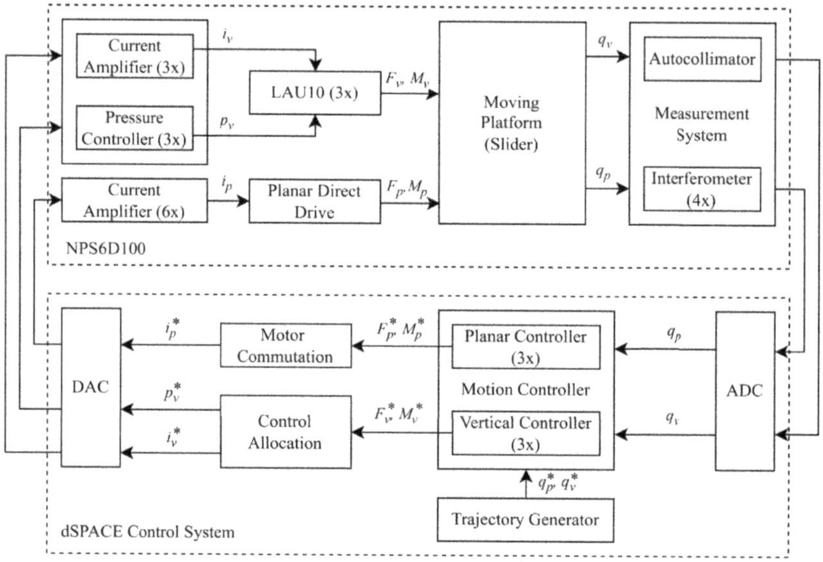


**Figure 3.** Control structure of the NPS6D100. Signals with subscripts $(\cdot)_p$ and $(\cdot)_v$ refer to planar and vertical quantities, respectively. Signals with superscript $(\cdot)^*$ indicate target values [30].

Individual feedback loops, one per axis, are implemented to achieve motion control across all six DOF. Figure 4 depicts the closed-loop block diagram for an individual axis. Each control loop follows a conventional PID structure, which is widely used for nanopositioning systems because of its outstanding balance between implementation simplicity and performance effectiveness [35,36]. To further improve the precision and minimize oscillations caused by mechanical resonances in the high-frequency range, additional filters have been integrated into each control loop. Among the wide variety of available filters, low-pass filters and band-stop filters (also referred to as notch filters) were implemented.

The core of the motion controller is the PID-type structure that includes a filtered derivative term. In the time domain, the controller is described by

$$\begin{aligned} v_{\mathrm{d}}(t) &= -\omega_{\mathrm{d}} \int_0^t v_{\mathrm{d}}(\tau)\,\mathrm{d}\tau + \omega_{\mathrm{d}}\, e(t), \\ u(t) &= K_{\mathrm{P}}\Big(e(t) + K_{\mathrm{I}} \int_0^t e(\tau)\,\mathrm{d}\tau + K_{\mathrm{D}}\, v_{\mathrm{d}}(t)\Big). \end{aligned} \tag{3}$$

Equivalently, in the Laplace domain, the controller (3) is expressed as

$$G_{\mathrm{c}}(s) = \frac{u(s)}{e(s)} = K_{\mathrm{P}}\left(1 + \frac{K_{\mathrm{I}}}{s} + K_{\mathrm{D}}\left(\frac{\omega_{\mathrm{d}} s}{s + \omega_{\mathrm{d}}}\right)\right), \tag{4}$$

where $K_P$, $K_I$, and $K_D$ are the proportional, integral, and derivative gains, respectively, while $\omega_d$ determines how aggressively the derivative term is filtered. Each axis is tuned independently to ensure reliable performance, with sufficient robust stability margins in terms of gain margin, phase margin, and bandwidth. The control parameters are listed in Table 4. Note that the PID parameters for all translational axes $(x, y, z)$ share the same values, which is a mere coincidence. As introduced in Section 2.3, the overactuated nature of the $z$-axis is handled by means of a well-suited control allocation method, where the overactuated problem of the $z$-axis is reduced into a SISO formulation similar to the $x$ and $y$ axes. The reader is referred to [30–32] to complement the discussion on the control allocation criteria.

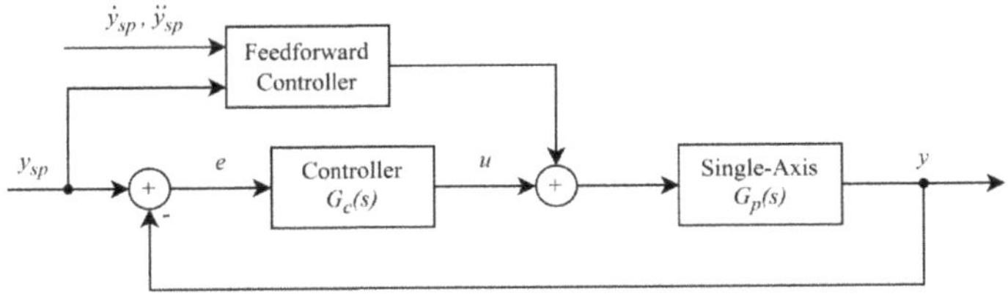


**Figure 4.** Control strategy for a single axis, where $y_{sp}$ is the setpoint [mm or arcsec], $y$ is the measured output [mm or arcsec], $e$ is the error [mm or arcsec], and $u$ is the control signal [N or Nm].

**Table 4.** Controller gains for all six closed-loop systems.

| Gain | $x$-Axis | $y$-Axis | $\varphi_z$-Axis | $z$-Axis | $\varphi_x$-Axis | $\varphi_y$-Axis |
|---|---|---|---|---|---|---|
| $K_p$ [N/mm, N m/″] | 663.63 | 663.63 | 0.0056 | 663.63 | 0.0038 | 0.0038 |
| $K_i$ [rad/s] | 201.06 | 201.06 | 62.832 | 201.06 | 62.832 | 62.832 |
| $K_d$ [s/rad] | 0.0145 | 0.0145 | 0.0159 | 0.0145 | 0.0159 | 0.0159 |
| $\omega_d$ [rad/s] | 2513.3 | 2513.3 | 1570.8 | 2513.3 | 1570.8 | 1570.8 |

Figure 5 depicts the Bode plot showing the measured frequency response of the translational motion axes $(x, y, z)$. The mechanical designs of the planar (Section 2.2) and vertical (Section 2.3) drive subsystems resemble the dynamics of a rigid body with mass and damping. Direct inspection reveals a decay of 40 dB per decade, reflecting the double integrator nature of the motion axes and the high stiffness in the actuation chain; see Figure 5. Additional resonance modes in the high frequency band confirm the presence of structural flexibilities in the motion platform which must be attenuated through the controller. For completeness, the measured frequency response of the tilting angles $(\varphi_x, \varphi_y, \varphi_z)$ are shown in Figure A1 (Appendix A).

Figure 6 presents the open-loop response with clearly visible robust stability characterized by sufficient gain margins and phase margins. To reach these outstanding robust stability margins, the tuning procedure was performed offline based on the measured frequency response of the motion axes. Despite only the translational motion axes $(x, y, z)$ being shown, the same tuning procedure was applied to the other rotational $(\varphi_x, \varphi_y, \varphi_z)$ DOFs. A summary of the control loop performance for all six motion axes is gathered in Table 5. For completeness, the computed open-loop response of the tilting angles $(\varphi_x, \varphi_y, \varphi_z)$ are shown in Figure A2 (Appendix A).

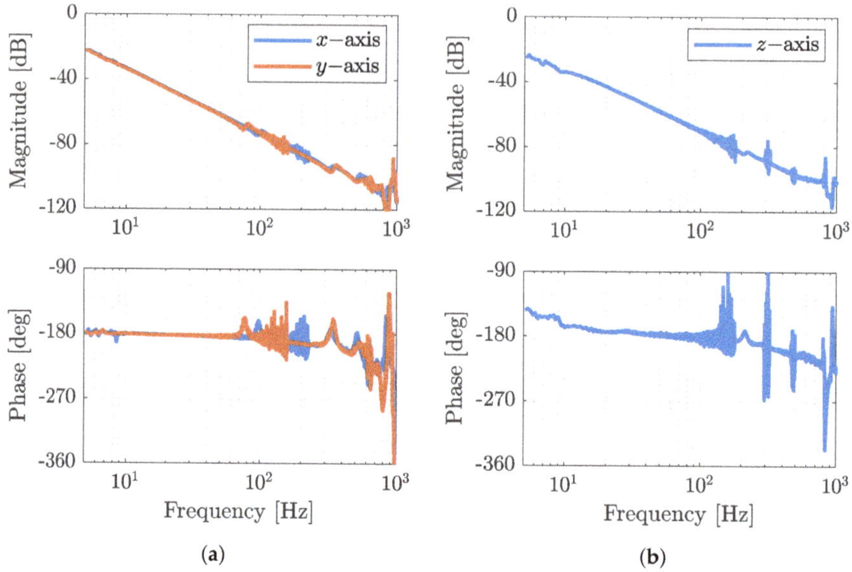


**Figure 5.** Measured FRFs of the translational DOFs $(x, y, z)$. (**a**) FRF of the $x$ and $y$ axes. (**b**) FRF of the $z$-axis.

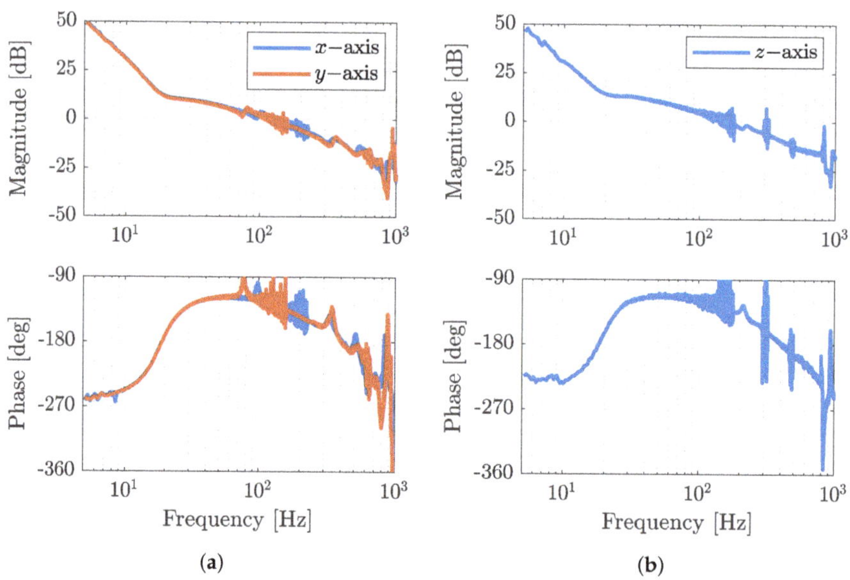


**Figure 6.** Open-loop FRFs of the translational DOFs $(x, y, z)$. (**a**) Open-loop FRF of the $x$ and $y$ axes. (**b**) Open-loop FRF of the $z$-axis.

Despite the overall robustness of the PID controller (4), high-frequency resonances in the motion axes may still disturb the precision performance of the NPS6D100. Therefore, the control design integrates a series connection of notch filters and low-pass filters, specifically tuned to attenuate targeted resonances and preserve precise positioning. These filters reduce oscillations by attenuating undesired high-frequency components in the control signal. The low-pass filter suppresses high-frequency noise (usually above 1 kHz), while the notch filter eliminates specific resonance frequencies that may produce vibration.

Even though the PID-based controller (4) performs well, especially in steady-state and low-frequency motion tasks (as shown in Section 4), it is not ideal for dealing with rapidly changing external disturbances or variations in system parameters. Such disturbing scenarios may require advanced control techniques. Several studies have explored advanced methods for handling periodic input disturbances and parametric uncertainty, as referenced in [32–34,37]. Furthermore, the interested reader may consult [30], which describes the design and real-time validation of an advanced control concept in a large-scale motion system with nanometer precision, also referred to as NPS6D200.

**Table 5.** Robustness metrics for all six closed-loop systems.

| Specifications | $x$-Axis | $y$-Axis | $\varphi_z$-Axis | $z$-Axis | $\varphi_x$-Axis | $\varphi_y$-Axis |
|---|---|---|---|---|---|---|
| Gain margin | 12.5 dB | 12.1 dB | 10.2 dB | 15.6 dB | 11.1 dB | 11.7 dB |
| Phase margin | 56.5° | 56.4° | 64.9° | 53.3° | 54.6° | 56.6° |
| Bandwidth | 121.8 Hz | 113.2 Hz | 104.9 Hz | 139.4 Hz | 101.2 Hz | 102.3 Hz |

## 4. Positioning Performance in 6D Closed Loop Control

This section presents the performance evaluation of the NPS6D100 nanopositioning system. Stationary and tracking exercises were performed to demonstrate that the motion system consistently achieves nanometer-level precision across the entire travel range.

Initial experiments in closed loop operation evaluate the capability of the NPS6D100 to maintain a fixed position. Figure 7 displays the time-domain responses of all six DOFs with the motion platform held at a fixed position in the center of the workspace. In this exercise, the position root mean square (RMS) errors are very low, 0.54 nm in $x$, 0.59 nm in $y$, and 0.47 nm in $z$, and the angular RMS errors are also small, with 13.65 m″ in $\varphi_x$, 17.97 m″ in $\varphi_y$, and 3.01 m″ in $\varphi_z$. These results validate that the system meets the performance expectations during steady-state operation, achieving sub-10 nm precision under controlled laboratory conditions. It is worth noting that the RMS level observed in Figure 7 reflects the influence of the control action but also contains the actual measurement signal noise. With a measurement resolution of 20 pm, the noise level of the measurement signal at the NPS6D100 is expected to be in the range of 0.2 nm standard deviation. In earlier investigations for a similar setup, the standard deviation of the fixed environment noise (with the controller switched off) was measured at $\sigma_x = 0.41$ nm, $\sigma_y = 0.33$ nm, and $\sigma_z = 0.36$ nm; see [6] for further discussion.

The subsequent analysis focuses on evaluating the positioning performance across the entire travel range. The measurement process was carried out by forming a grid with 1 mm spacing. At each point on the grid, the set of sensors measured the time series for one second and the RMS error was subsequently calculated and stored. As shown in Figure 8, the planar $(x, y)$ RMS error remained below 1 nm across the full ⌀100 mm planar workspace. The vertical positioning achieves comparable RMS error levels, that is, below 1 nm.

Figure 9 presents the servo errors along the 10 mm vertical range. The positioning errors in $x$, $y$, and $z$ remain in the nanometer-level and stable, even as the air gap between the planar coils and the permanent magnet arrays increases as height increases. These experiments show that the design without a coil lift system works effectively and, consequently, the direct force transfer behavior in the planar electromagnetic drivers remains stable over the whole vertical operating range.

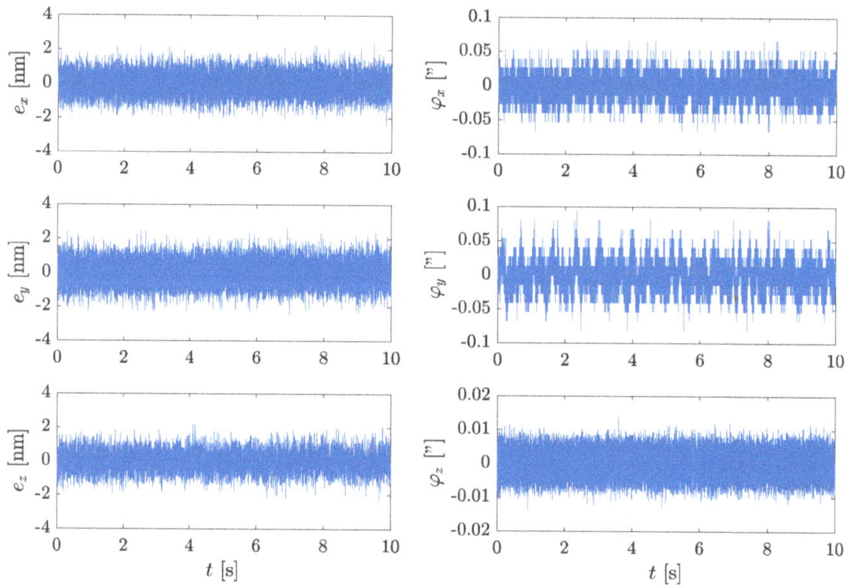


**Figure 7.** Time series in closed loop control at constant setpoint. RMS servo error: $e_x = 0.54\,\text{nm}$, $e_y = 0.59\,\text{nm}$, $e_z = 0.47\,\text{nm}$, $\varphi_x = 13.65\,\text{m}''$, $\varphi_y = 17.97\,\text{m}''$, and $\varphi_z = 3.01\,\text{m}''$.

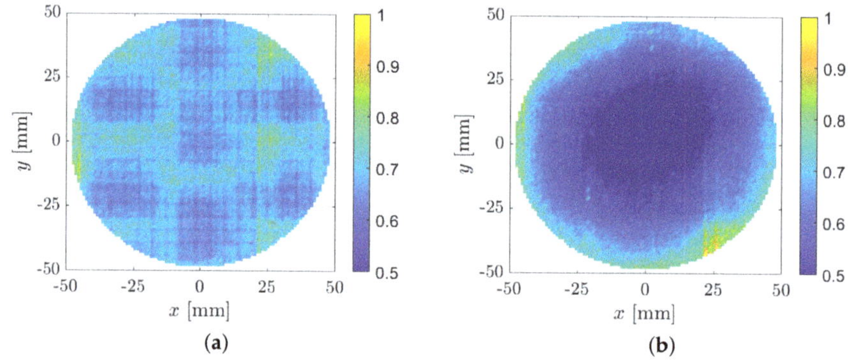


**Figure 8.** RMS servo error variation within the planar travel range at $z = 5\,\text{mm}$. (**a**) $xy$ servo error (RMS) in nm. (**b**) $z$ servo error. (RMS) in nm.

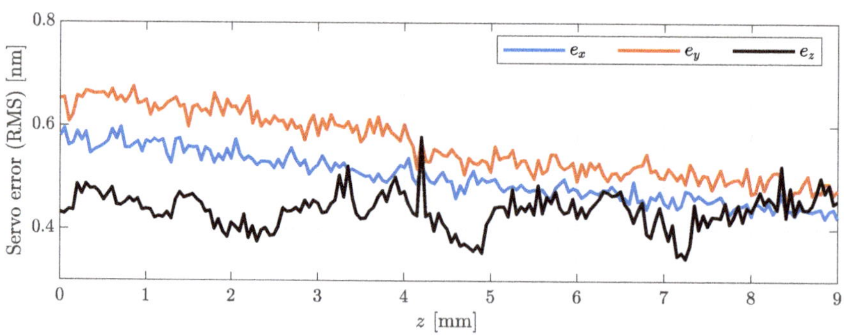


**Figure 9.** RMS servo error variation within vertical travel range measured at $(x, y) = (0, 0)$ mm. (blue)—$e_x$, (red)—$e_y$, (black)—$e_z$.

Then, step-wise motion experiments were carried out, which are fundamental in the field of precision engineering. Towards this end, the nanopositioning system NPS6D100 was commanded to perform nanometer-scale steps along the $x$, $y$, and $z$ axes, whereas the tilting angles ($\varphi_x$, $\varphi_y$, $\varphi_z$) were regulated to zero. Figure 10 illustrates the measured time series over a time span of 11 s. The steps are clearly visible, featuring minimum noise amplification and no visible significant overshoots. The raw signal (in blue lines) and the filtered signal (in red lines) both confirm that the NPS6D100 can reliably execute nanometer-level steps in all three spatial position coordinates.

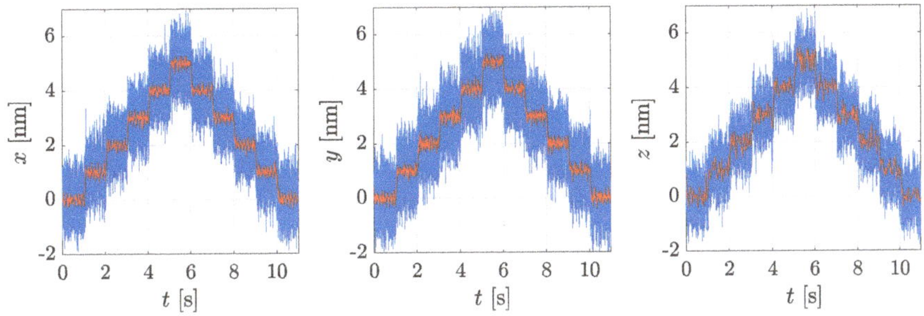


**Figure 10.** Nanometer staircase exercise in $x$-, $y$-, and $z$-directions (from left to right). Blue—raw data traced with 10 kHz; red—filtered data (4th order Chebyshev low pass, 40 Hz cutoff frequency). (**a**) Steps in $x$-axis. (**b**) Steps in $y$-axis. (**c**) Steps in $z$-axis.

The capability to precisely follow complex 3D trajectories was assessed through helix motions of different dimensions ranging from the nanometer-level to the millimeter-level. Figure 11 depicts a helix of dimensions ⌀40 nm × 9 nm traced with a velocity of 14.5 nm/s. The trajectory-tracking exercise was executed with sub-10 nm RMS servo errors. Furthermore, Table 6 summarizes these results for three helix sizes, including RMS errors and RMS currents. The nanopositioning system achieves RMS servo errors less than 3 nm across the macroscopic travel range, spanning from the nanometer to the millimeter scale. The low currents observed confirm the high efficiency of the drive units (planar and vertical) and the effectiveness of the low-dynamic weight force compensation mechanism integrated into the lifting modules.

**Table 6.** Summarized results of helix trajectory-tracking exercises.

| Commanded Helix Path | RMS Servo Errors | RMS Currents (Planar Coils) | RMS Currents (Lifting Modules) |
|---|---|---|---|
| ⌀40 nm × 9 nm<br>$v = 14.5$ nm/s | $e_x = 0.59$ nm<br>$e_y = 0.63$ nm<br>$e_z = 0.40$ nm | $i_A = 32.44$ mA<br>$i_B = 46.45$ mA<br>$i_C = 1.77$ mA | $i_1 = 2.41$ mA<br>$i_2 = 2.39$ mA<br>$i_3 = 2.09$ mA |
| ⌀40 µm × 9 µm<br>$v = 14.5$ µm/s | $e_x = 0.89$ nm<br>$e_y = 1.44$ nm<br>$e_z = 0.44$ nm | $i_A = 35.22$ mA<br>$i_B = 47.17$ mA<br>$i_C = 3.27$ mA | $i_1 = 2.56$ mA<br>$i_2 = 2.41$ mA<br>$i_3 = 2.12$ mA |
| ⌀40 mm × 9 mm<br>$v = 14.5$ mm/s | $e_x = 1.54$ nm<br>$e_y = 1.89$ nm<br>$e_z = 2.71$ nm | $i_A = 47.39$ mA<br>$i_B = 62.54$ mA<br>$i_C = 71.59$ mA | $i_1 = 70.54$ mA<br>$i_2 = 67.60$ mA<br>$i_3 = 73.99$ mA |

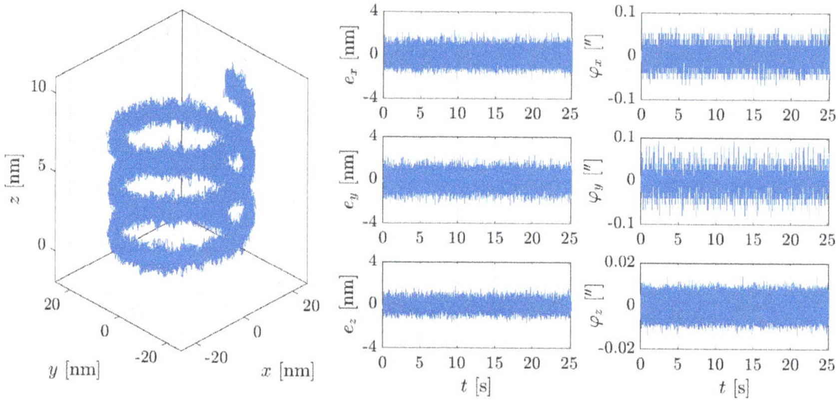


(a) (b)

**Figure 11.** The 3D motion exercise with a ⌀40 nm × 9 nm helix. (**Left**)—plot of the 3D position signals; (**right**)—servo error time series in the 6 DOFs during trajectory motion. (**a**) Isometric view. (**b**) Time series of each DOF.

## 5. Conclusions and Outlook

The NPS6D100 represents a solution where the coil lift system is not required and thus completes the range of positioning solutions that combine an integrated planar direct drive with direct vertical actuation. In direct comparison to the NPS6D200 [28,29], the working range and the coil lift system are the main differences, resulting in disturbance reduction and improved positioning performance.

The NPS6D100 is a high-precision nanopositioning system that enables 6-DOF motion within a ⌀100 mm × 10 mm workspace. The motion system was designed for sub-10 nm applications; it combines a 3D planar Lorentz-based direct drive, aerostatic guidings, and vertical actuation to achieve precise and reproducible motion across all axes.

A key feature of the NPS6D100 is the integration of three custom-developed LAUs, each combining a voice coil (electromagnetic) direct drive, a low-dynamic pneumatic weight force compensation, and an aerostatic bushing into a compact lifting module. This design ensures precise vertical positioning with minimum heat dissipation, making it suitable for long-term precision tasks without disturbing the measurement space. Each LAU also includes an internal vertical encoder for standalone initialization and control.

The measurement system uses a Cartesian mirror arrangement with three LIFs and a 2D autocollimator to provide high-resolution feedback in all six DOFs. This setup supports precise, closed-loop control and eliminates first-order Abbe errors. The system demonstrates sub-nanometer RMS servo errors under stationary setpoint conditions and maintains nanometer-level tracking performance during dynamic synchronous multiaxial motion.

The mechanical structure is compact and dynamically stable, and its performance under different motion conditions shows minimum variation in positioning precision. The NPS6D100 operates efficiently within the complete travel range, with vertical displacements of up to 10 mm, without introducing significant cross-coupling between axis or affecting the direct force transmission between the planar coils and the permanent magnet array.

The combination of stiffness, compactness, and high-precision measurement instruments makes the NPS6D100 well-suited for demanding applications such as sub-10 nm fabrication, long-range optical probing, and sample handling in nanoscale inspection or lithography.

Future developments will focus on investigating in situ calibration routines and integrating tool-specific probes to expand the functionality of the NPS6D100 within the nanotechnology field.

**Author Contributions:** Conceptualization, S.H., A.H. and M.K.; methodology, S.H., A.H. and M.K.; validation, S.H., A.H. and M.K.; investigation, S.H., A.H. and M.K.; writing—original draft preparation, S.H. and A.H.; visualization, S.H. and A.H.; writing—review and editing, S.H., A.H., M.K. and L.H. All authors have read and agreed to the published version of the manuscript.

**Funding:** This research was supported by the German "Land" of Thüringen in the scope of the internal research group "Next Generation Positioning".

**Institutional Review Board Statement:** Not applicable.

**Informed Consent Statement:** Not applicable.

**Data Availability Statement:** Research data are available upon request to the authors.

**Acknowledgments:** The authors gratefully acknowledge the fruitful cooperation with the Institute of Process Measurement and Sensor Technology and the Control Engineering Group at TU Ilmenau, as well as with SIOS GmbH and AeroLas GmbH.

**Conflicts of Interest:** All authors are employees of the IMMS, which is a (public) research institute owned by the German Land of Thüringen, and have no potential conflicts of interest. See also: https://www.imms.de/en/about-us.html (accessed on 13 July 2025).

## Abbreviations

The following abbreviations are used in this manuscript:

| | |
|---|---|
| ADC | Analog-to-Digital Converter |
| CMM | Coordinate Measurement Machine |
| DAC | Digital-to-Analog Converter |
| DOF | Degree of Freedom |
| FRF | Frequency Response Function |
| LAU | Lifting and Actuation Unit |
| LIF | Laser Interferometer |
| NPS | Nanopositioning System |
| PID | Proportional Integral Derivative (Controller) |
| RMS | Root Mean Square |

## Appendix A

Figure A1 depicts the measured frequency response function (FRF) of the tilting angles ($\varphi_x$, $\varphi_y$, $\varphi_z$). Moreover, the figure also shows the computed open-loop response of the tilting angles ($\varphi_x$, $\varphi_y$, $\varphi_z$) in order to complement the dynamic description of the NPS6D100.

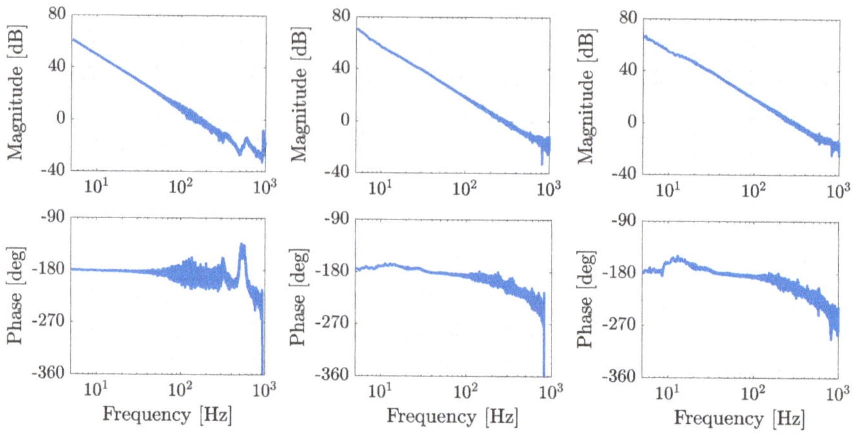


(**a**) (**b**) (**c**)

**Figure A1.** Measured FRFs of the tilting angles ($\varphi_x$, $\varphi_y$, $\varphi_z$). (**a**) FRF of the $\varphi_z$-axis. (**b**) FRF of the $\varphi_x$-axis. (**c**) FRF of the $\varphi_y$-axis.

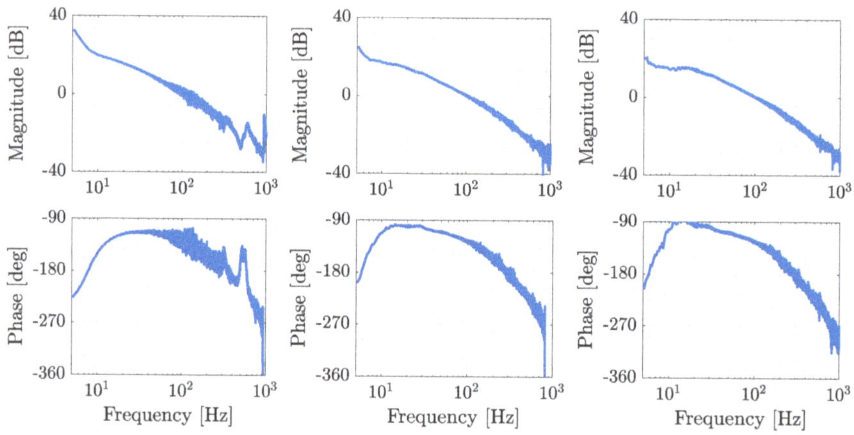


(**a**) (**b**) (**c**)

**Figure A2.** Open-loop FRFs of the tilting angles ($\varphi_x$, $\varphi_y$, $\varphi_z$). (**a**) FRF of the $\varphi_z$-axis. (**b**) FRF of the $\varphi_x$-axis. (**c**) FRF of the $\varphi_y$-axis.

## References

1. Oh, D.K.; Jeong, H.; Kim, J.; Kim, Y.; Kim, I.; Ok, J.G.; Rho, J. Top-down nanofabrication approaches toward single-digit-nanometer scale structures. *J. Mech. Sci. Technol.* **2021**, *35*, 837–859. [CrossRef]
2. Gour, J.; Beer, S.; Paul, P.; Alberucci, A.; Steinert, M.; Szeghalmi, A.; Siefke, T.; Peschel, U.; Nolte, S.; Zeitner, U.D. Wafer-scale nanofabrication of sub-5 nm gaps in plasmonic metasurfaces. *Nanophotonics* **2024**, *13*, 4191–4202. [CrossRef] [PubMed]
3. Chen, Y.; Shu, Z.; Zhang, S.; Zeng, P.; Liang, H.; Zheng, M.; Duan, H. Sub-10 nm fabrication: Methods and applications. *Int. J. Extrem. Manuf.* **2021**, *3*, 032002. [CrossRef]
4. Perego, M. Focus on sub-10 nm nanofabrication. *Nanotechnology* **2018**, *29*, 260201. [CrossRef] [PubMed]
5. Gorges, S.; Hesse, S.; Schäffel, C.; Ortlepp, I.; Manske, E.; Langlotz, E.; Dontsov, D. Integrated planar 6-DOF nanopositioning system. *IFAC-PapersOnLine* **2019**, *52*, 313–318. [CrossRef]
6. Hesse, S.; Schäffel, C.; Mohr, H.U.; Katzschmann, M.; Büchner, H.J. Design and performance evaluation of an interferometric controlled planar nanopositioning system. *Meas. Sci. Technol.* **2012**, *23*, 074011. [CrossRef]
7. Sreeja, C.; Godwinraj, D. Robust Control of Precision Nano-Positioning System for Microsurgical Applications. In Proceedings of the International Conference on ICT for Digital, Smart, and Sustainable Development, New Delhi, India, 22–23 April 2024; Springer: Singapore, 2024; pp. 493–505.

8. Basson, N.; Peng, C.H.S.; Geoghegan, P.; van der Lecq, T.; Steven, D.; Williams, S.; Lim, A.E.; Ho, W.H. A computational fluid dynamics investigation of endothelial cell damage from glaucoma drainage devices. *Sci. Rep.* **2024**, *14*, 3777. [CrossRef] [PubMed]
9. Gorges, S.; Leistritz, B.; Hesse, S.; Ortlepp, I.; Slotta, G.; Schäffel, C. Development of an integrated guiding and actuation element for high dynamic nanopositioning systems. In Proceedings of the 59th Ilmenau Scientific Colloquium, Ilmenau, Germany, 11–15 September 2017.
10. Gorges, S. A Lifting and Actuating Unit for a Planar Nanoprecision Drive System. Ph.D. Thesis, Technische Universität Ilmenau, Ilmenau, Germany, 2020.
11. He, Y.X.; Lin, R.W.; Li, R.J.; Li, J.; Cheng, Z.Y.; Pan, Q.S.; Huang, Q.X.; Fan, K.C. Nanopositioning X–Y stage with an embedded Six-DOF error compensation system based on Abbe and Bryan principles. *Measurement* **2024**, *227*, 114218. [CrossRef]
12. Moers, A.; van Riel, M.C.; Bos, E.J. *Design and Verification of the Trinano Ultra Precision CMM*; Universitätsbibliothek: Ilmenau, Germany, 2011.
13. Ro, S.K.; Park, J.K. A compact ultra-precision air bearing stage with 3-DOF planar motions using electromagnetic motors. *Int. J. Precis. Eng. Manuf.* **2011**, *12*, 115–119. [CrossRef]
14. Yagüe-Fabra, J.A.; Gao, W.; Archenti, A.; Morse, E.; Donmez, A. Scalability of precision design principles for machines and instruments. *CIRP Ann.* **2021**, *70*, 659–680. [CrossRef]
15. Schellekens, P.; Rosielle, N.; Vermeulen, H.; Vermeulen, M.; Wetzels, S.; Pril, W. Design for precision: Current status and trends. *Cirp Ann.* **1998**, *47*, 557–586. [CrossRef]
16. Bryan, J.B. The Abbe principle revisited: An updated interpretation. *Precis. Eng.* **1979**, *1*, 129–132. [CrossRef]
17. Seggelen, J. NanoCMM; A 3D Coordinate Measuring Machine with Low Moving Mass for Measuring Small Products in Array with Nanometer Uncertainty. Ph.D. Thesis, Eindhoven University of Technology, Eindhoven, The Netherlands, 2007.
18. Vermeulen, M.M.P.A.; Rosielle, P.; Schellekens, P. Design of a high-precision 3D-coordinate measuring machine. *Cirp Ann.* **1998**, *47*, 447–450. [CrossRef]
19. Hesse, S.; Schäffel, C.; Katzschmann, M.; Maaß, T.; Mohr, H. Planar motor concept for positioning with nanometer position uncertainty. In Proceedings of the 10th International Conference of the European Society for Precision Engineering and Nanotechnology, Zürich, Switzerland, 18–22 May 2008; pp. 150–154.
20. Manske, E.; Fröhlich, T.; Füßl, R.; Ortlepp, I.; Mastylo, R.; Blumröder, U.; Dontsov, D.; Kühnel, M.; Köchert, P. Progress of nanopositioning and nanomeasuring machines for cross-scale measurement with sub-nanometre precision. *Meas. Sci. Technol.* **2020**, *31*, 085005. [CrossRef]
21. Jäger, G.; Hausotte, T.; Manske, E.; Büchner, H.J.; Mastylo, R.; Dorozhovets, N.; Hofmann, N. Nanomeasuring and nanopositioning engineering. *Measurement* **2010**, *43*, 1099–1105. [CrossRef]
22. Manske, E.; Jäger, G.; Hausotte, T.; Füßl, R. Recent developments and challenges of nanopositioning and nanomeasuring technology. *Meas. Sci. Technol.* **2012**, *23*, 074001. [CrossRef]
23. Ortlepp, I.; Fröhlich, T.; Füßl, R.; Reger, J.; Schäffel, C.; Sinzinger, S.; Strehle, S.; Theska, R.; Zentner, L.; Zöllner, J.P.; et al. Tip-and laser-based 3D nanofabrication in extended macroscopic working areas. *Nanomanuf. Metrol.* **2021**, *4*, 132–148. [CrossRef]
24. Fern, F.; Fuessl, R.; Manske, E.; Schienbein, R.; Theska, R.; Ortlepp, I.; Leineweber, J. Measurement uncertainty analysis on a five-axis nano coordinate measuring machine NMM-5D following a vectorial approach. *TM-Tech. Mess.* **2021**, *88*, 61–70. [CrossRef]
25. Stauffenberg, J.; Ortlepp, I.; Belkner, J.; Dontsov, D.; Langlotz, E.; Hesse, S.; Rangelow, I.; Manske, E. Measurement precision of a planar nanopositioning machine with a range of motion of Ø100 mm. *Appl. Sci.* **2022**, *12*, 7843. [CrossRef]
26. Stauffenberg, J.; Reibe, M.; Krötschl, A.; Reuter, C.; Ortlepp, I.; Dontsov, D.; Hesse, S.; Rangelow, I.W.; Strehle, S.; Manske, E. Tip-based nanofabrication below 40 nm combined with a nanopositioning machine with a movement range of Ø100 mm. *Micro Nano Eng.* **2023**, *19*, 100201. [CrossRef]
27. SIOS GmbH. 2025. Available online: https://www.sios-precision.com/en/products/nanopositioning-and-nanomeasuring-machine/nanopositioning-platform (accessed on 7 May 2025).
28. Hesse, S.; Huaman, A.; Katzschmann, M.; Leistritz, B.; Herzog, L. NPS6D200—A Long Range Nanopositioning Stage with 6D Closed Loop Control. *Appl. Sci.* **2024**, *14*, 6972. [CrossRef]
29. Hesse, S.; Katzschmann, M.; Huaman, A.; Brasil, D.; Herzog, L. NPS6D200-Realization of high precision long stroke vertical actuation in conjunction with a planar direct drive. In Proceedings of the EUSPEN Special Interest Group Meeting on Precision Motion Systems & Control, Hertogenbosch, The Netherlands, 12–13 November 2024.
30. Huaman, A.S. Highly Dynamic Nanopositioning: A Robust Adaptive Control Approach. Ph.D. Thesis, Technische Universität Ilmenau, Ilmenau, Germany, 2024.
31. Seminario, R.; Schmitt, C.; Weise, C.; Reger, J. Control of an overactuated nanopositioning system with hysteresis by means of control allocation. In Proceedings of the 2022 IEEE 17th International Conference on Advanced Motion Control (AMC), Padova, Italy, 18–20 February 2022; pp. 280–287.

32. Huaman, A.S.; Reger, J. Robust Tracking Control with $\mathcal{L}_1$ Adaptive Augmentation for a Long-Stroke Vertical Nanopositioning System: Part I. In Proceedings of the 2022 IEEE Conference on Control Technology and Applications (CCTA), Trieste, Italy, 23–25 August 2022; pp. 614–620.
33. Huaman, A.S.; Reger, J. Robust Tracking Control with $\mathcal{L}_1$ Adaptive Augmentation for a Long-Stroke Vertical Nanopositioning System: Part II. In Proceedings of the 2022 IEEE Conference on Control Technology and Applications (CCTA), Trieste, Italy, 23–25 August 2022; pp. 621–627.
34. Huaman, A.S.; Reger, J. Robust Adaptive Tracking Control of a 3D Vertical Motion System for Nanometer Precision Applications. *IFAC-PapersOnLine* **2023**, *56*, 5332–5339. [CrossRef]
35. Fleming, A.J.; Leang, K.K. *Design, Modeling and Control of Nanopositioning Systems*; Springer: Cham, Switzerland, 2014; Volume 15.
36. Ru, C.; Liu, X.; Sun, Y. Nanopositioning technologies. In *Fundamentals and Applications*; Springer: Cham, Switzerland, 2016.
37. Huaman, A.S.; Reger, J. Disturbance Rejection in Periodic Motion with Enhanced Constant Velocity Transitions via Adaptive Augmentation. In Proceedings of the 2025 IEEE International Conference on Mechatronics (ICM), Wollongong, Australia, 28 February–2 March 2025; pp. 1–6.

actuators

*Article*

# A Study on Reducing Loss in PCB Motor Stator Using Multi-Via Structure

**Su-Bin Jeon [1], Do-Hyeon Choi [1], Hyung-Sub Han [1], Yun-Ha Song [1] and Won-Ho Kim [2,*]**

[1] Next Generation Energy System Convergence, Gachon University, Seongnam 13120, Republic of Korea; harry89155@naver.com (S.-B.J.); dino0403@naver.com (D.-H.C.); brosubi@naver.com (H.-S.H.); 6228140@naver.com (Y.-H.S.)
[2] Electrical Engineering, Gachon University, Seongnam 13120, Republic of Korea
* Correspondence: wh15@gachon.ac.kr

**Abstract:** This study proposes a multi-via structure as a loss-reduction design technique to mitigate current crowding in a slotless axial flux permanent magnet motor (AFPM) equipped with printed circuit board (PCB) stators. The PCB stator enables high current density operation through parallel copper-foil stacking and supports an ultra-compact structural configuration. However, current concentration in the via regions can increase copper loss and phase resistance. In this work, the via position and diameter were defined as design variables to perform a sensitivity analysis of current distribution and phase resistance variation. The effects of current density dispersion and the potential for copper loss reduction were evaluated using three-dimensional finite-element analysis (FEA). The results confirm that adopting a multi-via structure improves current path uniformity and reduces electrical losses, thereby enhancing overall efficiency. Furthermore, the analysis shows that excessive via enlargement or overuse does not necessarily yield optimal results and, in certain cases, may lead to localized current peaks. These findings demonstrate that the multi-via structure is an effective and appropriate design strategy for PCB stators and highlight the importance of optimized via placement tailored to each stator configuration.

**Keywords:** actuator; axial flux permanent magnet motor; PCB motor; multi-via; via

## 1. Introduction

The global mobility market, including shared, electric, and autonomous services, is expected to grow at a compound annual rate exceeding 12% through 2030 [1]. An academic survey of shared-mobility business models likewise confirms that double-digit compound growth will persist into the mid-2020s [2]. This expansion sharply increases the demand for compact actuators that must be installed in large numbers inside vehicles [3].

An anti-lock braking system (ABS) requires one miniature hydraulic pump motor per wheel—four per vehicle—and must deliver counter-torque within a few milliseconds after slip is detected [4]. Experiments published in *Actuators* report torque-response times below 15 ms for electric actuators [5]. To meet such high-speed control requirements, axial flux permanent magnet (AFPM) machines have attracted considerable interest. By generating torque in the radial direction, AFPMs achieve both high torque density and a short axial length, enabling efficient operation in severely space-constrained environments [6,7]. Laminated core AFPMs, however, suffer from intricate coil insertion, localized heat concentration, and tight manufacturing tolerances [8].

Printed circuit board (PCB) coreless AFPMs have emerged to overcome these limitations. Removing the magnetic core eliminates iron loss, while multilayer parallel copper traces provide a high conductor fill factor. Since the experimental demonstration of an ultra-thin AFPM employing dual-stator PCB windings [9], electromagnetic and thermal performance have been continuously improved through composite coil architectures [10] and refined PCB winding designs [11]. A slotless PCB-wound electric/hydraulic brake (EHB) motor has further shown that a flat form factor and rapid dynamic response can be achieved simultaneously [12]. A comparative optimization study on brake-by-wire (BBW) actuators has reinforced the suitability of PCB-AFPMs for high-response braking systems [13].

The parallel layer structure of PCB stators accommodates high current while distributing it laterally, thereby reducing copper loss. When driven by GaN-FET power electronics, PCB windings have achieved substantial gains in power density [14]; eddy current and circulating current losses associated with variations in the number of winding layers have also been quantified [15]. Optimizing thermal-via placement alone has been reported to lower the winding temperature of a PCB-AFPM by more than 15 °C [12].

Nonetheless, the vias that interconnect the layers of a multilayer PCB winding concentrate current locally, increasing the current density within the via and raising phase resistance [16]. Multiphysics analysis shows that, under pulsed excitation, temperature rise at the winding–via junction of a PCB-wound AFPM is pronounced, making thermal management essential [17]. A comparative study of PCB winding patterns in high-speed coreless AFPMs further reports that via layout markedly affects circulating and eddy current losses, which in turn play a decisive role in overall efficiency [18]. Although electric machines generally incur multiple loss components, including iron, mechanical, and stray losses, their contribution is negligible in the structure considered. The rotor, composed of solid S45C carbon steel, rotates synchronously with the stator field, thereby preventing relative flux variation and suppressing eddy current formation. The stator, built from laminated non-oriented electrical steel (35PN230), limits circulating currents through its lamination layers, which minimizes eddy current losses and leaves hysteresis loss as the dominant core component [19]. Consequently, copper loss emerges as the principal loss mechanism in coreless PCB stators under high current density, and this study therefore focuses on copper loss minimization through multi-via optimization. The quantitative relationships between via geometry (diameter, number, and placement) and phase resistance or peak current density remain insufficiently characterized [20]. A systematic review of 38 papers published between 2019 and 2024 highlights the absence of predictive via-loss models as a major research gap [21].

Previous studies have mainly focused on qualitative comparisons of PCB winding structures in terms of thermal and loss characteristics, while systematic investigations into the quantitative relationship between via diameter, number, layout, and the resulting phase resistance and current density remain limited. Therefore, the originality of this study lies in conducting a comparative FEM analysis of single-via, inline multi-via, and grid multi-via structures under identical conditions, clarifying the trade-off between phase resistance, current density, and copper loss, and proposing an optimized via layout design strategy tailored for actuator-oriented PCB AFPM. Using a three-phase, two-turn PCB stator rated at 114.67 W for an ABS actuator as a case study, the effects of via diameter, number, and placement on phase resistance and peak current density are systematically compared. The results show that simply enlarging or increasing the number of vias does not guarantee optimal performance; in certain cases, local current spikes cause a non-linear rise in phase resistance. The proposed grid multi-via layout reduces phase resistance and peak current density by 20.76% and 20.72%, respectively, relative to the single-via baseline, lowering copper loss by 10.4% while remaining compatible with standard PCB fabrication.

The remainder of this paper is organized as follows. Section 2 introduces the basic design of the PCB distributed-winding motor for actuators. Section 3 presents the multi-via design of the PCB stator. Section 4 describes the FEM methodology and performance evaluation. Finally, Section 5 concludes the paper.

## 2. Basic Design of PCB Distributed Winding Motor for Actuators

### 2.1. Distributed Winding PCB Axial Flux Permanent Magnet Motor

Figure 1 shows an axial flux permanent magnet motor (AFPM) equipped with a PCB stator. Replacing the conventional laminated iron core with a slotless PCB stator greatly simplifies the design and markedly shortens the axial length [10,11]. In a traditional AFPM, windings must be wrapped around stator teeth, complicating manufacturing and increasing stack height [6]. In contrast, printing conductors directly onto the PCB substrate eliminates tooth geometry and enables a much thinner profile [12]. Formed from ultra-thin copper foil and supported by a parallel layer architecture, the PCB windings can withstand high current densities—indeed, PCB stators have been reported to operate at current densities three to four times higher than those of conventional wound machines without thermal issues [15]. Once the circuit pattern is finalized, automated printing accelerates production and lowers cost; in addition, the PCB fabrication process offers high repeatability and minimal quality variation, providing clear advantages for high-volume applications such as compact precision actuators [20].

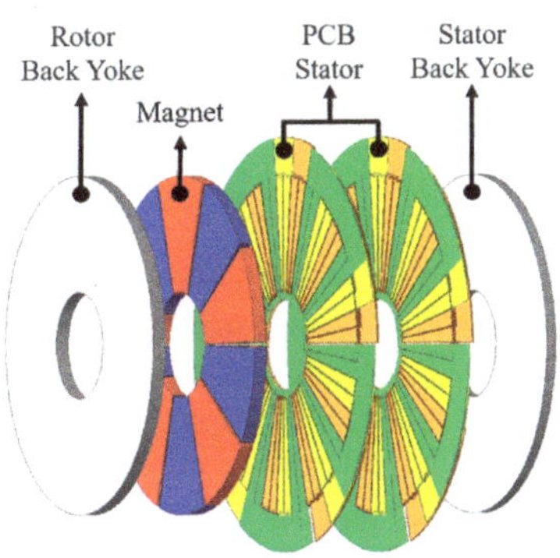


**Figure 1.** Distributed winding PCB axial flux permanent magnet motor (AFPM).

Table 1 summarizes the key design parameters of the proposed PCB AFPM motor model, including the material specifications of the magnet, rotor, and stator, as well as the principal geometric dimensions (outer diameter, inner diameter, axial length, and back-yoke thickness). This information defines the baseline configuration used in the FEM simulations and clarifies the structural constraints of the analyzed model.

Table 2 summarizes the key performance metrics of the proposed motor, obtained through three-dimensional finite-element analysis (3D FEA) under rated operating conditions. At a rated speed of 3650 rpm, the motor delivers an output power of 114.67 W with a corresponding shaft torque of 0.3 Nm. The phase current is set to 12.37 Arms, resulting in a peak current density of 29.46 A/mm$^2$ in the conductor region. Copper and ac losses are estimated at 48.52 W and 5.2 W, respectively, leading to a total input power of approximately 168.4 W. Under these conditions, the overall efficiency is calculated to be 68.10%. The back-EMF at rated speed under no-load conditions is 3.23 Vrms.

**Table 1.** Design parameters of the PCB AFPM motor model.

| Parameter | Unit | Value |
| --- | --- | --- |
| Magnet | - | N45UH |
| Rotor | - | S45C |
| Stator | - | 35PN230 |
| Number of phases | - | 3 |
| Magnet thickness | mm | 2.7 |
| Rotor backyoke thickness | mm | 2.7 |
| Stator backyoke thickness | mm | 2 |
| Outer diameter | mm | 65.4 |
| Inner diameter | mm | 20.6 |
| Axial length | mm | 10.1 |

**Table 2.** 3D Finite element analysis (3D FEA) data.

| Parameter | Unit | Conventional Model |
| --- | --- | --- |
| Back EMF | Vrms | 3.23 |
| Rated Speed | Rpm | 3650 |
| Load Torque | Nm | 0.3 |
| Current | Arms | 12.37 |
| Current density | $A/mm^2$ | 29.46 |
| Copper Loss | W | 48.52 |
| AC Loss | W | 5.2 |
| Output Power | W | 114.67 |
| Efficiency | % | 68.10 |

### 2.2. *Trace Configuration of the Distributed Winding PCB Motor*

Figure 2a shows the single-phase PCB substrate, whereas Figure 2b shows the three-phase PCB substrate. The stator integrates all three phases on a single board, adopting a distributed winding layout with a pole-to-slot ratio of 1:3 [11]. Six copper layers are stacked to obtain two turns per slot for each phase, and the upper and lower layers are electrically linked through vias, guaranteeing uninterrupted current flow in the multilayer PCB [11]. As illustrated in Figure 2a, each phase appears to split into two separate conductor shapes, but these traces are connected through end-turns to form one continuous current path; the two shapes are, therefore, segments of the same phase rather than independent phases. Figure 2b consequently represents a complete three-phase, two-turn (3Φ 2T) winding configuration. Compared with conventional winding techniques, this PCB stator greatly simplifies manufacturing and is well suited to scalable mass production [12,20]. The design is especially advantageous for ultra-compact precision-actuator applications, where severe space constraints demand a thin, low-profile structure that still meets strict packaging requirements [3].

Figure 3 presents a cross-sectional view of the effective copper traces interconnected by via holes. The effective copper trace serves as the primary current-carrying path that directly contributes to torque generation. The effective traces for each phase are distributed across different PCB layers and electrically connected through vias to form a closed-loop circuit. These traces function as active conductors that interact magnetically with the rotor to deliver driving force.

Figure 4 shows the cross-section of the PCB winding layers. The end-turn traces provide return paths between effective copper traces within the same phase, and, positioned along the inner and outer boundaries of the PCB stator, the current begins at the inner end-turns (Layers 3 and 6), passes sequentially through each layer via the vias, and exits through the outer end-turns [11]. Although the end-turn traces themselves do not directly contribute to torque generation, they are essential for maintaining current continuity and circuit completeness [11]. Consequently, the structure consists of six parallel effective copper traces and two parallel end-turns on the inner and outer sides, establishing a stable two-turn current loop per phase. This parallel conductor arrangement lowers both resistance and copper loss, thereby enhancing overall efficiency [12].

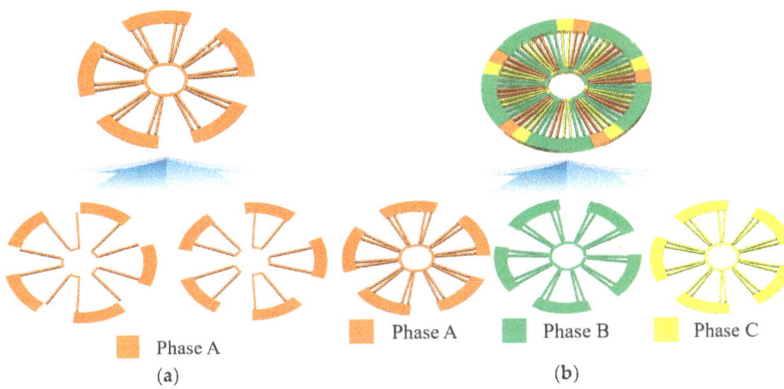


**Figure 2.** (**a**) Single-phase PCB stator structure. (**b**) Three-phase PCB stator.

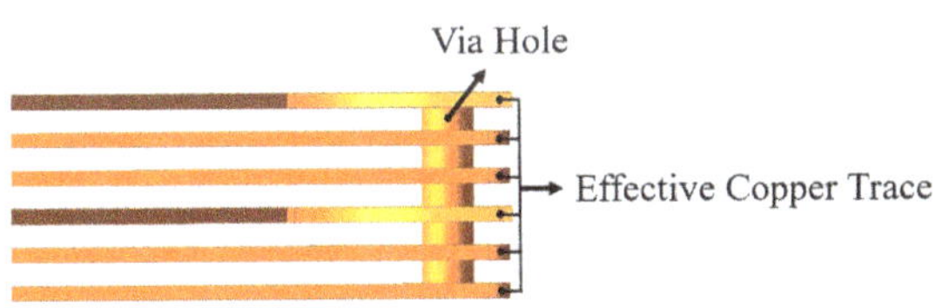


**Figure 3.** Cross-sectional view of effective copper traces connected by via holes.

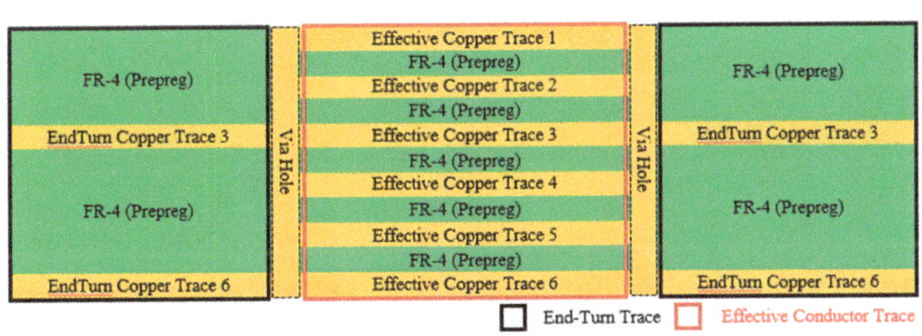


**Figure 4.** Cross-sectional structure of the PCB stator.

The conductor resistance $R$ is defined by Equation (1):

$$R = \rho \frac{L}{A} \tag{1}$$

where $\rho$ is the electrical resistivity, $L$ is the current-path length, and $A$ is the conductor cross-sectional area.

$$A_{Total} = N \cdot A \tag{2}$$

When the number of parallel conductors is $N$, the total effective cross-sectional area increases as shown in (2), which directly leads to a reduction in the overall conductor resistance.

Figure 5 presents a three-dimensional exploded view of the plated via structure in the multilayer PCB stator. The inner wall of each drilled via is coated with a 25 μm copper layer, forming a cylindrical conductor [11]. Via pads on every layer act as bonding sites for the plating, providing dependable electrical connections to the upper- and lower-layer copper traces, while multiple FR-4 prepreg layers insulate adjacent copper sheets and preserve mechanical rigidity [11]. These vias link the upper and lower conductor traces, completing the current path across the six-layer PCB stack.

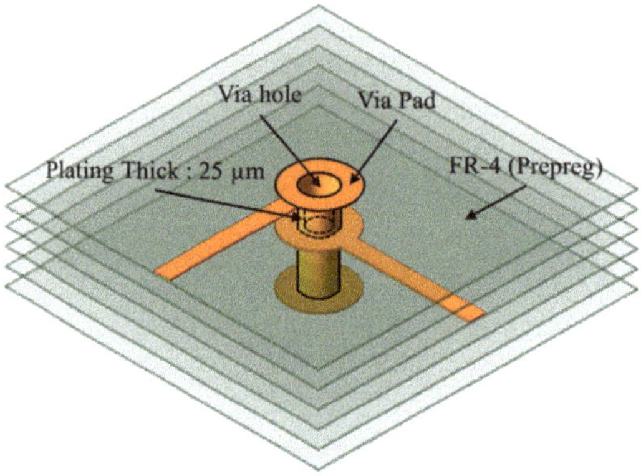


**Figure 5.** Detailed structure of plated via and conductor layers in PCB stator.

Effective conductor traces constitute the primary copper paths that carry current and generate torque; traces distributed across different layers are electrically interconnected through the vias to produce a single closed circuit with two turns per phase [11]. End-turn traces offer return paths within each phase, running between the inner (Layers 3 and 6) and outer end-turns and positioned along the stator boundary to maintain continuity of the inter-layer conductor network [11]. Fabricated with the standard 25 μm plating thickness, the vias align precisely with the pads of both effective and end-turn traces, thereby reducing electrical resistance and mitigating current crowding in the multilayer structure [16,17]; this ensures stable, well-distributed current paths even under high current density conditions [17].

The integrated architecture enables straightforward manufacturing while preserving an ultra-thin, lightweight form factor [12], delivering high electromagnetic performance and electrical reliability for space-constrained applications such as compact precision actuators [3].

### 2.3. Distributed Winding End-Turn Structure of the PCB Stator

Figure 6 shows a three-dimensional enlarged view of the plated via network in the three-phase, two-turn PCB stator. Black arrows mark the current direction along each layer's effective trace, and yellow columns depict inter-layer current flow through the vias. The stator integrates three phases with two dedicated end-turn layers per phase; this separation minimizes inter-phase interference and physically isolates each phase's current path [11]. The total number of end-turn layers is determined by the turns per slot and the number of parallel PCB layers, which together close the current loop [11]. The parallel layer arrangement was tuned within the feasible design range to optimize end-turn placement, yielding a stable current distribution across all traces. As Figure 6 illustrates, the stator employs a distributed winding strategy that exploits the end-turn regions, providing magnetic force balancing and current distribution benefits comparable to those of conventional slotted distributed windings—even in a slotless structure [6].

End-turns interconnect the effective traces of each phase through vias, so current flows in a distributed manner across the parallel conductors, lowering current density and conductor resistance and thereby reducing copper loss [15]. Finally, the via-based inter-layer connections keep current path lengths among the effective traces precisely aligned, contributing to balanced torque generation across all three phases [11].

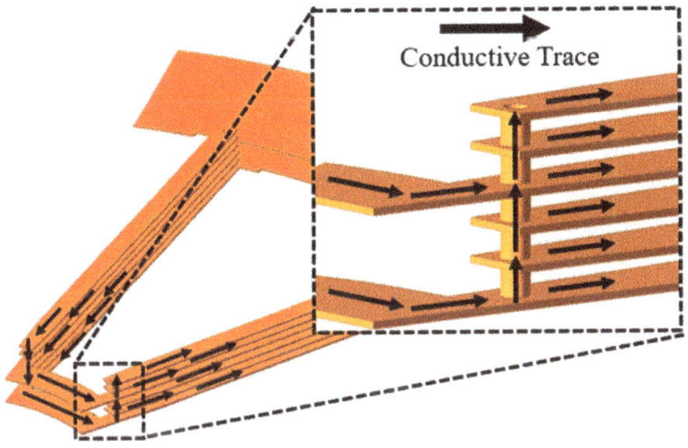


**Figure 6.** Current flow path through multilayer conductive traces and vias in PCB stator.

## 3. Multi-Via Design of the PCB Stator

### *3.1. Limitations of Conventional Single-Via Structures*

Figure 7 shows the current density distribution concentrated inside the via of the conventional single-via model. The simulation results indicate peak current densities exceeding $5 \times 10^5$ A mm$^{-2}$ along the via wall (red region), whereas the surrounding copper traces exhibit much lower values (cyan–green regions). Although a single via links the inter-layer current paths in a multilayer PCB, the actuator-grade AFPM PCB stator investigated here must carry 12.37 A to deliver its rated output of 114.67 W; consequently, a single-via configuration cannot distribute the current effectively. Under the limited copper plating thickness of 25 μm, local heating and thermal saturation risks increase, the effective cross-sectional area for current flow is restricted, phase resistance rises, and both power loss and efficiency degradation occur [17]. Over prolonged operation, such thermal and electrical concentration can weaken via reliability and elevate the likelihood of failure [17].

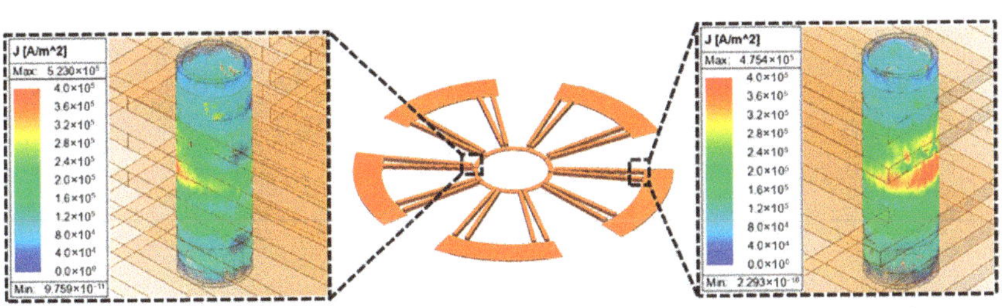


**Figure 7.** Current density distribution concentrated in the via hole of the conventional model.

Figure 8 illustrates the cross-sectional layout of the via hole, plated copper barrel, and surrounding pad area. A PCB via is a cylindrical conductor formed by plating the inner wall of a drilled hole; its conductive cross-section is determined by the plating thickness and the drill diameter. The effective current-carrying area of the via is represented by an annular cross-section, as expressed in Equation (3).

$$A_{Via} = \pi\left(r^2 - (r-t)^2\right) = \pi\left(2rt - t^2\right) \tag{3}$$

where $r$ is the radius of the drilled hole, and $t$ is the plating thickness of the via wall.

$$t \ll r \tag{4}$$

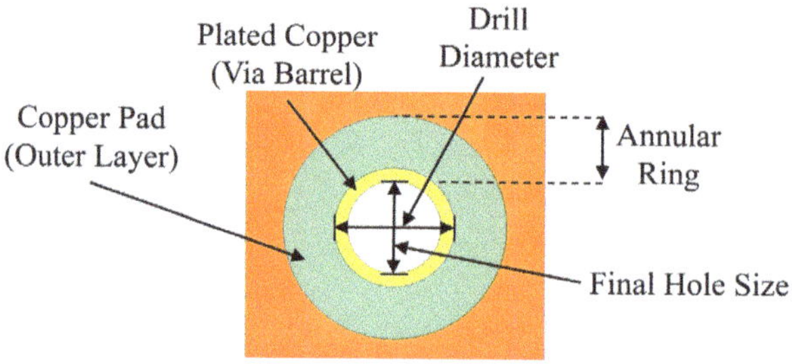


**Figure 8.** Cross-sectional layout of the via hole, plated barrel.

In most cases, as shown in Equation (4), the expression can be approximated and simplified as in Equation (5).

$$A_{Via} \approx 2\pi rt \tag{5}$$

Accordingly, the electrical resistance of the via is given by the general conductor-resistance formula, as shown in Equation (6).

$$R_{Via} = \rho \cdot \frac{l}{A_{Via}} \approx \rho \cdot \frac{l}{2\pi rt} \tag{6}$$

where $\rho$ is the electrical resistivity of copper, and $l$ is the PCB thickness (i.e., the via length). Equation (6) shows that the via resistance $R_{Via}$ decreases as the via radius $r$ increases. Nevertheless, from a practical design standpoint, arbitrarily enlarging the via radius is undesirable for three reasons. First, because the conductive copper is only a thin plated layer of thickness $t$, increasing the drill radius reduces the proportion of conductive plating in the via's cross-sectional area [17]. Second, an excessively large via diameter restricts routing space on the PCB and undermines the mechanical and thermal robustness of the current path [16]. Third, since the plating thickness $t$ is fixed by process constraints, increasing the radius alone does not meaningfully improve current-carrying capacity [17].

$$\pi\left(2rt - t^2\right) \tag{7}$$

The effective cross-sectional area of a via is given by Equation (7) when the plating thickness $t$ is much smaller than the radius $r$, the area can be approximated as $2\pi rt$. Accordingly, the via resistance is defined by Equation (8) and, in theory, decreases as the via radius increases.

$$R = \rho l/(2\pi rt) \tag{8}$$

However, as the drill diameter grows, the proportion of the actual copper conductive layer within the cross-section diminishes, indicating that enlarging a single via's diameter is not an effective way to reduce resistance.

### 3.2. Multi-Via Structure

Figure 9 compares the single-via and multi-via configurations in the inner and outer end-turn regions. The single-via configuration (left) employs one plated through hole for each layer-to-layer connection, whereas the multi-via configuration (right) incorporates multiple parallel vias to reduce local current crowding and lower effective resistance. The top images show enlarged views of the inner end-turn region, while the bottom images illustrate the outer end-turn region. All conductor layers are implemented on a multilayer PCB with 2 oz copper plating. In the single-via structure, each effective conductor layer is interconnected by only one via, producing several disadvantages. First, the current is forced through a narrow path, resulting in a high current density within the via [17]. Second, the limited cross-sectional area raises electrical resistance and, consequently, copper loss [17]. Third, the associated thermal concentration can degrade long-term reliability [16]. Because the via lies directly on the main current path, its resistance adds to the total equivalent resistance, and the elevated local current density concentrates heat in the via, accelerating aging and reducing durability [16,17].

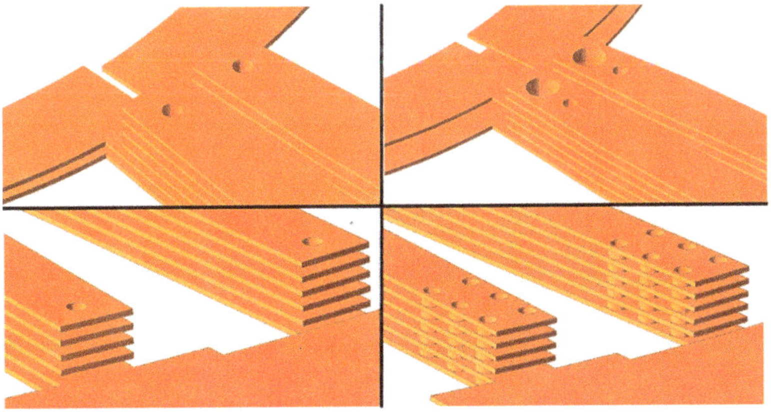

**Figure 9.** Comparison of single-via and multi-via configurations at the inner and outer end-turn regions.

Therefore, a multi-via design must do more than simply increase the via count; the vias must be placed strategically along the current path to minimize loss and provide an optimized, reliable layout [16]. To systematically evaluate and optimize the multi-via design, this study employed a step-by-step procedure. First, a two-turn PCB stator consisting of two 6-layer boards was chosen as the baseline configuration, considering the trade-off between back-EMF requirements and AC loss. This selection reflects a limitation inherent to the chosen motor topology, which may restrict the direct applicability of the results to other winding patterns. Step 1: The single-via structure was analyzed by gradually varying the via diameter to establish reference trends in phase resistance and current density. Step 2: Inline multi-via structures were introduced, and the effects of sequential via placement at the inner and outer end-turns were examined. Step 3: Grid multi-via structures were applied, distributing multiple smaller vias across the conductor segment to improve current sharing and reduce resistance. Step 4: Finally, the inline and grid configurations were quantitatively compared, and the optimal layout was identified by balancing current distribution effectiveness with conductor area utilization.

### 3.3. Inline-Type and Grid-Multi-Via Structures

The top row shows the inline configuration, while the bottom row presents the grid configuration. (a) and (b) display the inner and outer end-turns of the inline design, where vias are sequentially inserted from the conductor tip. (c) and (d) illustrate the grid design, in which multiple parallel vias are distributed throughout the conductor segment to equalize current density. Compared with the inline configuration, the grid structure more effectively mitigates current crowding, reduces resistance, and enhances thermal dissipation, while maintaining full compatibility with standard PCB fabrication processes.

Figure 10a shows the inline multi-via structure at the inner end-turn. Vias are arranged in a single row and inserted in the sequence $1 \rightarrow 2 \rightarrow 3$, starting from the terminal end of the conductor. This layout directs incoming current from the inner side through the conductor tip first, mitigating localized current concentration and reducing heat generation.

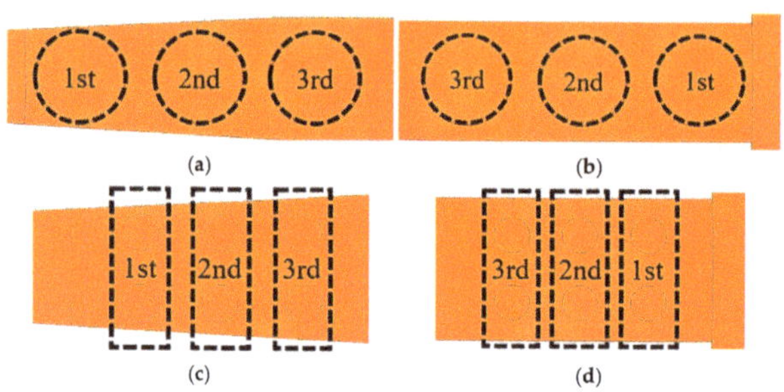


**Figure 10.** Inline and grid multi-via placements for PCB stators. Top row: inline configuration (**a**) inner end turn with sequential vias; (**b**) outer end turn with sequential vias. Bottom row: grid configuration (**c**) inner end turn with distributed parallel vias; (**d**) outer end turn with distributed parallel vias.

Figure 10b presents the inline multi-via structure at the outer end-turn. Although the physical order is $3 \rightarrow 2 \rightarrow 1$, the insertion sequence remains $1 \rightarrow 2 \rightarrow 3$. This arrangement distributes current entering from the outer side toward the conductor tip first, optimizing the current path and enhancing both electrical and thermal performance.

Figure 10c illustrates the grid multi-via structure at the inner end-turn. Each row consists of a pair of vertical vias, added sequentially in the order $3 \rightarrow 2 \rightarrow 1$. Consequently, two, four, and six vias are distributed across the conductor segment, effectively reducing current density and electrical resistance.

Figure 10d shows the grid multi-via structure at the outer end-turn. While the physical arrangement is $3 \rightarrow 2 \rightarrow 1$, the insertion sequence again follows $1 \rightarrow 2 \rightarrow 3$. Because current enters from the outer side, prioritizing via connections near the conductor tip minimizes thermal loss and electrical resistance.

In the two-turn conductor pattern used in this study, the grid multi-via approach offers superior electrical performance compared with the inline configuration. In inline designs, if space is insufficient for multiple vias at the same location, enlarging the via diameter can lower internal current density; however, the resulting larger plated area occupies more of the effective conductor, reducing copper cross-section and increasing phase resistance. When adequate spacing between conductors is available—as in the present two-turn layout, the grid method, which distributes multiple small vias, more effectively mitigates current density and lowers resistance. Considering factors such as number of turns, conductor width, via arrangement, parallel current paths, and manufacturability, the grid multi-via configuration is optimal for a two-turn PCB stator. This design provides superior resistance reduction and thermal dissipation performance while remaining compatible with standard PCB fabrication processes.

### 3.4. Simulation Conditions

In this study, the rotor and stator structures were kept identical; only the via configuration was varied to compare and analyze electrical characteristics such as phase resistance and current density. The analysis conditions were as follows. First, for the single-via structure, the via diameter was gradually increased, and the resulting changes in phase resistance and current density distribution were evaluated. This analysis considered both the reduction in current density due to the enlarged via diameter and the opposing effect of decreased effective conductor area—caused by the enlarged plated via region within the same conductor width—on resistance behavior. Second, for the multi-via structures, two configurations were investigated: inline multi-via and grid multi-via. For both configurations, the via diameter was incrementally increased within the same effective conductor width. The current distribution effect and conductor loss characteristics of each structure were then compared. Through this evaluation, an electrical performance optimization strategy was established by balancing the benefit of an increased plated via area against the reduction in effective conductor cross-section.

In addition, the finite-element simulations were performed using ANSYS Maxwell 3D, Release 2025 R2 with the Eddy Current Solver for resistance and current density evaluation. Adaptive meshing with a minimum element size of 200 μm was applied to ensure accuracy, and representative mesh distributions are shown in Figure 11. Specifically, the mesh applied to the PCB stator and via regions for eddy current analysis is presented in Figure 11a, while Figure 11b,c illustrates the mesh discretization of the multilayer PCB circuit and the refined airgap/airband region used in transient electromagnetic simulations for torque and dynamic field evaluation. No special boundary conditions were applied beyond the default solver settings.

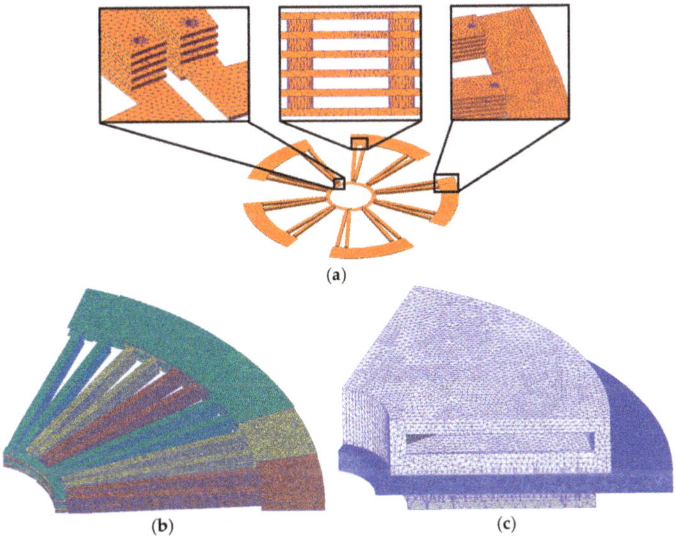

**Figure 11.** Representative mesh plots for FEM simulations: (**a**) mesh of PCB stator and via regions used in eddy current analysis for resistance and current density evaluation, (**b**) mesh discretization of the multilayer PCB circuit for transient electromagnetic simulations, and (**c**) refined airgap and surrounding airband mesh for transient analysis.

## 4. FEM Methodology and Performance Evaluation

### 4.1. Analysis Items

The stator employs a parallel layer configuration with two vias—inner and outer—allocated to each slot. These vias electrically connect the upper and lower effective conductors, enabling current distribution across layers. In this study, the diameters of the inner and outer vias were varied independently to evaluate their effects on phase resistance and current density, with the aim of identifying the optimal combination of via diameter and placement. The analysis began with the single-via structure, uniformly increasing both inner and outer via diameters in successive FEM simulations. Building on these results, inline multi-via (IM-via) and grid multi-via (GM-via) configurations were then applied to the inner and outer regions. After determining the inner via condition that minimized phase resistance, the same procedure was applied to the outer via to obtain the overall optimal layout. This step-by-step approach allowed separate sensitivity studies of phase resistance and current density with respect to via diameter and layout, enabling a quantitative optimization that balances current distribution effectiveness against conductor area utilization.

### 4.2. FEM-Based Performance Analysis

Figure 12 shows that, in the single-via configuration, a via diameter of 0.60 mm produces the lowest values—0.1008 Ω in phase resistance and 425 A/mm$^2$ in peak current density—corresponding to reductions of approximately 4.63% and 22.73%, respectively, relative to the 0.30 mm baseline. These improvements stem from the enlarged effective current path and enhanced current spreading that accompany an increase in via diameter, which mitigate loss and local heating caused by current crowding. Once the via diameter surpasses this optimal value, however, both phase resistance and peak current density rise again because the larger via removes copper from the effective conductor, narrowing the remaining conductive path. Simply enlarging a single via, therefore, does not always reduce resistance; beyond a certain point, it can have the opposite effect. Optimal via design must balance current spreading effectiveness against retention of a sufficient copper cross-section.

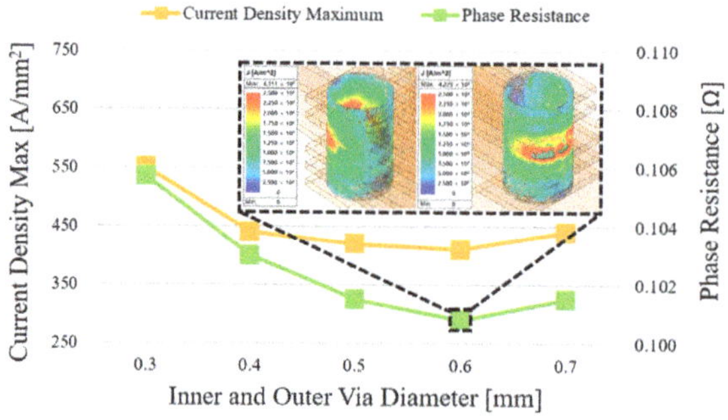


**Figure 12.** Phase resistance and peak current density vs. uniform inner/outer via diameter (single-via structure).

Based on the preceding single-via analysis, Figure 13 evaluates phase resistance and current density characteristics by incrementally applying the IM-via structure to the inner via while keeping the outer via fixed at the baseline diameter of 0.30 mm. For each via position (1st, 2nd, and 3rd), the diameter was varied from 0.30 mm to 0.70 mm, and the resulting phase resistance values were quantitatively compared. The lowest phase resistance occurs with a 0.50 mm diameter for the 1st via and 0.30 mm for the 2nd via, identifying this pair as the optimal IM-via configuration. Adding a 3rd via provides only marginal resistance reduction, and in some cases even increases resistance, because additional vias reduce the effective conductor area and distort the current path, leading to local crowding and extra loss. The graph in Figure 13 plots the minimum phase resistance value obtained for every via-diameter combination (0.30–0.70 mm) at the 1st, 2nd, and 3rd positions, making the optimal design point easy to identify. In summary, for the IM-via structure, the 1st-via 0.50 mm and 2nd-via 0.30 mm combination is optimal. These results confirm that precise coordination of via position and diameter, rather than simply increasing the via count, is essential for improving electrical performance.

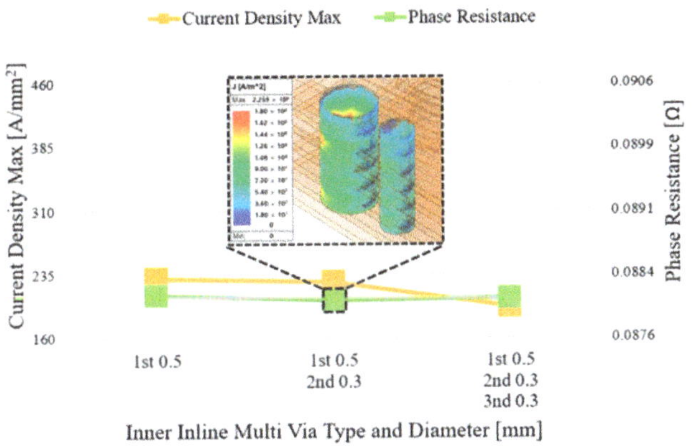


**Figure 13.** Effect of inner inline multi-via diameter combinations on phase resistance and peak current density.

Figure 13, under the same conditions as the IM-via analysis in Figure 12, evaluates phase resistance and current density characteristics by fixing the outer via at the baseline diameter of 0.30 mm and applying a GM-via configuration to the inner via. In the GM-via scheme, two vias are placed in parallel across the conductor width at each position. Considering space constraints and conductor interference, the via diameter was varied from 0.30 mm to 0.40 mm. The analysis proceeded in three stages: two vias at the 1st position, four vias at the 1st + 2nd positions, and six vias at the 1st + 2nd + 3rd positions. Phase resistance and current density performance were compared for every combination.

The best result was obtained with two 0.30 mm diameter vias at the 1st position, yielding a minimum phase resistance of 0.0878 Ω. Relative to the baseline model (inner and outer vias both 0.30 mm, phase resistance 0.1057 Ω), this represents a 16.96% reduction. The peak current density at the inner via also decreased from 410 A/mm$^2$ to 212 A/mm$^2$, a 48.29% reduction. Adding vias at the 2nd and 3rd positions, thereby increasing the via count, often raised phase resistance. Excessive vias reduce the effective conductor area and distribute current inefficiently, which degrades electrical performance. In the GM-via layout, horizontally paired vias further increase copper area loss and conductor interference as diameter grows, raising resistance.

Figure 14 shows the lowest phase resistance value for each combination obtained from FEM simulations over the 0.30–0.40 mm diameter range at the 1st, 2nd, and 3rd stages. Overall, the GM-via scheme achieves lower resistance than the IM-via approach, and the simplest configuration (two 0.30 mm vias at the 1st position) proves most effective. This option is also the easiest to manufacture and confirms that adding more vias does not necessarily improve performance.

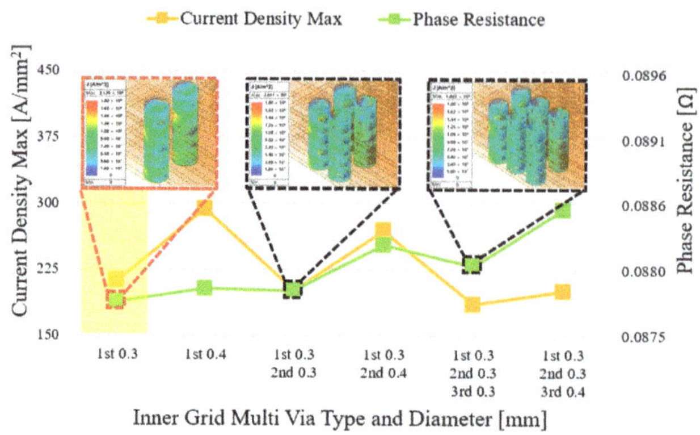


**Figure 14.** Effect of Inner Grid Multi-Via Diameter Combinations on Phase Resistance and Peak Current Density.

Figure 15 builds on Figure 14, in which the inner via uses the GM-via scheme, and the configuration with two 0.30 mm vias at the 1st position proves optimal. With that condition fixed, the present stage applies the IM-via structure to the outer via to evaluate additional phase resistance and current density characteristics. The arrangement is the same as before, except that the outer vias are stacked inward from the stator edge in the order 1st → 2nd → 3rd. For each position, the via diameter is varied stepwise from 0.30 mm to 0.70 mm, and the resulting performance differences attributable to current spreading are quantified.

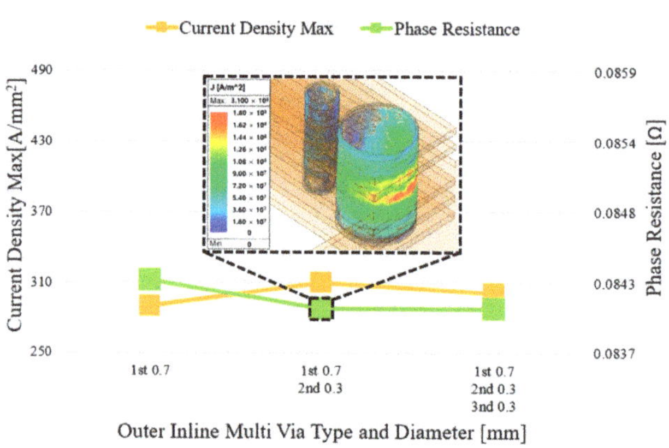


**Figure 15.** Effect of outer inline multi-via diameter combinations on phase resistance and peak current density.

The lowest phase resistance is obtained with a 0.70 mm via at the 1st position and a 0.30 mm via at the 2nd position, indicating that a large-diameter via at the outermost location secures the main current path, while smaller vias in subsequent positions provide auxiliary current dispersion. Adding a 3rd via yields no significant additional reduction, offering limited benefit compared with the added design complexity.

Figure 15 plots the minimum phase resistance value for every position–diameter combination obtained from the FEM simulations, confirming that strategic selection of via positions and diameters, rather than merely increasing the via count, is decisive for improving electrical performance.

Figure 16 summarizes the FEM results for all GM-via combinations at the 1st, 2nd, and 3rd positions, graphing the lowest phase resistance value obtained in each set. For the inner via, the GM-via arrangement identified in Figure 14—two 0.30 mm vias at the 1st position—gives the best electrical performance, yielding a phase resistance of 0.0878 Ω and a peak current density of 212 A/mm$^2$. With this inner via condition fixed, the same GM-via scheme is applied to the outer via, and the resulting phase resistance and current-density characteristics are evaluated.

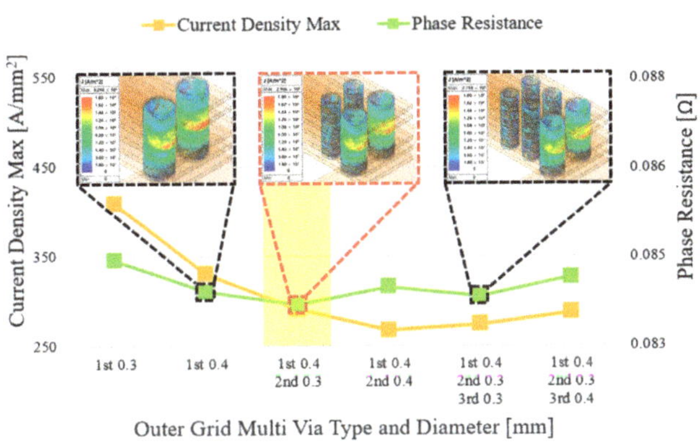


**Figure 16.** Effect of outer grid multi-via diameter combinations on phase resistance and peak current density.

For the outer GM-via, two side-by-side vias are placed at each position (1st, 2nd, and 3rd) with diameters swept from 0.30 mm to 0.40 mm, resulting in two, four, and six total vias for the 1st, 1st + 2nd, and 1st + 2nd + 3rd stages, respectively. FEM analysis shows that a 0.40 mm via at the 1st position combined with a 0.30 mm via at the 2nd position (1st 0.4, 2nd 0.3) provides the best performance, with a phase resistance of 0.0837 Ω and a peak current density of 290 A/mm$^2$. Relative to the optimal inner GM-via case with a single 0.30 mm outer via (phase resistance 0.0878 Ω), this configuration lowers phase resistance by 4.67%. Compared with the baseline single-via design (inner and outer vias both 0.30 mm, peak outer via current density 422 A/mm$^2$), the peak current density decreases by 31.28%. Even against the optimal IM-via outer arrangement (1st 0.7 mm, 2nd 0.3 mm; phase resistance 0.0843 Ω, current density 310 A/mm$^2$), the GM-via layout achieves lower phase resistance and current density. Overall, it reduces phase resistance by 20.76% relative to the single-via baseline.

These results confirm that the GM-via approach provides a more efficient current-distribution path for outer via design. Applying GM vias strategically to both inner and outer locations yields the greatest performance improvement over the single-via and IM-via schemes, demonstrating that precise selection of via position and diameter, rather than simply increasing via count or size, is essential for optimal PCB stator design.

According to the three-dimensional finite-element results in Table 3, the grid multi-via (GM-via) design evaluated under the same operating conditions as the baseline single-via model (identical back-EMF, rated speed of 3650 rpm, load torque of 0.3 Nm, line current of 12.57 A rms, and output power of 114.67 W) reduces the peak current density from 42.19 A·mm$^{-2}$ to 33.45 A·mm$^{-2}$ (≈20.7%) and lowers the phase resistance from 0.106 Ω to 0.083 Ω (≈21.7%). This resistance reduction directly decreases $I^2R$ losses, with copper loss falling from 50.08 W to 39.70 W (≈20.7%), which in turn raises overall efficiency from 66.72% to 71.01%, that is, +4.29 percentage points (≈6.4%). These results confirm that the

GM-via layout equalizes current paths, mitigates localized current crowding, and effectively increases the usable conductor cross-section under identical mechanical specifications.

**Table 3.** Comparison of 3D finite-element analysis (FEA) results between the conventional and improvement through FEM approach model.

| Parameter | Unit | Conventional Model | Improvement Through FEM Approach Model | Change |
|---|---|---|---|---|
| Back EMF | Vrms | 3.2 | 3.23 | - |
| Load Torque | Nm | 0.3 | 0.3 | - |
| Current | Arms | 12.57 | 12.57 | - |
| Current density | A/mm$^2$ | 42.19 | 33.45 | 20.7% |
| Phase Resistance | Ohm | 0.106 | 0.083 | 21.7% |
| Copper Loss | W | 50.08 | 39.70 | 20.7% |
| AC Loss | W | 7.11 | 7.11 | - |
| Output Power | W | 114.67 | 114.67 | - |
| Efficiency | % | 66.72 | 71.01 | 6.4% |

In Figure 17, (a) Copper loss, increasing approximately with the square of load current (torque), is ≈20% lower in the GM-via design at rated operation, consistent with the resistance reduction. (b) AC loss, dominated by speed (frequency-dependent effects), shows minor sensitivity to torque and peaks near the highest rpm. (c) Output power, following $P_{out}$-T$\omega$ reaches 114.67 W at the rated point, indicating that loss reduction does not compromise deliverable power. (d) Efficiency, reflecting the combined influence of losses and output, exceeds 70% near the rated region, with a maximum of 71.01% compared to 66.72% for the single-via baseline.

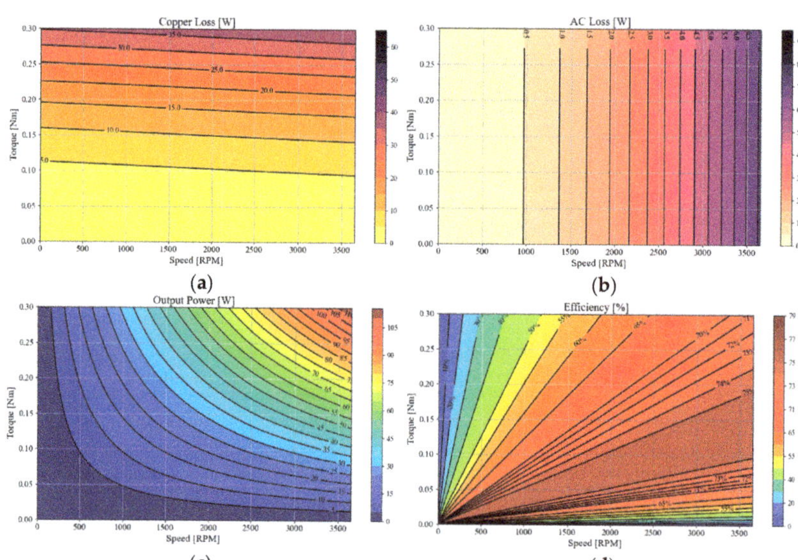


**Figure 17.** FEM-based performance maps of the improved grid multi-via PCB stator over the torque–speed operating range.

## 5. Conclusions

This study presents an optimal design for a compact axial flux permanent magnet motor (AFPM) that employs an ultra-thin, high-density PCB stator. A multi-via configuration was introduced to reduce phase resistance and peak current density, thereby improving overall electrical performance. Step-by-step FEM analyses were carried out for both in-line and grid multi-via schemes to determine the optimal via arrangement. Using the grid multi-via structure, phase resistance decreased from 0.106 Ω to 0.084 Ω (≈20.8%), and peak current density fell from 42.19 A/mm$^2$ to 33.45 A/mm$^2$ (≈20.7%) relative to the single-via baseline. Copper loss was reduced from 50.08 W to 39.7 W, while overall efficiency increased from 66.72% to 71.01%, with output power and other operating conditions held constant. These results demonstrate that carefully selecting via position and diameter, rather than simply increasing the via count, provides the best balance between design complexity and electrical performance. Unlike earlier PCB-stator motor studies, which lacked systematic optimization of via structures, this work offers a practical design methodology that simultaneously accounts for via diameter, placement, and the number of parallel layers. Real manufacturing constraints, such as minimum drill diameter and copper clearance limits, are also included, ensuring both practicality and reproducibility.

**Author Contributions:** Conceptualization and design, W.-H.K. (including study concept, critical revision, and final approval); methodology development and data curation, S.-B.J. (including experimental design, data management, drafting the initial manuscript, and approval of the final version); software development and resource provision, D.-H.C. (including technical implementation, critical review, and final accountability for the software components); validation and overall research oversight, Y.-H.S. (ensuring data integrity and accountability throughout the study); formal analysis and investigation, H.-S.H. (performing comprehensive data analysis, interpretation, and figure visualization); writing—original draft preparation, S.-B.J. (responsible for drafting the manuscript followed by critical revision by all authors); writing—review and editing, S.-B.J. and D.-H.C. (providing in-depth manuscript reviews, revisions, and final manuscript approval); visualization, Y.-H.S. and H.-S.H. (developing data visualizations, ensuring clarity in presentation); supervision and project management, W.-H.K. (overseeing the entire research process and endorsing the final submission). All authors have read and agreed to the published version of the manuscript.

**Funding:** This work was partly supported by the Korea Institute of Energy Technology Evaluation and Planning (KETEP) grant funded by the Korea government (MOTIE) (20214000000060, Department of Next Generation Energy System Convergence based on Techno-Economics-STEP) and in part by the Korea Institute of Energy Technology Evaluation and Planning (KETEP) grant funded by the Korea government (MOTIE) (RS-2024-00419152, Development of 5 kW industrial logistics electric platform technology).

**Data Availability Statement:** The original contributions presented in the study are included in the article; further inquiries can be directed to the corresponding author.

**Conflicts of Interest:** The authors declare no conflicts of interest.

## References

1. Hsieh, F.-S. Emerging Research Issues and Directions on MaaS, Sustainability and Shared Mobility in Smart Cities with Multi-Modal Transport Systems. *Appl. Sci.* **2025**, *15*, 5709. [CrossRef]
2. Turoń, K.; Tóth, J. Innovations in Shared Mobility—Review of Scientific Works. *Smart Cities* **2023**, *6*, 1545–1559. [CrossRef]
3. Mittal, V.; Lotwin, M.; Shah, R. A Review of Bio-Inspired Actuators and Their Potential for Adaptive Vehicle Control. *Actuators* **2025**, *14*, 303. [CrossRef]
4. Yu, P.; Sun, Z.; Xu, H.; Ren, Y.; Tan, C. Design and Analysis of Brake-by-Wire Unit Based on Direct Drive Pump–Valve Cooperative. *Actuators* **2023**, *12*, 360. [CrossRef]
5. Gong, X.; Qian, L.; Ge, W.; Wang, L. Research on the Anti-Disturbance Control Method of Brake-by-Wire Unit for Electric Vehicles. *World Electr. Veh. J.* **2019**, *10*, 44. [CrossRef]
6. Hao, Z.; Ma, Y.; Wang, P.; Luo, G.; Chen, Y. A Review of Axial-Flux Permanent-Magnet Motors: Topological Structures, Design, Optimization and Control Techniques. *Machines* **2022**, *10*, 1178. [CrossRef]
7. Zhao, X.-K.; Zou, X.-P.; Guo, Q.-C.; Zhu, L.-K. Axial Flux Permanent Magnet Synchronous Motor Cogging Torque Calculation Method Based on Harmonic Screening. *Energies* **2025**, *18*, 3779. [CrossRef]

8. Liu, W.; Yang, J.; Pei, L.; Rafiei, M.; Yang, Y.; Wang, Y.; Cui, J.; Guo, Y.; Zhang, B. Structural Optimization and Electromagnetic Performance Research of Axial Magnetic Field Tidal Current Generators. *Energies* **2025**, *18*, 2520. [CrossRef]
9. Gao, P.; Gu, Y.; Wang, X. The Design of a Permanent Magnet In-Wheel Motor with Dual-Stator and Dual-Field-Excitation Used in Electric Vehicles. *Energies* **2018**, *11*, 424. [CrossRef]
10. Wang, X.; Li, T.; Cui, X.; Zhao, X. Design and Analysis of Coreless Axial Flux Permanent Magnet Machine with Novel Composite Structure Coils. *Energies* **2022**, *15*, 5162. [CrossRef]
11. Wang, X.; Lu, H.; Li, X. Winding Design and Analysis for a Disc-Type Permanent-Magnet Synchronous Motor with a PCB Stator. *Energies* **2018**, *11*, 3383. [CrossRef]
12. Choi, D.-H.; Han, H.-S.; Hong, M.-K.; Jung, D.-H.; Kim, W.-H. Design for Loss Reduction in a Compact AFPM Electric Water Pump with a PCB Motor. *Energies* **2025**, *18*, 2538. [CrossRef]
13. Arasteh, E.; Assadian, F. A Comparative Analysis of Brake-by-Wire Smart Actuators Using Optimization Strategies. *Energies* **2022**, *15*, 634. [CrossRef]
14. Musumeci, S.; Barba, V. Gallium Nitride Power Devices in Power Electronics Applications: State of Art and Perspectives. *Energies* **2023**, *16*, 3894. [CrossRef]
15. Yang, L.; Zhao, J.; Fu, W.; Liu, X.; Zhu, J.; Ai, C. A Comprehensive Investigation of Winding Eddy and Circulating Current Losses of Stator Iron Coreless PMBLDC Motors. *Energies* **2023**, *16*, 5523. [CrossRef]
16. Bauer, A.; Dieterich, D.; Urschel, S. Comparison of Eddy Current Loss Calculation Techniques for Axial Flux Motors with Printed Circuit Board Windings. *Energies* **2025**, *18*, 2603. [CrossRef]
17. Salim, N.; Nikam, S.P.; Pal, S.; Wankhede, A.K.; Fernandes, B.G. Multiphysics analysis of printed circuit board winding for high-speed axial flux permanent magnet motor. *IET Electr. Power Appl.* **2019**, *13*, 805–811. [CrossRef]
18. Chulaee, Y.; Lewis, D.; Heins, G.; Patterson, D.; Ionel, D.M. Winding Losses in Coreless Axial Flux PM Machines with Wave and Spiral PCB Stator Topologies. In Proceedings of the 2022 IEEE Energy Conversion Congress and Exposition (ECCE), Detroit, MI, USA, 9–13 October 2022; pp. 1–6. [CrossRef]
19. Mahmouditabar, F.; Baker, N. A Review on the Effect of Electrical Steel Manufacturing Processes on the Performance of Electric Machines. *Energies* **2023**, *16*, 7954. [CrossRef]
20. Chulaee, Y.; Lewis, D.; Mohammadi, A.; Heins, G.; Patterson, D.; Ionel, D.M. Circulating and Eddy Current Losses in Coreless Axial Flux PM Machine Stators with PCB Windings. *IEEE Trans. Ind. Appl.* **2023**, *59*, 4010–4020. [CrossRef]
21. Shuaibu, I.; Wei, E.H.T.; Kannan, R.; Samaila, Y.A. The Utilization of Printed Circuit Boards (PCBs) in Axial Flux Machines: A Systematic Review. *Eng. Proc.* **2025**, *87*, 13. [CrossRef]

MDPI AG
Grosspeteranlage 5
4052 Basel
Switzerland
Tel.: +41 61 683 77 34

*Actuators* Editorial Office
E-mail: actuators@mdpi.com
www.mdpi.com/journal/actuators

www.ingramcontent.com/pod-product-compliance
Lightning Source LLC
Chambersburg PA
CBHW040811250926
51850CB00007BA/1142
*9783725860951*